Logos and Máthēma

Polish Contemporary Philosophy and Philosophical Humanities

Edited by Jan Hartman

Volume 1

PETER LANG

Frankfurt am Main · Berlin · Bern · Bruxelles · New York · Oxford · Warszawa · Wien

Roman Murawski

Logos and Máthēma
Studies in the Philosophy of Mathematics
and History of Logic

PETER LANG
Internationaler Verlag der Wissenschaften

Bibliographic Information published by the Deutsche Nationalbibliothek
The Deutsche Nationalbibliothek lists this publication in the Deutsche Nationalbibliografie; detailed bibliographic data is available in the internet at http://dnb.d-nb.de.

Cover Design:
© Olaf Gloeckler, Atelier Platen, Friedberg

The publication was financially supported
by the Faculty of Mathematics and Computer Science
of Adam Mickiewicz University in Poznań.

ISSN 2191-1878
ISBN 978-3-631-61804-2
© Peter Lang GmbH
Internationaler Verlag der Wissenschaften
Frankfurt am Main 2011
All rights reserved.

All parts of this publication are protected by copyright. Any utilisation outside the strict limits of the copyright law, without the permission of the publisher, is forbidden and liable to prosecution. This applies in particular to reproductions, translations, microfilming, and storage and processing in electronic retrieval systems.

www.peterlang.de

*Dedicated to
my wife Hania and daughter Zosia*

Foreword

The volume contains 20 essays devoted to the philosophy of mathematics and the history of logic. They have been divided into four parts. Part 1 contains papers considering general philosophical problems of mathematics. In the essay "Mathematical Knowledge" basic epistemological problems of mathematics are considered. Main doctrines in the epistemology of mathematics have been presented and analyzed. The interrelations between logic and philosophy of mathematics as well as some current tendencies in the philosophy of mathematics have been studied. The essay "On the Power and Weaknesses of the Axiomatic Method" discusses the meaning of the axiomatic method for the methodology of mathematics. In "Remarks on the Mathematical Universe" the problem of the existence and the nature of mathematical entities is considered. Various conceptions that appeared in the history are presented and examples of mathematicians and logicians declaring for them are indicated. Consequences of those conceptions for doing mathematics are considered. In the last essay in this part "Structuralism and Category Theory in the Contemporary Philosophy of Mathematics" set-theoretical (Bourbaki-style) and category-theoretical approaches to structuralism in the philosophy of mathematics are compared. Advantages and disadvantages of them are indicated.

Part 2 is devoted to problems concerning Hilbert's program and the influence on it of the discovery of the incompleteness phenomenon. In the essay "Hilbert's Program: Incompleteness Theorems vs. Partial Realizations" the question whether Gödels' incompleteness theorems did reject Hilbert's program is studied. Generalizations and strengthenings of Gödel's results as well as generalized and relativized Hilbert's programs and their meaning for the philosophy of mathematics are considered. The essay "On the Distinction Proof–Truth in Mathematics" contains some historical, philosophical and logical considerations connected with the distinction between proof and truth in mathematics. The crucial rôle of Gödel's incompleteness theorems as well as of the undefinability of truth vs. definability of provability and the rôle of finitary vs. infinitary methods are stressed. The problem of the necessity of extending the available methods by new rules of inference and new axioms is also considered. The discovery of the incompleteness phenomenon destroyed the old conviction that axiomatic method is the ideal method for mathematics. Therefore it was not immediately accepted by logicians. Reactions to this discovery are considered in the next paper. The essay "Gödel's Incompleteness Theorems and Computer Science" indicates some applications of Gödel's results to the discussion of problems of computer science. In particular the problem of relations between the mind and machine (arguments by J. J. C. Smart and J. R. Lucas), Gödel's opinion on this issue and some interpretations of the incompleteness theorems from the point of view of the information theory are presented. Though it seems that human mind is not fully equivalent to a machine (computer), nevertheless some of its functions can be mechanized. Next essay "The Present State of Mechanized Deduction, and the Present Knowledge of Its Limitations" tells about attempts to develop procedures of mechanized deduction. It indicates also various limitations of them.

One of the aims of Hilbert's program was to show that the classical mathematics (referring to the actual infinity) is safe and free of any inconsistencies. Hence the attempts to prove the consistency of basic mathematical theories by finitistic methods undertaken in Hilbert's school (Ackermann, von Neumann). Gödel's results show that new non-finitistic methods must be applied here. Those problems are considered in the paper "On Proofs of the Consistency of Arithmetic". The paper "Decidability vs. Undecidability. Logico-Philosophico-Historical Remarks" presents the decidability problem from a philosophical and historical perspective. It indicates also basic mathematical and logical results concerning (un)decidability of particular theories and problems. In the paper "Undefinability of Truth. The Problem of Priority: Tarski vs. Gödel" it is argued that Tarski obtained the theorem on the undefinability independently from Gödel though he made clear his indebtedness to Gödel's methods. On the other hand Gödel was aware of the formal undefinability of truth in 1931 but he did not publish this result – reasons for that are considered. The problem of definability and undefinability of the concepts of satisfaction and truth from a more technical point of view is considered in the essay "Troubles With (the Concept of) Truth in Mathematics" closing Part 2.

Parts 3 and 4 are devoted to the work of Polish logicians and mathematicians in the philosophy of mathematics and in logic. Part 3 begins by an essay presenting the philosophical system of deep and interesting but unfortunately rather forgotten and underestimated Polish philosopher and mathematician Józef Maria Hoene-Wroński. In the essays "Philosophical Reflection on Mathematics in Poland in the Interwar Period" and "Philosophy of Mathematics in the Warsaw Mathematical School" the views and tendencies of the most outstanding representatives of Lvov-Warsaw Philosophical School and of the founders of Polish Mathematical School are presented and analyzed. The problem whether those views had any influence on logical and mathematical research is considered. Philosophical views concerning mathematics of Andrzej Mostowski, an outstanding representative of the second generation of the Lvov-Warsaw School, are presented and discussed in the next essay.

The last Part 4 contains three essays devoted to Polish mathematical logic. The first one "Stanisław Piątkiewicz and the Beginnings of Mathematical Logic in Poland" presents information on the life and work of Stanisław Piątkiewicz. His *Algebra w logice* (1888) contained an exposition of the algebra of logic and its use in representing syllogisms. This was the first original Polish publication on symbolic logic (it appeared 20 years before analogues works by Łukasiewicz and Stamm). In the next essay contribution of Polish logicians to the recursion theory is presented. The final essay is devoted to logical investigations at the University of Poznań in the period 1945–1955. It is considered whether there was any continuation of the Lvov-Warsaw Logical School in Poznań.

I would like to thank all who helped me in the work on this book. First of all I thank the co-authors who agreed to include into the volume our joint papers, in particular Professor Tadeusz Batóg, Professor Jerzy Pogonowski, Profesor Jan Woleński and Doctor Izabela Bondecka-Krzykowska. I thank also Faculty of Mathematics and Computer Science of Adam Mickiewicz University in Poznań for the financial support and Doctor Paweł Mleczko from that university for the help in the preparing the camera ready version of the book. Last but not least I thank Mr. Łukasz Gałecki from Peter Lang Verlag for his patient and helpful assistance.

Poznań, March 2011 *Roman Murawski*

Contents

Part I: Philosophy of Mathematics (in General)

Mathematical Knowledge 15

On the Power and Weaknesses of the Axiomatic Method 53

Remarks on the Mathematical Universe 63

Structuralism and Category Theory in the Contemporary Philosophy of Mathematics (with Izabela Bondecka-Krzykowska) 73

Part II: Hilbert's Program vs. Incompleteness Phenomenon

Hilbert's Program: Incompleteness Theorems vs. Partial Realizations . 83

On the Distinction Proof-Truth in Mathematics 101

Reactions to the Discovery of the Incompleteness Phenomenon 115

Gödel's Incompleteness Theorems and Computer Science 127

The Present State of Mechanized Deduction, and the Present Knowledge of Its Limitations 137

On Proofs of the Consistency of Arithmetic 159

Decidability vs. Undecidability. Logico-Philosophico-Historical Remarks 165

Undefinability of Truth. The Problem of the Priority: Tarski vs. Gödel . 177

Troubles With (the Concept of) Truth in Mathematics 187

Part III: Philosophy of Mathematics in Poland

The Philosophy of Hoene-Wroński 205

Philosophical Reflection on Mathematics in Poland in the Interwar Period 215

Philosophy of Mathematics in the Warsaw Mathematical School 227

Andrzej Mostowski on the Foundations and Philosophy of Mathematics (with Jan Woleński) 243

Part IV: Mathematical Logic in Poland

Stanisław Piątkiewicz and the Beginnings of Mathematical Logic in Poland (with Tadeusz Batóg) 261

Contribution of Polish Logicians to Recursion Theory 267

Logical Investigations at the University of Poznań in 1945–1955 (with Jerzy Pogonowski) 283

References 295
Editorial Note 327
Index of Names 331

Part I
Philosophy of Mathematics (in General)

Mathematical Knowledge

Since its very beginnings mathematics played a special and distinguished rôle in the human knowledge. It was close to the ideal of a scientific theory, even more, it established such an ideal and served as a pattern of a theory. It has played an important rôle also in the development of the epistemology. In fact mathematics has been through ages a pattern of any rational knowledge and the paradigm of a priori knowledge. Hence the importance and meaning of philosophical and methodological reflections on mathematics as a science. Such reflections have accompanied mathematics since ancient Greece.

In philosophical reflections on mathematics one can distinguish two principal groups of problems: ontological and epistemological. Among main questions of the first group are the following ones: what is the subject of mathematics, in particular what is the nature of mathematical objects, where and how do they exist, what are the criteria of their existence, what is the source and origin of mathematical objects, what is the nature and properties of the mathematical infinity.

Epistemological problems concerning mathematics (which are the main subject of the present article) can be divided into four groups:

– the problem of cognitive methods used and accepted in mathematics. In particular one considers here the problem of sources and origin of mathematical knowledge, the problem of the process of arriving at new results, the problem of methods of justifying mathematical statements and theorems, the problem of the validity of such methods, the problem of the nature of mathematical proofs, of criteria of distinguishing correct and incorrect proofs, of the justification of the axiomatic-deductive method and of the range of its applicability as well as the problem of the status of the axioms and of their origin, problems of decidability and the question whether there are any (and what) limits or bounds of mathematical knowledge. Here belong also the problems whether deduction is the only legitimate method in mathematics or should it be combined with induction and generally with empirical methods? Or is the method of proofs and refutations the proper method of establishing new results? What is the rôle of intuition in mathematical knowledge? Should nonconstructive methods be allowed in mathematics or one should restrict mathematics to constructive methods only?

– the problem of the type of mathematical knowledge. One asks here in particular whether mathematical knowledge is a priori or an empirical knowledge, whether mathematical theorems are analytic or synthetic statements, what is the value of mathematical statements, does a mathematician discover or create mathematical reality and its properties, and consequently mathematical knowledge. If one has to do with discovering in mathematics then which methods can be used here, and similarly in the case of the alternative answer. One also considers here

the problem of the relations between pure and applied mathematics, in particular the fundamental question why abstract mathematical theorems can be applied to the description of physical phenomena of the external world.
– the problem of a systematization of mathematical knowledge, and in particular the problem of the unification of mathematics,
– the problem of the dynamics and the development of mathematics as well as the problem of the place and the rôle of mathematics in the whole (system) of culture and especially in relation to other domains of human scientific knowledge.

Note that this list of problems and questions is not complete and particular items of it overlap each other.

Both aspects of the philosophical reflection on mathematical knowledge distinguished above – ontological and epistemological – are interconnected. Answers to some questions induce and imply (or even force) solutions to other problems. Nevertheless – to accord with the subject of this article – we shall concentrate here on epistemological problems (being conscious the whole time of the fact that one cannot escape some ontological solutions and decisions).

Philosophy of mathematics and in particular the epistemology of mathematics are of course connected with other branches of philosophy and with mathematics itself. The development of mathematics, the development and changes of the subject of study and of methods of mathematics lead to the development of the philosophical reflection on mathematics and to the change and revision of previous doctrines. As an example, perhaps the most striking one, can serve the introduction and development of the non-Euclidean geometries in the nineteenth century. On the other hand new mathematical problems induce new philosophical questions and problems (for example the recent use of computers in proving mathematical theorems). Hence the importance of the history of mathematics to the philosophy of mathematics (one can even say, paraphrasing I. Kant, that the history of mathematics without philosophy is blind and the philosophy of mathematics without history is empty). Also the development of logic, especially of the mathematical logic at the turn of the nineteenth century, as well as of the mathematical studies of mathematics as a science (metamathematics) played a great rôle for the epistemology of mathematics making possible the precise formulation of various problems and notions (such as proof, truth, consistency) as well as their solution (indicating for example some limitations and bounds of the axiomatic-deductive method) – cf. Gödel's incompleteness theorems, Löwenheim–Skolem theorems or Tarski's theorem on the undefinability of truth).

Philosophy (and in particular the epistemology) of mathematics plays a double rôle with respect to mathematics: on the one hand it describes and codifies the methods actually used in mathematics (one should distinguish here of course between the context of a discovery and the context of justification) and on the other it plays a normative rôle establishing and justifying the legitimate and correct methods of mathematics.

From various possible ways of presenting the main doctrines in the epistemology of mathematics we have chosen the historical one, because, as L. Kołakowski wrote, "All that is really important in the philosophy, is being discovered by learning its history; great philosophers sensibilize us to the plurality of perspectives from which the world can be considered as well as to the plurality of languages in which it can be described".[1] Hence the essay is organized in the following way: At the beginning the predecessors of the contemporary doctrines are presented. Next the main modern conceptions in the epistemology of mathematics (connected with logicism, intuitionism and formalism) are considered. Finally recent trends in the philosophy of mathematical knowledge are described.

1. Predecessors of the Contemporary Doctrines

The real reflection on mathematics as a science began by Plato (427–347 B.C.). His philosophy of mathematics grew out of his theory of ideas. He claimed that the subject of mathematics are mathematical (arithmetical and geometrical) ideas (or forms).[2] They are real entities conceived as being independent of perception and being apprehended, as being capable of absolutely precise definition and as being absolutely permanent, that is to say timeless or eternal. Hence a mathematician does not create mathematical objects and their properties but does discover them. Consequently the mathematical knowledge is based on the reason and the proper method of mathematics is the axiomatic method – Plato was probably the first who introduced it. Mathematics is very close to Plato's ideal of knowledge because it abstracts from changeable phenomena and concentrates on unchangeable, timeless, mind-independent and definite objects and relations between them. Plato admitted that a mathematician uses in his research practice observations and drawings or performing constructions but they serve only the process of remembering the proper mathematical objects (ideas) and not the creation of them (Plato refers here to his theory of *anamnezis*). Hence mathematics is a science whose aim is the description of timeless, mind-independent and definite mathematical objects (ideas) and their mutual relations. Consequently all mathematical propositions are necessarily true. Their necessity is independent of their being apprehended by a mathematician, independent of any formulation and thus of any rules governing a natural or artificial language. Mathematical theorems can be applied to the description of the objects of sense-experience because

---

1   Cf. L. Kołakowski, Zawód błazna jest mi bliższy. Z Leszkiem Kołakowskim rozmawia (korespondencyjnie) Paweł Śpiewak, *Res Publica* 9 (1988), 30.
2   According to Aristotle, Plato distinguished between the arithmetical and geometrical ideas (forms) and the so-called mathematicals, each of which is an instance of some unique form – each form having many such instances.

the latter are to a certain degree similar to, or better, approximate the ideas (Plato says here that, for example, one apple participates in the arithmetical idea One).

Aristotle (384–322 B.C.), the disciple of Plato, developed his philosophy of mathematics partly in opposition to that of Plato and partly independently of it. He rejected Plato's theory of ideas claiming that mathematical objects are forms of things, are idealizations obtained by the process of abstraction. Hence they do not exist timelessly and independently of things but are in a sense in things. Consequently mathematical propositions as being only idealizations cannot be necessarily true. The necessity cannot be found in any single statement about mathematical objects but in hypothetical statements saying that if a certain proposition is true then a certain other proposition is also true. Hence using today's terminology we can say that for Aristotle the necessity of mathematics was that of logically necessary hypothetical propositions. Aristotle paid much more attention to the structure of whole theories in mathematics than to isolated propositions (cf. *Physics* II, 9, 200a, 15–19; *Metaphysics* 1051a, 24–26). According to him the base of any knowledge is formed by general notions which do not need to be defined and by general propositions which do not need to be proved. All other notions should be defined and all other statements should be proved. He distinguished in any theory four basic components (cf. *The Posterior Analytics* I, 10, 76a, 44–77a, 3): (1) the principles which are common to all sciences (Aristotle called them axioms, they correspond to logical axioms and axioms of identity in today's terminology), (2) the specific principles which are taken for granted by the mathematician engaged in the demonstration of theorems (Aristotle called them postulates, they correspond to non-logical axioms in the terminology of today's formal logic), (3) definitions (add that Aristotle did not assume that what is defined exists) and (4) existential hypotheses assuming that what has been defined exists independently of our perception and thought (note that according to Aristotle such hypotheses seem not to be required for pure mathematics). It is worth adding that Aristotle saw in mathematics also some aesthetic elements, even more, he claimed that they play an important rôle in the development of mathematical knowledge. In fact mathematics says, though not explicitly, about the beauty and reveals some of its elements (cf. *Metaphysics* 1078a, 52–1078b, 4). Note that similar ideas can be found also by Proclus, a neoplatonic philosopher living in the 5th century, or by Henri Poincaré, French mathematician and philosopher living in the nineteenth century.

Plato's philosophy of mathematics and Aristotle's ideas concerning the structure of a scientific theory and in particular of a mathematical theory found their deepest application and realization in *Elements* by Euclid (365 (?)–300 (?) B.C.). In fact this work established a paradigm in mathematics prevailing up until the end of the nineteenth century called today Euclidean paradigm.

The *Elements* was on the one hand the presentation of results obtained by Greek mathematicians in the last 300 years before Euclid and on the other it gave a firm basis for the future development of mathematics. They consisted of 13 books: books I–IV were devoted to the plane geometry, book V to Eudoxus' theory of proportions in its purely geometrical form, book VI to the similarity of plane figures, books VII–IX to arithmetic (the ancient number theory), book X to incommensurable magnitudes and books XI–XIII to solid geometry. Every book began by definitions of new notions and by a list of axioms and postulates (one can see here the influence of Aristotle). Note that the *Elements* contained no list of primitive undefined terms, but, on the contrary, Euclid attempted to define all the terms he used (eventually those "definitions" were rather explanations of notions than proper definitions in the strict sense). It was possible because he, as Aristotle, did not distinguish between the language of the considered theory and the colloquial language – in fact the language of a theory was not separated from the natural language.

The postulates, axioms and definitions supplied the starting point for Euclid's proofs. His aim was to prove all principles by showing that they follow necessarily from the basic assumptions. In this way he wanted on the one hand to strengthen the mathematical knowledge by increasing the rigor with which already known laws could be proved and on the other to extend this knowledge by proving new and hitherto unknown laws. He wanted to organize mathematics (first of all geometry) in a systematic deductive form. The exact analysis of Euclid's proofs however indicates that there were certain gaps in them.

Nevertheless the *Elements* established a pattern of a scientific theory and in particular a paradigm in mathematics. Since Euclid till the end of the nineteenth century mathematics was developed as an axiomatic (in fact rather a quasi-axiomatic) theory based on axioms and postulates. Proofs of theorems contained several gaps – in fact the lists of axioms and postulates were not complete, one freely used in proofs various "obvious" truths or referred to the intuition. Consequently proofs were only partially based on axioms and postulates. Almost no attention was paid to the precization and specification of the language of theories – in fact the language of the theories was simply the unprecise colloquial language.

Add that the Euclid's approach (connected with Platonic idealism) to the problem of the development of mathematics and the justification of its statements (which found its fulfilment in the Euclidean paradigm), i.e. justification by deduction (by proofs) from explicitly stated axioms and postulates, was not the only approach and method which was used in the ancient Greek (and later). The other one (call it heuristic) was connected with Democritean materialism. It was applied for example by Archimedes who used not only deduction but any methods, such as intuition or even experiments (not only mental ones), to solve problems. Though the Euclidean approach won and dominated in the history one should note that it formed rather an ideal and not the real scientific practice of mathematicians. In fact rigorous, deductive mathematics was a rather rare phenomenon. On the contrary, intuition and heuristic reasoning were the animating forces of mathematical research practice. The vigorous but rarely rigorous mathematical activity produced "crises" (for example the pythagoreans' discovery of the incommensurability of the diagonal and side of a square, Leibniz's and Newton's problems with the explanation of the nature of infinitesimals, Fourier's "proof" that any function is representable in a Fourier series, antinomies connected with Cantor's imprecise and intuitive notion of a set).

There was a significant problem connected with the axiomatic-deductive method, namely the problem of the choice of postulates and axioms. For Plato they were simply necessary truth. Aristotle spoke as though he felt that every science had its own definite principles (which should function as postulates) and its own definite primitive terms (for every definite term there was just one correct way of defining it). Euclid expressed no opinion on such questions. Proclus (410–485), the neoplatonic author of *Commentary to the First Book of Euclid's "Elements"* claimed that the common feature of axioms and postulates is the fact that they need no justification or proof, they can be accepted as known. The difference between them is similar to that between theorems and problems. Axioms contain facts which are immediately obvious and do not make any trouble for our thought; in postulates on the other hand one tries to find facts and properties which can be easily established and by which no sophisticated procedures or constructions are needed. He accepted Plato's ideas of the origin of mathematical notions and said that they have their source in the soul which contains their patterns. This induces also the method of mathematics – in fact the proper method of mathematics is not intuition but the discursive method consisting of deduction from the premises.

It is worth noting here (anticipating the development of the events in the epistemology of mathematics and in the mathematics itself) that in fact till the end of the nineteenth century mathematicians were convinced that axioms and postulates should be simply true statements, hence sentences describing the real state of affairs in the mathematical reality. Only the development of non-Euclidean geometries in the nineteenth century called the attention to the possibility that this is not necessary, that one can develop theories based on any consistent set of axioms. But the way to the full consciousness of this was long and not direct.

Middle Ages did not bring new important ideas into the philosophy of mathematics. The views and theses of Plato, Aristotle and Euclid were developed and commented and mathematics was developed along the lines established by the *Elements*. Only in the seventeenth century some new ideas appeared. The intensive development of natural sciences and of mathematics brought new problems which should be solved. One looked also for some principles which would unify the whole edifice of human scientific knowledge.

As a founder of modern philosophy one usually considers René Descartes (1596–1650). He was the first philosopher whose outlook was profoundly affected by the new physics and astronomy. He was a philosopher and a mathematician. As a mathematician he is known first of all as the inventor of the analytic (coordinate) geometry (though not quite in its final form) – it was in fact based on the application of algebra to geometry and contributed very much to the unification of those two branches of mathematics which since antiquity were developed as two separate parts of mathematics. Descartes contributed also significantly to the methodology of mathematics. To explain it one should start from his principle claiming that all things that we conceive very clearly and very distinctly are true. Hence the criterion of certainty in science is based on clear and distinct ideas. Descartes proclaimed a program of universal rational knowledge build along the principles similar to those of mathematics. According to him only mathematicians are constructing proofs and therefore only mathematics provides an unfailing and secure knowledge. It has its sources in the fact that only quantitative properties are considered in mathematics. Hence Descartes' idea of bounding every scientific theory to such considerations and his idea of creating a universal analytical and mathematical theory called by him *mathesis universalis*. In mathematics itself – being the pattern of any other science – only analytic methods should be applied. Descartes allowed in it only intuition and deduction. Axioms of mathematics were for him infallible and indubitable truths.

The analytic method should enable us to discover the simple components of thoughts. And this what was simple, was for Descartes clear and distinct, hence certain. In *Discours de la méthode* (1637) he enumerates some rules which suffice in every scientific theory. According to them one should not accept any statement which is not clear and distinct, one should apply the analytic method and "decompose" any problem into so many components that are enough to find a solution and finally to deduce more complex truths from simple truths, i.e., from axioms.

Blaise Pascal (1623–1662), French philosopher and mathematician one generation younger than Descartes, was not so "bewitched" by the power of reason. He distinguished two parts, two realms: the realm of the reason and that of a heart. Reason cannot help us to solve existential problems. Descartes' clear and distinct ideas are of no help here. He wrote: "*Le coeur a ses raisons, que la raison ne connait pas*". There is of course a question what should be understood under "le coeur"? Various interpretations have been provided. One of them identifies it with the human ability to know the supernatural things, other one with the intelectual intuition.

In the realm of reason mathematics, and in particular geometry, was – according to Pascal – the pattern and ideal. Geometry is the only domain of human knowledge which provides infallible and indubitable proofs. The new ideal scientific method based on geometry should be founded on two rules: (1) one should not use any term whose meaning has been not exactly explained, and (2) all statements should be proved. Pascal was of course conscious of the fact that in scientific practice one cannot define all terms and prove all statements. Hence he allowed to accept without definition some terms which are clear by the "natural light" and to accept some clear initial principles (axioms) on which proofs can be based. Among such clear primitive terms are the notions of space, time, movement, number or equality. They are clear because the very nature gave us the understanding of them. Similarly for the case of axioms – they are clear by *le coeur*. But the deduction of theorems from the axioms proceeds in the realm of reason and according to its rules. Both are certain and secure though they take place on two different levels.

The common feature of Descartes' and Pascal's ideas was the conviction of the universal character of mathematics as a pattern of a scientific theory – just mathematics was for them an ideal of human knowledge and its methods could (and should) be applied in all domains. The reason for that was the fact that only mathematics and its methods can lead to a secure and infallible knowledge (only mathematics can give a real justification of its statements and claims). On the other hand they proposed several conditions which should be fulfilled to obtain a valuable theory.

Descartes' idea of constructing a universal science based on the patterns of mathematics and giving a frame for any scientific knowledge was further developed by Gottfried Wilhelm Leibniz (1646–1716). He proclaimed the idea of a universal logical calculus. The latter was connected with his idea of treating logic in a mathematical way,. i.e., by methods characteristic for mathematics. Leibniz attempted first of all to design a universal symbolic language, *characteristica universalis*. It was supposed to be a system of signs fulfilling the following conditions: (i) there is to be a one-one correspondence between the signs of the system (provided they are not signs of empty places for variables) and ideas or concepts (in the broadest sense), ii) the signs must be chosen in such a way that if an idea (thought) can be decomposed into components then the sign for this idea will have a parallel decomposition, (iii) one must devise a system of rules for operating on the signs such that if an idea N is a logical consequence of an idea M, then the 'picture' of N can be interpreted as a consequence of the 'picture' of M (this is a sort of completeness condition).

According to these conditions all simple concepts corresponding to simple properties ought to be expressed by single graphical signs, complex concepts by combinations of signs. This was based on a fundamental general assumption that the whole possible vocabulary of science can be obtained by combinations of some simple concepts. The method of constructing concepts was called by Leibniz *ars combinatoria*. It was a part of a more general method – a calculus – which should enable people to solve all problems in a universal language. It was called *mathesis universalis*, *calculus universalis*, *logica mathematica*, *logistica*. Leibniz hoped that *characteristica universalis* would, in particular, help to decide any philosophical problem.[3] He claimed that "he owed all his discoveries in mathematics exclusively to his perfect way of applying symbols, and the invention of the differential calculus was just an example of it" (cf. L. Couturat 1901, pp. 84–85). Note that Leibniz's idea was in fact an idea of a universal logic (based on mathematics) and the idea of mechanization of reasonings (which should be reduced to manipulations of symbols according to certain rules referring only to the form and not to the contents of statements written in the appropriate symbolic language). He proposed using mechanical calculations in aid of deductive reasoning (which meant in particular the introduction of calculations into logic).

Leibniz did not succeed in realizing his idea of *characteristica universalis*. One of the reasons was that he treated logical forms intensionally rather than extensionally. This could not be reconciled with the attempt to formalize logic completely and transform it into a universal mathematics of utterly unqualified generality. Another source of difficulty was his conviction that the combination of symbols must be a necessary result of a detailed analysis of the whole of human knowledge. Hence he did not treat the choice of primitive fundamental notions as a matter of convention. His general metaphysical conceptions induced a tendency to search for absolutely simple and primitive concepts (an analogue of monads), the combinations of which would lead to the rich variety of notions. As a partial realization of Leibniz's idea of *characteristica universalis* one can treat mathematical logic developed on the turn of the nineteenth century (see below).

The idea of a universal logical calculus which should form a framework for any valid reasoning in mathematics and generally in any scientific theory was not his only contribution to the epistemology of mathematics. The other one was his distinction between truths of reasoning and truths of fact on the one hand and the distinction between primitive and derived truths on the other. Primitive truths are truths known by intuition, they do not need any justification because they are clear by themselves and cannot be deduced from anything simpler and more sure. Derived truths are truths which can be reduced to primitive ones; they form the demonstrative knowledge. Truths of reasoning and truths of fact were characterized by Leibniz in the following way: "Truths of reasoning are necessary and their opposite is impossible: truths of fact are contingent and their opposite is possible. When a truth is necessary, its reason can be found by analysis, resolving it into more simple ideas and truths, until we come to those which have primacy ..." (cf. *Monadology*). Truths of reason are grounded in the 'principle of contradiction' (which Leibniz took as covering the principle of identity and of the excluded middle). Facts can neither justify nor refute them. They are not based on facts and do not concern facts – they concern only the possibility. Hence

---

3   He wrote: "And when this comes [i.e., when the idea of universal language is realized – R.M.] then two philosophers wanting to decide something will proceed as two calculators do. It will be enough for them to take pencils, go to their tablets and say: *Calculemus!* (Let us calculate!)" (cf. G.W. Leibniz 1975–1890, pp. 198–201).

they are true not only in the actual world but in all possible worlds, similarly as the logical laws. They do not say about any specific type of objects. Truths of fact are based on facts which can either justify or reject them. They are true only in the actual world.

Leibniz claimed that not only trivial tautologies but all the axioms, postulates, definitions and theorems of mathematics are truths of reason. This means that they are necessary and eternal, they are not based on facts or experience and are true in all possible worlds.

Leibniz's distinction between truths of reason and truths of fact found its development and in a sense a fulfilment by Immanuel Kant (1724–1804). All later analyzes of the problem referred to him.

Starting from the assumption that the general form of a judgement is "A is B" Kant distinguished in *Kritik der reinen Vernunft* between analytic and synthetic judgements. A judgement is analytic if and only if nothing but reflection upon the concepts in the judgement and upon the form of combination of these concepts is needed to enable us to know whether the judgement is true. A judgement is synthetic if and only if mere reflection upon the concepts in it and upon their combination is not sufficient to enable us to know whether it is true; to know that appeal to something further (experience or intuition or both is required). Consequently analytic judgements are uninformative, they tell us nothing we did not already have to know just to understand them. Observe that if one assumes additionally that analyticity of a judgement follows from the fact that its subject is included in its predicate then by definition all analytic judgements are true. On the other hand synthetic judgements can be either true or false, they are informative. Note that Kant's terminology was a bit psychologistic – but one can eliminate this difficulty and rephrase his definitions by replacing the term "statement" by "judgement". Kant claimed that all laws of formal logic are analytic but there are also other analytic statements.

Kant as the first linked up the analyticity and syntheticity with the property of being a priori or a posteriori. Roughly speaking one can say that a statement is a posteriori if it is empirical, i.e., based on experience and requiring justification from experience, and it is a priori if it is attainable prior to experience and its justification does not need any reference to experience. Kant formulated the following famous problem: how are synthetic a priori judgements possible? He claimed that there are such statements in our knowledge, for example theorems of arithmetic or geometry. His answer to this fundamental question was revolutionary for the epistemology. He claimed that synthetic a priori statements are possible because there are a priori components in our knowledge. In fact space and time as forms of our intuition (*Anschauung*) and categories as forms of reason[4] are such

---

4   Kant distinguished twelve categories divided into four sets of three: (1) of quantity: unity, plurality, totality; (2) of quality: reality, negation, limitation; (3) of relation: substance-and--accident, cause-and-effect, reciprocity; (4) of modality: possibility, existence, necessity.

a priori elements. They are necessary conditions of any knowledge. In particular the a priori intuition of time is a basis for arithmetic and the intuition of space – for geometry.

Kant divided synthetic a priori statements into two classes: intuitive and discursive. The former are connected with the structure of perception and perceptual judgements, the latter with the ordering function of general notions. An example of a discursive synthetic a priori proposition is the principle of causality. Kant claimed (contrary to Leibniz!) that all propositions of pure mathematics belong to the intuitive class of synthetic a priori statements. They cannot be empirical because they are necessary. On the other hand they are synthetic because they are about the structure of space and time as revealed by what can be constructed in them. They are a priori because pure space and time are a priori conditions of any perception of physical objects. On the other hand propositions of applied mathematics are either synthetic a posteriori (if they concern the empirical contents of sense perceptions) or synthetic a priori (if they concern the structure of space and time). Pure mathematics considers the structure of space and time free from empirical material and applied mathematics has for its subject matter the structure of space and time together with the material filling them.

It is necessary to stress here the distinction which Kant made between the thought of a mathematical concept and its construction. The former requires merely internal consistency while the latter requires that perceptual space should have a certain structure. Consequently one can postulate the existence of, for example, 5-dimensional sphere but one cannot construct it. Hence Kant did not deny the possibility of consistent geometries other than Euclidean one. So the development of non-Euclidean geometries in the nineteenth century did not refute Kant's philosophy of mathematics.

There is still one point in Kant's philosophical ideas concerning mathematics which should be indicated here. It is his theory of the actual infinity. Following Aristotle he distinguished between potential infinity and actual infinity. But he did not claim, as Aristotle did, that the latter is logically impossible. His idea was to treat it as an Idea of reason, i.e., as an internally consistent notion which is however inapplicable to sense experience since instances of it can be neither perceived nor constructed. On the other hand it is needed in mathematics. This approach to the actual infinity will be later used by David Hilbert in his formalistic program (see below).

Kant's aprioristic thesis concerning mathematics has been criticized in the nineteenth century by empiricists. One of them was John Stuart Mill (1806–1873). He developed the methodological version of the empiricism and attempted to justify it using logic. He applied it also to mathematics and attempted to argue that mathematics is in fact an empirical science. In particular he claimed that the source of mathematics is the reality perceived by senses. Mathematical concepts have been simply abstracted from the reality by omitting some properties of real objects and by generalizing and idealizing other properties. Hence mathematical propositions are not necessarily true. Their necessity can be reduced only to the fact that they are logical consequences of assumptions. But the very assumptions are far from being necessary and certain, on the contrary, they are only hypotheses and in fact one can adopt any propositions as assumptions. Hence in mathematics the necessity can be attributed only to logical connections between propositions and not to very propositions themselves. Consequently mathematical propositions are necessary truths only in such a degree as the axioms are. But the latter can be any sentences, even more, in practice axioms are usually false statements because they do not describe the real world but are only idealizations and generalizations of its properties. One can easily see here a similarity of Mill's views and the views of Aristotle (though the latter did not claim that axioms of mathematics can be any sentences).

2. Modern Doctrines in the Epistemology of Mathematics

Modern philosophy of mathematics was dominated by three schools: logicism, intuitionism and formalism. They emerged in the last quarter of the nineteenth century and the first thirty years of the twentieth century. They referred of course to earlier doctrines, e.g., to Plato, Aristotle, Leibniz and Kant. Their origin just in the period between 1870 and 1930 was connected with the origin and the intensive development of mathematical logic on the one hand and with the crisis in the foundations of mathematics at the end of the nineteenth century on the other.[5] It was connected with the discovery of antinomies in Cantor's set theory, i.e., with the discovery of pairs of mutually inconsistent propositions each of which could be justified with the same degree of certainty. They are called today logical antinomies (to distinguish them from semantical antinomies in which the concepts of meaning and reference are involved). Their source was the imprecise notion of a set used by Cantor. Examples of those antinomies are Burali-Forti antinomy of the greatest ordinal (it was known to Cantor), Cantor's antinomy of

---

5   It is usually called the second crisis. As the first crisis one means the discovery of the incommensurable magnitudes by the Pythagoreans in the ancient Greece. This led to the change of the notion of a number and to replacing arithmetic by a geometrical algebra. Sometimes one adds here also the seventeenth century crisis connected with the development of the differential and integral calculus by W. G. Leibniz and I. Newton (then the crisis of the nineteenth century is called the third crisis). In fact basic notions of this calculus such as, e.g., the notion of a differential (an infinitely small magnitude) in the form introduced by Leibniz, were simply inconsistent. Nevertheless the calculus has been successfully developed and applied. Only in the twentieth century the consistent basis for the Leibniz's calculus has been developed by the nonstandard analysis of A. Robinson. This example indicates that mathematical theories can be (and have been) often successfully developed and applied without a satisfactory consistent basis and that such a basis has been provided many years later.

the set of all sets and Russell's antinomy of the irreflexive classes (called today simply Russell's antinomy). Attempts to overcome difficulties revealed by the antinomies stimulated the researches also on the philosophical level and led to the development of logicism, intuitionism and formalism. This would not be possible without the modern mathematical logic developed at the end of the nineteenth century (this is true especially in the case of logicism and formalism).

Mathematical logic was a partial realization of Leibniz's idea of *characteristica universalis*. Its development was connected with the process of mathematization of logic and with the extention of Aristotelian logic (which was in fact restricted to the logic of names). Among the pioneers of the modern logic one should mention August de Morgan, George Boole, Charles Sanders Peirce and Ernst Schröder. They developed the so called algebra of logic. Beside it there has been also developed the non-algebraic form of logic by Gottlob Frege and Bertrand Russell. Both trends provided the necessary technical background which enabled the development of logicism and formalism. In particular Frege's fundamental logical work *Begriffsschrift, eine der arithmetischen nachgebildete Formelsprache des reinen Denkens* (1879) opened the new period in the history of formal logic (though its reception by the contemporaries, especially by the representatives of the algebraic trend, for example by E. Schröder or J. Venn, was not enthusiastic, on the contrary, Frege's work was not accepted and quickly forgotten and the reasons for that concerned not only the merits – for example Frege used a very complicated and "hard" symbolism – but also of personal nature). *Begriffsschrift* contained the (first in the history of logic) formal axiomatic system, more exactly the formal system of propositional calculus with implication and negation as the only connectives, and provided for the first time full analysis of the notion of a quantifier by giving a suitable system of axioms for them. Frege's formal system of propositional calculus was a system in which the deduction of theorems from axioms was complete and without gaps, i.e., rules of inference and the very notion of a formal proof have been precisely defined at the very beginning. In this way Frege's work contributed in an essential manner to the clarification and making precise of the basic notion of the epistemology of mathematics, i.e., of the notion of proof. Recall that since Plato, Aristotle and Euclid the axiomatic-deductive method was treated as the best method of mathematics, but the very notion of a proof was in fact understood rather intuitively and no precise definition of it was given. Having fixed and described in a precise way the rules of inference, Frege defined a proof as a sequence of sentences in a fixed precisely described formal language such that every element of this sequence is a result of applying one of the rules of inference to formulas appearing earlier in this sequence and the rules of inference could refer only to the form of formulas and not to their meaning or sense. In this way the intuitive notion of a consequence (being in fact of a rather psychological than of mathematical or logical character) has been replaced by a formal logical notion of a proof and consequence based on the notion of a rule of inference. This approach has been developed later (among others by Bertrand Russell, David Hilbert and Kurt Gödel) and contributed very much to the establishing of a new paradigm in mathematics, namely of the logico-set-theoretical paradigm.

Discussing the sources of modern doctrines in the philosophy of mathematics one should mention also the development of the non-Euclidean geometries in the thirties of the nineteenth century and the axiomatization of geometry at the end of the nineteenth century. Their impact can be seen first of all in the philosophy of geometry but they influenced the epistemology of mathematics in general as well. In particular they contributed to the discussion of the problem whether mathematical knowledge is an *a priori* or *a posteriori* knowledge, whether mathematics is an analytic or synthetic science. The fact that non-Euclidean geometries have been shown to be as logically consistent as Euclidean geometry shook the conviction that a mathematical theory should be based on true axioms, that geometry is a description of properties of the real space and that consequently the choice of primitive concepts and axioms is in a sense determined by the physical reality.

The idea of axiomatizing geometry in a complete way appeared simultaneously in works of several mathematicians. The most important were here the contributions of Moritz Pasch, Giuseppe Peano and the Italian school (G. Veronese, M. Pieri, F. Enriques) and David Hilbert. Pasch in his *Vorlesungen über neuere Geoemetrie* (1882) formulated the idea that if geometry is to be really a deductive science then the process of deducing theorems in it must be independent of any meaning of geometrical notions not determined by axioms. The final separation of geometry from the empirical reality was done in Hilbert's *Grundlagen der Geometrie* (1899). In this way geometry became pure mathematical theory. Axioms were not treated any longer as evident and necessary statements. The question about their truth lost its meaning and sense. As axioms any sentences could be adopted. The main problem was now the problem of consistency of given axioms. Geometrical deductive systems became uninterpreted axiomatic systems various interpretations of which are possible. In this way the traditional philosophical view which regarded geometrical knowledge as synthetic a priori knowledge of our world has been decisively refuted.

2.1. Logicism

The main thesis of logicism states that mathematics can be reduced to logic, i.e., mathematics is a part of logic and logic is an epistemic ground of all mathematics. This thesis can be formulated as the conjunction of the following three theses: (1) all mathematical concepts (in particular all primitive notions of mathematical theories) can be explicitly defined by purely logical notions, (2) all mathematical theorems can be deduced (by logical deduction) from logical axioms and definitions, (3) this deduction is based on a logic common for all mathematical theories, i.e., the justification of theorems in all mathematical theories refers to the same basic principles which form one logic common for the whole of mathematics (here is also included a thesis that any argumentation in mathematics should be formalized). Hence theorems of mathematics have uniquely determined contents and it is a logical contents. Moreover, mathematical theorems are analytic (similarly as logical theorems). Note that the founders of logicism did not state precisely what is the character of logical laws. Frege for example claimed only that they are not rules of nature but "rules of rules of nature" and that they are not laws of thinking but "laws of truth".

The genesis of logicism can be seen in the philosophical controversy between rationalism and empiricism concerning the character of mathematical propositions (judgements). Logicism was connected with Locke's and Leibniz's view that mathematical propositions have tautological character and with the Leibniz's view about the possibility of algorithmizing all mathematical reasonings (and generally all scientific reasonings).

Logicism can be treated as an extension of the nineteenth century tendency to the unification of the classical mathematics (in particular of analysis) by reducing it to the arithmetic. Pioneers of this trend were Karl Weierstrass and Richard Dedekind. It consisted in showing that the theory of real numbers which forms the foundations for analysis can be developed on the basis of the arithmetic of natural numbers, i.e., that the notion of a real number can be defined in terms of natural numbers and some concepts of set theory and that all properties of reals can be deduced from theorems of the arithmetic of natural numbers. This task has been in fact done by R. Dedekind in his paper *Stetigkeit und irrationale Zahlen* (1872).

In this situation the problem of reducing the arithmetic of natural numbers to a simpler theory arose. It was solved by Gottlob Frege (1848–1925), the founder of logicism, in two of his books: *Grundlagen der Arithmetik* (1884) and *Grundgesetze der Arithmetik* (vol. I – 1893, vol. II – 1903). Frege opposed there both empiricism claiming that arithmetical laws are simply inductive generalizations and formalism which treated them as rules of operating on symbols as well as kantianism which viewed them as synthetic a priori propositions. Frege claimed that arithmetic (and consequently the whole of mathematics) is a part of logic and that its theorems are analytic and *a fortiori* they are a priori.[6] On the other hand Frege disagreed with Kant who stated that mathematical theorems do not extend our knowledge, on the contrary, Frege claimed that they are in fact informative. Frege's definition of analyticity said that a sentence A is analytic if and

---

6   One should note here that Frege treated geometry in a different way than arithmetic. In fact he claimed that geometry is synthetic (and not analytic as arithmetic) because it says about one particular domain and that it is a priori because its axioms do not need to be proved.

only if *A* is provable on the basis of logic and definitions alone. He claimed that all arithmetical notions can be defined by logical concepts and all arithmetical theorems can be deduced from laws of logic and definitions. He showed in the indicated works how this can be done. He applied here the notion of the equipollence of sets introduced by Cantor. In fact Frege did not use sets (which are not logical notions) but was talking about concepts and their extensions and about the equipollence of extensions. This enabled him to remain on the ground of logic only. In modern terminology one would say that Frege first defined a cardinal number as a class of equipollent sets and then defined the natural numbers in terms of the successor function. Note that concepts were understood by Frege in an absolute, platonic way as existing independently of time, space and human knowledge. Mathematicians do not create them and their properties but are only discovering them.

Having defined the notion of a natural number in terms of logical notions Frege proved in *Grundgesetze* several properties of them. Unfortunately it turned out that the system of logic used by him was inconsistent. This was discovered by Bertrand Russell (1872–1970) in 1901 and in 1902 communicated to Frege. In fact Russell observed that one can construct in Frege's system an antinomy of irreflexive classes called today Russell's antinomy.[7] This antinomy was published and analysed by him in the book *The Principles of Mathematics* (1903) where he expressed (and defended) several views on the foundations of mathematics close to those of Frege.

Russell undertook the task of reducing mathematics to logic in the monumental work *Principia Mathematica* written together with Alfred North Whitehead (1861–1947) (vol. I – 1910, vol. II – 1912, vol. III – 1913). One finds there a new completely reconstructed system of logic called the ramified theory of types. It was based on the general assumption that the totality of properties which can be considered forms an infinite hierarchy of types: properties of the first type are properties of individuals, properties of the second type are properties of properties of the first type, etc. This hierarchy does not contain properties which could hold of objects of different levels. To avoid the vicious circle of impredicative definitions Russell and Whitehead introduced not only types but also orders of properties (orders depended on the form of a formula describing a considered object or property). By using those means they could eliminate the antinomy of irreflexive classes.

---

7   It is today formulated in the following way: Given a set X one can ask whether it is its own element or not. So consider a set Z of all such sets X that X is not its own element, i.e., $Z = \{X : X \notin X\}$. What are the properties of the set Z, in particular is Z its own element or not? If one answers YES, i.e., $Z \in Z$, then $Z \notin Z$ because in Z are only sets being not their own elements. On the contrary, if the answer is NO, i.e., $Z \notin Z$, then $Z \in Z$ because Z does not have the property of the elements of Z. Consequently $Z \in Z$ if and only if $Z \notin Z$ which is a contradiction.

Properties, called by Russell propositional functions, played in the theory of types the same rôle as concepts and their extensions by Frege. Hence Russell and Whitehead could simply adopt his definition of natural numbers. Some new problems appeared while proving properties of natural numbers. In particular it has turned out that to prove that for every natural number there exists a successor of it one needs an additional assumption being of a non-logical character, in fact the axiom of infinity stating that there exist infinitely many individuals is needed. Russell proposed a solution to this problem by suggesting to consider in the case of any such theorem not the theorem itself but rather an implication the antecedent of which is just the axiom of infinity and the succedent the considered theorem (by deduction theorem this implication can be proved in logic). In a similar way one can treat also other theorems whose proofs require additional assumptions such as for example the axiom of choice.

It has been shown in *Principia* how to reduce not only the arithmetic but the whole of mathematics to logic, i.e., to the ramified theory of types. It is worth noting here that – contrary to Frege – Russell and Whitehead treated concepts not in a platonic way but in a nominalistic way. Hence they did not postulate the independent existence of sets and all symbols for sets treated only as signs denoting nothing. Sets were reduced by them to propositional functions.

Russell claimed that the whole of mathematics is analytic.[8] His notion of analyticity can be given by the equivalence that a sentence *A* is analytic if and only if *A* is a tautology and a tautology can be characterized by three properties: (i) it is a priori, (ii) its negation is inconsistent and (iii) it is invariant with respect to logical constants (note the similarity of the latter to Bolzano's characterization of analyticity). On the other hand it is not clear whether Russell admitted that tautologies are "empty" or, on the contrary, they are informative – in fact he wrote in *Introduction to Mathematical Philosophy* that "logic ic concerned with the real world just as truly as zoology, though with its more abstract and general features" (1919, p. 169).

Russell's thesis about the analyticity of the whole of mathematics concerned also geometry (compare this with the views of Frege). In fact Russell distinguished two types of geometry: pure geometry in which one deduces consequences from the adopted axioms and does not ask whether they are true or not – this geometry is a priori and is not synthetic, hence it is analytic; and applied geometry being in fact a part of physics – it is an empirical science and is synthetic but not a priori.

The system of *Principia Mathematica* has been changed and modified later, in particular by L. Chwistek and F. P. Ramsey who introduced the so called simple theory of types. This was an extensional theory (in contrast with the original Whitehead and Russell's theory which was in fact intensional). The theory of

---

8   One should note here that Russell's views evolved. In particular before 1910 he claimed that logic and mathematics are synthetic. The thesis about the analyticity of mathematics has been proclaimed by him since 1910, i.e., since the publication of *Principia Mathematica*.

types has been accepted as the best system for the foundations of mathematics, it became a basis to which other researches referred (cf. for example Tarski's theory of satisfaction and truth (1933) or Gödel's work on the incompleteness of first-order systems (1931a)). It played this rôle till the fifties when its functions were taken up by the axiomatic set theory. The system of *Principia Mathematica* (and its later simplifications) was the first complete, consistent and natural system of logic. It was in a sense a synthesis of all earlier conceptions in the field of logic and the foundations of mathematics. It indicated the power and meaning of formal methods in logic and mathematics, in particular it showed that the formal principles of logic provide a sufficient tool for deduction of theorems from any given axioms.

Logicism played an important rôle in the development of the foundations of mathematics and of the philosophy of mathematics. Though various critical remarks concerning it (and in particular the theory of types) has been formulated one of its fundamental merits is to indicate the elegant way of a systematization of the mathematical knowledge and of making precise the intuitive notion of a mathematical proof. The logicists used here the results of the mathematical logic (and on the other hand contributed very much to its development). They underlined the universal and simultaneously fundamental character and rôle of logic in mathematics. Since it has turned that the reduction of mathematics to logic done by logicists was in fact the reduction to logic in a broader sense (several non purely logical assumptions were necessary in this reduction), the today's version of it says that the whole of mathematics is reducible to logic and set theory. There exists also another version of logicism, a methodological one, called if-thenism. It is based on the finitistic character of the operation of logical consequence and on the deduction theorem and claims that theorems proved in mathematical theories should be understood as implications whose antecedents are finite conjunctions of axioms; such implications are theses of logic.

Add at the end that logicism is not necessarily connected with the thesis about the analyticity of mathematical statements. In fact logicism claims only that there is a certain "homogeneity" of logic and mathematics with respect to the partition of propositions into synthetic and analytic. One can of course imagine a version of logicism claiming that logic and mathematics are both synthetic a priori or a posteriori.

2.2. Intuitionism

Intuitionism as a doctrine in the philosophy of mathematics has been founded by the Dutch mathematician Luitzen Egbertus Jan Brouwer (1881–1966). Intuitionists saw as their predecessors those philosophers and mathematicians who claimed that mathematics is a science possessing a definite contents and that the human mind is able to perceive directly mathematical objects and to formulate about them synthetic a priori judgements. Hence they willingly referred to I. Kant and to Paul Natorp. It seems that the ideas expressed later explicitly in works of Brouwer were in a sense in the air at the end of the nineteenth century.

For the first time the intuitionistic ideas appeared by German mathematician Leopold Kronecker (1823–1891) and his students in the seventies and eighties of the nineteenth century. In *Über den Zahlbegriff* (1887) Kronecker formulated a program of "arithmetization" of algebra and analysis, i.e., the program of founding those domains of mathematics on the most fundamental notion of a number.[9] He developed a unified theory of various types of numbers based on the primitive intuition of a natural number. His scientific credo has been summarized in the best way in his sentence: *"Die ganzen Zahlen hat der lieber Gott gemacht, alles andere ist Menschenwerk"*.[10] A consequence of this attitude was for example the fact that Kronecker admitted only those definitions of numbers which give a procedure of deciding whether a given number satisfies it or not. He accepted only "pure" existence proofs, i.e., proofs of existential theses giving constructions of the postulated objects.

Philosophical and methodological theses of Kronecker were a reaction to K. Weierstrass' attempts of applying methods of Cantor's set theory (in particular the theory of infinite sets) to the number theory and to the theory of functions. Similar were the sources of the ideas of the group of French mathematicians called today the Paris school of intuitionism or French semi-intuitionists: R. L. Baire, E. Borel, H. L. Lebesgue and the Russian mathematician N. N. Luzin. Their considerations on the foundations of mathematics were mainly connected with the study of the rôle of the axiom of choice. They did not create a compact philosophy doctrine but formulated several general remarks on the margin of their mathematical investigations in the theory of functions. Their common feature is certain constructivistic tendency. Many of their views were later adopted by Brouwer.

Discussing the problem of forerunners of Brouwer's intuitionism one must mention the French mathematician Henri Poincaré (1854–1912). His philosophical attitude can be characterized by saying that he was an apriorist, intuitionist and constructivist as well as a founder of the conventionalism. According to Poincaré the main rôle in the human mathematical knowledge is played by the creative activity of the mind and by its ability to construct concepts. This creative activity is manifested in various ways. One of its manifestation is intuition which appears both in the unconscious as well as conscious work. It has spontaneous and rational character, it gives the consciousness of clarity and evidence. It does not

---

9   The restriction of mathematics to algebra and analysis was the consequence of a thesis (Kronecker referred here to C. F. Gauss) that, for example, geometry or mechanics are independent of human mind because they refer to the external reality.
10 "The integer numbers were made by God, everything else is the work of man."

need the evidence of senses. There are various kinds of intuition: generalization by induction, the intuition of pure number, reference to senses or imagination, preexisting ability to construct the concept of a group as a pure, and not sensory, form of knowledge (this idea was especially important in his philosophy of geometry). Poincaré claimed that intuition should be supplemented by a discursive knowledge, i.e., it should be preceded and concluded by the conscious work in which the intuitive "revelations" are verified in a rational way.

One of the kinds of intuition is mathematical induction. It forms the base of number theory and of the whole of mathematics. According to Poincaré mathematical induction is a synthetic a priori judgement which extends our knowledge, it is the archetype of mathematical reasoning. Therefore theorems of mathematics (with the exception of geometry!) based on it are synthetic. He considered reasonings using induction as "the exact type of the a priori synthetic intuition" (cf. *La science et l'hypothèse*, English translation, p. 388). Induction is not empirical, it cannot be reduced to logic. Logic is tautological and cognitively empty, it enables us only to build analytic judgements, therefore mathematics cannot be reduced to logic only (as logicists maintain). Our trust in mathematical induction comes from the fact that "it is only the affirmation of the power of the mind which knows it can conceive of the indefinite repetition of the same act, when the act is once possible" (cf. *La science et l'hypothèse*, English translation p. 388).

Poincaré claimed that mathematical objects are being constructed by human mind, that there is no domain of mathematical knowledge which would be independent of the knowing subject. One of the consequences of this anti-realistic attitude was the rejection of the actual infinity and the restriction of mathematics to objects which can be defined by finitely many words only.

Discussing Poincaré's philosophy of mathematics one must mention also his conventionalism. It was manifested mainly in his philosophy of natural science but also in his philosophy of geometry. In the latter it contributed to the overcoming of the traditional views according to which geometry is the description of the real empirical space. Poincaré claimed that axioms and postulates of geometry are (contrary to statements of arithmetic!) neither synthetic a priori judgements nor empirical facts. They are conventions and implicit definitions. The choice between various possible conventions is based on experimental facts but we are free in this choice – the unique restriction is to avoid inconsistency. The question which geometry is a true geometry (this problem was especially important after the introduction and development of the non-Euclidean geometries) is, according to Poincaré a wrong question. He wrote: "One geometry cannot be more true than the other; it can be only more convenient" (*loc. cit.*). And the Euclidean geometry turns out to be the most convenient one. Poincaré used the idea of F. Klein to characterize geometries as theories of invariants of appropriate groups of transformations and the very concept of a group, more exactly the possibility of constructing a concept of a group as a pure form of knowledge preexisting in our mind (it was treated by him as a part of intuition). In this way geometry was not any longer "*la science de la verité*" but it became "*la science de la consequence*" (Poincaré).

Brouwer presented his philosophical views for the first time in his doctoral dissertation *Over de Grondslagen de Wiskunde* (1907).[11] The main aim of him (and of the intuitionism) was to avoid inconsistencies in mathematics. Brouwer proposed means to do that which turned out to be very radical and led in the effect to the deep reconstruction of the whole of mathematics.

The main fundamental thesis of Brouwer's intuitionism is the rejection of platonism in the philosophy of mathematics, i.e., of the thesis about the existence of mathematical objects which is independent of time, space and human mind. The proper ontological thesis is the conceptualism. According to Brouwer mathematics is a function of the human intellect and a free activity of the human mind, it is a creation of the mind and not a theory or a system of rules and theorems. Mathematical objects are mental constructions of an (idealized) mathematician. As a consequence one should reject the axiomatic-deductive method as a method of developing and founding mathematics. It is not sufficient to postulate only the existence of mathematical objects (as it is done in the axiomatic method) but one must first construct them. One must also reject the actual infinity. An infinite set can be understood only as a law or a rule of forming more and more of its elements, but they will never exist as forming an actual totality. Hence there are no uncountable sets and no cardinal numbers other than $\aleph_0$.

The conceptualistic thesis of intuitionism implies also the rejection of any nonconstructive proofs of the existential theses, i.e., of proofs giving no constructions of the postulated objects. In fact in intuitionism "to be" equals "to be constructed". Brouwer claimed that just nonconstructive proofs were the source of all troubles and mistakes in mathematics. Since such proofs are usually based on the law of the excluded middle ($p \vee \neg p$) as well as on the law of double negation ($p \equiv \neg\neg p$), intuitionists could not any longer use the classical logic. Moreover they claimed that logic is neither a basis for mathematics nor a starting point of it. Brouwer said that logic is based on mathematics, that it is secondary and dependent on mathematics and not vice versa, i.e., logic is a basis for mathematics as the logicists assert.

According to Brouwer and the intuitionism mathematics is based on the fundamental intuition of time. One sees here the connection with the philosophy of Kant. Indeed the intuitionism accepts Kant's thesis about the a priori time and rejects his thesis about the a priori space. The intuition of time is a basis for the mental construction of natural numbers. Moreover this construction is the basic mathematical activity from which all other mathematical activity springs. One of the consequences of this is that arithmetical statements are synthetic a priori judgements.

---

11 He developed them in his inaugural lecture at the University of Amsterdam *Intuitionisme en formalisme* (1912) and in the paper "Consciousness, Philosophy and Mathematics" (1949).

A mathematical theorem is a declaration that a certain mental construction has been completed. All mathematical constructions are independent of any language. Hence there is no language (formal or informal) which would be safe for mathematics and would protect it from inconsistencies. It is a mistake to analyze the language of mathematics instead of analyzing the mathematical thinking. Mathematics should be justified not "on paper" but "in the human mind". This intuitionists' thesis contradicts the thesis of the logicism and especially of the formalism about the meaning and importance of the formal reconstruction of mathematics and of the formal proof of the consistency of mathematical theories. It is also incomparable with views of all those philosophers and mathematicians since Plato who claimed that any abstract thinking is dependent of a language.

One of the consequences of Brouwer's theses described above and forming a base for the doctrine of intuitionism was the need for the reconstruction of the whole of mathematics according to the intuitionists' principles. Brouwer began to realize this task with his students in 1912. They reconstructed the concept of the continuum, the theory of point sets, theory of functions, theory of countable well orderings, the theory of complex functions, projective geometry, algebra, topology, measure theory, affine geometry and others. From the point of view of the epistemology of mathematics most important were the works of Brouwer's pupil Arend Heyting (1898–1980). Indeed he popularized the ideas of Brouwer and attempted to explain them in a language usually used in the reflection on mathematics – note that Brouwer expressed his ideas in a lnguage far from the standards accepted by mathematicians and logicians and therefore not always understandable. It seems that without those attempts of Heyting the ideas of Brouwer would soon disappear.

In particular Heyting constructed the first formalized system of the intuitionistic propositional calculus, i.e., of the propositional calculus satisfying principles of the intuitionism. This system has been never accepted by Brouwer. This sceptical attitude has its reasons in the thesis about the essential inexaustibility of the totality of mental processes which can be accepted as valid. Consequently there exists no formal system which would adequately represent the human mathematical activity. Indeed the latter is always dynamic and not closed while the former is static and closed. Hence no formal system can be a complete and adequate description of the intuitionistic mathematics. The latter is more poor than the classical mathematics and, what more, much more complicated and consequently not so useful in applications. On the other hand intuitionistic logic has turned out to be a very useful tool in various parts of mathematics, in particular in the topos theory or in the theoretical computer science.

2.3. Constructivism

Intuitionism is one of the constructivistic trends in the foundations and philosophy of mathematics. Constructivism is a common name for various doctrines the main thesis of which is the demand to restrict mathematics to the consideration of constructive objects and to constructive methods only. Hence constructivism is a normative attitude the aim of which is not to build appropriate foundations for and to justify the existing mathematics but rather to reconstruct the latter according to the accepted principles and to reject all the methods and results which do not fulfil them. The constructivistic tendencies appeared in the last quarter of the nineteenth century as a reaction against the intensive development of highly abstract methods and concepts in mathematics inspired by Cantor's set theory.

There are various constructivist programs and schools. They differ by their interpretation and understanding of the concept of constructivity. One of the most developed schools is intuitionism discussed above. Others are finitism, ultraintuitionism (called also ultrafinitism or actualism), predicativism, classical and constructive recursive mathematics, Bishop's constructivism. It is impossible to describe here all those doctrines in detail. Note that they are not based on one philosophical system, on the contrary, they accept different, not always compatible, philosophical, in particular ontological assumptions. In general one can distinguish four types of constructivism according to the accepted ontological basis: (1) objectivism claiming that objects of mathematics are objective results of constructive processes existing independently of the knowing subject which constructs them, (2) intentionalism which ascribes to mathematical objects being results of appropriate constructive processes the intentional existence (being characteristic for cultural entities), (3) mentalism claiming that objects of mathematics being products of mental acts exist only in those acts, (4) nominalism according to which mathematical objects are concrete and definite spatio-temporal objects.

Constructivistic tendencies in mathematics contributed very much (and are still contributing) to making precise various notions and ideas of mathematics. They are very important also for the computer science. On the other hand constructivistic mathematics is in fact much more poor than the classical one.

2.4. Formalism

The third main school in the philosophy of mathematics is formalism created by German mathematician David Hilbert (1862–1943). Hilbert was of the opinion that the attempts to justify and found mathematics undertaken hitherto, especially by the intuitionism, were unsatisfactory because they led to the restriction of mathematics and to the rejection of various parts of it, in particular those considering infinity. The aim of his program formulated for the first time in his famous lecture at the Second International Congress of Mathematicians in Paris in 1900 was to save the integrity of the existing classical mathematics (dealing with the actual infinity) by proving that it is secure. Among twenty three main problems which should be solved Hilbert mentioned there as Problem 2 the task of proving the consistency of axioms of arithmetic (under which he meant number theory and analysis) (cf. Hilbert 1901). He has been returning to the problem of justification of mathematics in his lectures and papers, especially in the twenties, where he proposed a method of solving it called today Hilbert's program. One should add that Hilbert saw the supra-mathematical significance of the whole issue writing that "the definite clarification of the nature of the infinite has become necessary, not merely for the special interests of the individual sciences but for the honor of human understanding itself" (cf. *Über das Unendliche*).

Hilbert's program of clarification and justification of mathematics was Kantian in character. One can see here a turn in the direction of idealism. In Kant's philosophy ideas of reason, or transcendental ideas, are concepts which transcend the possibility of experience but on the other hand are answer to a need in us to form our judgements into systems that are complete and unified. Therefore we form judgements concerning an external reality which are not uniquely determined by our cognition, judgements concerning things in themselves. To do that we need ideas of reason.

In likening the infinite to a Kantian idea of pure reason, Hilbert suggested that it is to be understood as a regulative rather than a descriptive device. Sentences concerning the infinite, and generally expressions which Hilbert called ideal propositions, should not be taken as sentences describing externally existing entities. We use ideas of reason and ideal elements in our thinking because they allow us to retain the patterns of classical logic in our reasonings.

Hilbert distinguished between the unproblematic, 'finitistic' part of mathematics and the 'infinitistic' part that needed justification. Finitistic mathematics deals with so called real propositions which are completely meaningful because they refer only to given concrete objects (add that real propositions play the rôle of Kant's judgements of the understanding (*Verstand*)). Infinitistic mathematics on the other hand deals with the so called ideal propositions that contain reference to infinite totalities (they play the rôle of Kant's ideas of pure reason). Ideal propositions play an auxiliary rôle in our thinking, they are used to extend our system of real judgements. Hilbert believed that every true finitary proposition had a finitary proof. Infinitistic objects and methods enabled us to give easier, shorter and more elegant proofs but every such proof could be replaced by a finitary one (this is the reflection of Kant's views of the relationship between the ideas of reason and the judgements of the understanding). Hilbert was also convinced that consistency implies existence and that every proof of existence not giving a construction of postulated objects is in fact a presage of such a construction.

Hilbert proposed to justify the infinitistic mathematics by finitistic methods because only they can give it security. He wanted to do it via proof theory (*Beweistheorie*). Its main goal was to show that proofs which use ideal elements in order to prove results in the real part of mathematics always yield correct results. One can distinguish here two aspects: consistency problem (prove by finitistic method that the infinitistic mathematics is consistent) and conservation problem (show by finitistic methods that any real sentence which can be proved in the infinitistic part of mathematics can be proved also in the finitistic part, even more, that there is a finitistic method of transforming infinitistic proofs of real sentences into finitistic ones).

Hilbert's proposal to carry out this program consisted in two steps. To be able to study seriously mathematics and mathematical proofs one should first of all define accurately the very concept of a proof. Hilbert used here the results of mathematical logic (G. Peano, G. Frege and B. Russell), in particular the idea of a formalized system in which a mathematical proof is reduced to a series of very simple and elementary steps, each of which consists in performing a purely formal transformation on the sentences which have been previously proved. Hence the first step proposed by Hilbert was to formalize mathematics, i.e., to reconstruct infinitistic mathematics as a big, elaborate formal system. The second step was to give a proof of the consistency and conservativeness of mathematics by considering formal proofs, i.e., strings of symbols of the appropriate artificial symbolic language. This was just the aim of the proof theory created by Hilbert (and called also metamathematics). It was a theory in which (formalized) mathematical theories and their properties are to be studied by mathematical methods.

One should note here that formalization was for Hilbert only an instrument used to prove the correctness of (infinitistic) mathematics. Hilbert did not treat mathematical theories as games on symbols or collections of formulas without any contents. Formalization was only a methodological tool in the process of studying the properties of the preexisting mathematical theories.[12] On the other hand Hilbert, contrary to the intuitionists, connected thinking with a language. He claimed that thinking, similarly to the process of speaking and writing, takes place by constructing and ordering sentences.

Hilbert represented a strongly anti-logicistic attitude. He maintained that mathematics cannot be deduced from logic alone, logic does not suffice to justify mathematics – hence the attempts of Frege and Russell were in his opinion fruitless.

Hilbert and his school had scored some successes in realization of the program of justifying infinitistic mathematics. In particular Wilhelm Ackermann showed by

---

12 Later various radical versions of formalism appeared, in particular the so called strict formalism of Haskel B. Curry (cf. Curry 1951). Mathematics was reduced in it to the study of formalized theories and nothing was assumed except the symbols constituting a given system.

finitistic methods the consistency of a fragment of arithmetic of natural numbers. But soon something was to happen that undermined Hilbert's program. In 1931 the Austrian mathematician Kurt Gödel (1906–1978) proved that arithmetic of natural numbers and all formal systems containing it are essentially incomplete provided they are consistent (and based on a recursive, i.e., effectively recognizable set of axioms).[13] He announced also a theorem stating that no such theory can prove its own consistency, i.e., to prove the consistency of a given theory T containing arithmetic one needs methods and assumptions stronger than those of the theory T. Hence in particular one cannot prove the consistency of an infinitistic theory by finitistic methods.[14]

Gödel's methods were used to indicate still another feature of axiomatic theories. The studies initiated by Alonzo Church and continued by others have shown that most theories interesting from the mathematical point of view are undecidable, i.e., there does not exist (and cannot exist) an effective method for deciding whether a given statement can be proved (justified) on the basis of a given system of axioms.[15]

Gödel's results indicated certain limitations of the axiomatic-deductive method considered since antiquity to be the best method for mathematics. They showed that one cannot include the whole of mathematics in a consistent formalized system based on the first order predicate calculus – what more, one cannot even include in such a system all truths about natural numbers. In this way it has been also shown that the concept of a formal proof which was supposed to be the precization of the imprecise notion of a mathematical proof is not adequate. In the research practice mathematicians are using any (correct) methods to solve problems and to answer questions. The scope of the admissible methods is not fixed or bounded beforehand. They are chosen according to the needs and problems that appear. On the other hand there is no precise definition of a correct method in mathematics. Therefore the hopes that the precise notion of a formal

---

13 The undecidable arithmetical sentence constructed by Gödel in his proof of the incompleteness theorem had a metamathematical contents rather than mathematical (it stated: "I am not a theorem"). Though interesting for logicians it was rather artificial from the mathematical point of view. Hence one could still cherish hopes that all sentences which are interesting and reasonable from the mathematical point of view (whatever it means) are decidable and that in the domain of such sentences the attempts to make precise the notion of a mathematical theory and a mathematical proof by using (first order) formal theories are successful. They were shuttered by results of J. Paris, L. Harrington and L. Kirby (1979–1982) indicating examples of undecidable sentences about natural numbers of the directly mathematical (in fact combinatorial or number-theoretic) contents (cf. Paris–Harrington 1977 and Kirby–Paris 1982; see also Murawski 1994).

14 It should be noted here that this is only a rough formulation of Gödel's theorem on the unprovability of consistency. One must take here into account also the way in which the metamathematical notion of consistency of a given theory has been formalized.

15 The ambiguous notion of being effective has been made precise by means of the theory of recursive (computable) functions.

proof based on (first order) logic (with fixed axioms and rules of inference) will provide such a definition. Gödel's incompleteness theorems indicated that this is not (and cannot be) the case. They revealed also the distinction between syntactical and semantical notions, in particular between provability and truth. Note that formalists considered formal provability to be an analysis of the concept of mathematical truth. Gödel showed that semantic truth cannot be adequately expressed by syntactical provability. In fact there is a "gap" between them, more exactly the notion of provability (for any first order formal theory) is definable in the language of arithmetic of natural numbers by a formula containing only one existential quantifier (such formulas a called Σ_1^0 formulas) while the notion of a true sentence of the arithmetic of natural numbers is not definable by an arithmetical formula (hence it is not arithmetic, indeed it is hyperarithmetic; similarly for other theories).

One can also prove that it is not only impossible to characterize mathematical structures, e.g., the structure of natural numbers, adequately by a (first order) formalized theory (because such theories are always incomplete), but that any description of a considered structure (such as the structure of natural numbers) by (first order) axioms is inadequate in the sense that the theory has a great variety of models, even models very different from the structure one wants to describe. Hence first order logic assumed to be the best tool in reconstructing mathematics is too weak. On the other hand higher order characterizations (e.g., by second order notions and second order logic) are not so regular and natural (from the logical, methodological and philosophical point of view).

Gödel's results struck Hilbert's program. Did they reject it? This question cannot be answered definitely for the simple reason – Hilbert's program was not formulated precisely enough, it used vague terms as finitistic, real, ideal which were never precisely defined. Both Hilbert and Gödel were ready after the incompleteness theorems to extend the scope of admissible methods by allowing some forms of infinitistic reasonings. Gödel doubted whether all correct proofs can be captured in a single formalized system.

The idea of extending the admissible methods and allowing general constructive methods instead of only finitistic ones was explicitly formulated by Paul Bernays. A motivation for this shift from the original Hilbert's program could be Gödel's reduction (found independently also by Gerhard Gentzen) of classical arithmetic to the intuitionistic arithmetic of Heyting and Gentzen's proof of the consistency of arithmetic by transfinite induction (which was apparently accepted by Hilbert). But there arises a problem: what is meant by constructivity? This concept is in general much less clear than that of finitism. Nevertheless the broadening of original Hilbert's proof theory postulated by Bernays has become an accepted paradigm (it is usually called the generalized Hilbert's program). Investigations were carried out in this direction and several results have been obtained.

Another consequence of Gödel's incompleteness results is the so called relativized Hilbert's program. If the entire infinitistic mathematics cannot be reduced to and justified by finitistic mathematics then one can ask for which part ot it is that possible. In other words: how much of infinitistic mathematics can be developed within formal systems which are conservative over finitistic mathematics with respect to real sentences? This question constitutes the relativized version of the program of Hilbert. Recently results of the so called reverse mathematics developed mainly by H. Friedman and S. G. Simpson contributed very much to this program.[16] In fact they showed that several interesting and significant parts of classical mathematics are finitistically reducible. This means that Hilbert's program can be partially realized.

2.5. Logico-Set-Theoretical Paradigm

Studies on the foundations of mathematics and on the philosophy of mathematics in the nineteenth century and in the first half of the twentieth century led to the establishing of the new paradigm of mathematics, called logico-set-theoretical paradigm. It replaced the Euclidean paradigm (described above) prevailing up until the end of the nineteenth century. Several events and achievements contributed to the establishing of the new paradigm. Among the most important are the origin and the development of set theory (G. Cantor), arithmetization of analysis (A. Cauchy and K. Weierstrass, R. Dedekind), axiomatization of the arithmetic of natural numbers (G. Peano), non-Euclidean geometries (N. I. Lobachevsky, J. Bolyai, C. F. Gauss), axiomatization of geometry (M. Pasch, D. Hilbert), the development of mathematical logic (G. Boole, A. De Morgan, G. Frege, B. Russell). Besides those "positive" factors there was also a "negative" factor, viz., the discovery of antinomies in the set theory (C. Burali-Forti, G. Cantor, B. Russell) and semantical antinomies (G. D. Berry, K. Grelling). They showed that the intuitive concept of a set is vague and a precise definition of it is needed. The latter was provided by axiomatizing set theory (E. Zermelo, T. Skolem). Semantical antinomies indicated the necessity of distinguishing between language and metalanguage.

The main features of the logico-set-theoretical paradigm can be characterized as follows: (1) set theory became the fundamental domain of mathematics, in particular some set-theoretical notions and methods are present in any mathe-

---

16 From the philosophical point of view reverse mathematics is a reductionist program. Its main aim is to study the rôle of the comprehension axiom (the axiom on the existence of sets) in the mathematics. In particular one considers there a problem which forms of the comprehension axiom are necessary and sufficient to prove various particular theorems of analysis, algebra, topology, etc. Detailed description of the results of the reverse mathematics and of their meaning for the philosophy of mathematics is given in (Murawski 1994) (one can also find there an extensive bibliography).

matical theory and set theory is the basis of mathematics in the sense that all mathematical notions can be defined by primitive notions of set theory and all theorems of mathematics can be deduced from axioms of set theory, (2) languages of mathematical theories are strictly separated from the natural language, they are artificial languages and the meaning of their terms is described exclusively by axioms; some primitive concepts are distinguished and all other notions are defined in terms of them according to precise rules of defining notions, (3) all mathematical theories have been axiomatized,[17] (4) there is a precise and strict distinction between a mathematical theory and its language on the one hand and metatheory and its metalanguage on the other (the distinction was explicitly made by A. Tarski), (5) two crucial concepts for mathematics, i.e., the concept of a consequence and the concept of a proof have been precisely defined.[18]

One should emphasize here the significant rôle played by the mathematical logic and the foundations of mathematics in the development of the philosophy of mathematics. This has been evident especially after 1930. Results of those domains contributed to the process of making precise various philosophical problems and explaining crucial methodological concepts (such as proof, truth, consistency, etc.) and indicated several important properties of axiomatic systems which implied the necessity of the revision of some ideas of the epistemology of mathematics. In particular one should mention here the precise definition of truth and model (Tarski) and the so called limitation theorems, i.e., Gödel's incompleteness theorems, Church's theorem on the undecidability, Tarski's theorem on the undefinability of truth and Skolem–Löwenheim's theorems on the cardinality of models and on nonstandard models (indicating the impossibility of a unique characterization of structures by first order axiomatic systems).

Studies on the foundations of geometry and arithmetic and especially the metamathematical studies on the set theory pointed out some problems connected with the axiomatization of mathematics. Gödel showed that all richer (i.e., containing arithmetic of natural numbers) theories are essentially incomplete. Axiomatization of geometry and the development of non-Euclidean geometries threw some light on the problem which system of geometry is true. On the other hand it has turned out that some interesting and important (for various branches of mathematics) hypotheses of set theory, i.e., the axiom of choice and the continuum hypothesis, are independent of other accepted axioms for sets (K. Gödel proved in 1939 that the axiom of choice and the continuum hypothesis are consistent

---

17 It does not mean that axioms of mathematical theories were fixed once and for ever. On the contrary, axiomatizations of theories are being developed. We mean here that in proving theorems one can use axioms and only axioms and it is not allowed to apply for example drawings or so called evident facts.
18 The concept of a consequence was defined by A. Tarski. To the process of formulating a precise definition of the concept of a proof contributed G. Frege, B. Russell, A. N. Whitehead, D. Hilbert, P. Bernays, W. Ackermann, S. Jaśkowski and G. Gentzen.

with the axioms of Zermelo–Fraenkel set theory and in 1963 P. Cohen proved that they are independent). Since set theory is a fundamental mathematical theory (in the sense explained above) this indicated that there is in fact no firm basis for mathematics fixed once and for ever, i.e., various set theories are possible (i.e., consistent) and can serve as a basis for mathematics. Which is the proper one? And what does it mean? Which axioms should and can be accepted in mathematics? What should decide of the acceptance or rejection of particular axioms? Which new axioms can be added to solve particular problems in mathematics, for example to solve the continuum hypothesis or the axiom of determinacy (which is inconsistent with the axiom of choice). What is the justification of axioms of large cardinals? Such problems were present in the philosophy of mathematics since its origins but now they are showed in a new light.

The indicated problems are especially pressing in the current set theory. Therefore we shall discuss them just on the example of this theory. One can distinguish three types of arguments used to justify the axioms: intrinsic, extrinsic and heuristic (those classes are not disjoint). The intrinsic arguments are based on the very notion of a set, they refer to the primitive intuition of a set (there is of course a problem what are the sources of this intuition). Extrinsic arguments are stated in terms of consequences or inter-theoretic connections. In particular they refer to the fact that a considered axiom (or theorem) has been confirmed in special particular cases, that it implies unknown results of the lower level, that it provides new proofs of old results, that it enables a unification of new and old results in such a way that the old results become special instances of the new ones, that it enables to extend the patterns known for weaker theories, that it provides new strong methods of solving problems unsolved so far, that it enables us to solve various hypothesis or to establish some connections between theories. Heuristic arguments of justification are of an a priori character and are not uniform. They refer to various principles such as for example Cantorian finitism (infinite sets are similar to finite ones), limitation of size (there exist only sets which are not too "big" in comparison with sets already accepted), maximization (everything that can be a set, is a set), realism (based on the distinction between existence and definability, it rejects the restriction of existing sets only to definable ones), uniformity (the universe of sets is in fact uniform, i.e., the same properties and situations appear anew at higher levels).

3. Current Trends in the Philosophy of Mathematics

The philosophy of mathematics after 1930 has been shaped by Gödel's incompleteness theorems and the consciousness of limitations of the axiomatic-deductive method revealed by them. It was characterized on the one hand by the dominance of the classical doctrines like logicism, intuitionism and formalism and on the other by the emergence of some new conceptions.

One should mention here Willard Van Orman Quine's holistic philosophy of mathematics and his indispensability argument according to which mathematics should be considered not in separation from other sciences but as an element of the collection of theories explaining the reality (cf. Quine 1951a, 1951b, 1953). Mathematics is indispensible there, in particular in physical theories, hence its objects do exist.

In this way Quine attacked the anti-realist and anti-empiricist approaches to the philosophy of mathematics. This cleared the way for empiricist approaches. One of them is the quasi-empiricism of Hilary Putnam who claimed (cf. Putnam 1975) that mathematical knowledge is not a priori, absolute and certain, that it is rather quasi-empirical, fallible and probable, much like natural sciences. He argues that quasi-empirical mathematics is logically possible and that ordinary mathematics has been quasi-empirical all along. In (1967) Putnam proposed a modal picture of mathematics according to which mathematics does not study any particular objects themselves but rather possibilities involving any objects whatsoever. Hence mathematics studies the consequences of axioms and asserts also the possibility of its axioms having models. The introduction of modalities opened the door to new epistemologies of mathematics.

Quine's–Putnam's indispensability argument was criticized by Hartry Field whose theory belongs to one of the most discussed proposals in the philosophy of mathematics in the recent years (cf. Field 1980 and 1989). Analyzing the rôle of mathematics in the natural sciences, especially in physics, Field comes to the conclusion that it is not true that mathematics is indispensable in them and that science uses mathematics merely as a theoretically dispensable descriptive and inferential short-cut only. Mathematical objects play there another rôle than abstract theoretical objects. In fact the latter extend the purely observational theories while in theories using abstract mathematical objects one cannot prove more than in a theory which does not refer to such objects. In other words any statement that does not refer to mathematical objects, which is a consequence of a mathematical extension of Field-style theory, is already a consequence of the nonmathematical part of the theory. Field illustrated his program of formulating versions of scientific theories that do not presuppose the existence of numbers and functions by developing an intrinsic version of Newtonian gravitation theory. This leads him to the nominalism – he claims that mathematics is only a useful auxiliary fiction, a set of propositions which enable us to formulate and to justify statements about the real world which itself has in fact no interpretation.

In contrast to Field, Charles Chihara and Philip Kitcher claim that natural sciences require something like the mathematical formalism to formulate and develop its theories. Chihara maintains (cf. Chihara 1990) that this formalism is not about mathematical objects but it concerns the possibility of taking open sentences. Unfortunately he says little about the epistemology of those possibilities. Kitcher views mathematics as an idealizing theory – it describes how we would segregate, arrange and collect physical objects if we lived in an infinite world and had perfect memories, etc. In (1983) he attempts to show that the growth of mathematical knowledge is far more similar to the growth of scientific knowledge than is usually appreciated. He offers a picture of mathematical knowledge which rejects mathematical apriorism. It is shown how early mathematical theories described empirically based idealizations and how theory gave birth to the study of even more remote idealizations.

Some interesting philosophical ideas concerning mathematics can be also found by K. Gödel who formulated them especially in connection with some problems of set theory (cf. Gödel 1944 and 1947). His philosophy of mathematics can be characterized as platonism. He claimed that mathematical objects exist in the reality independently of time, space and the knowing subject. He stressed the analogy between logic and mathematics on the one hand and natural sciences on the other. Mathematical objects are transcendental with respect to their representation in mathematical theories. The basic source of mathematical knowledge is intuition though it should not be understood as giving us the immediate knowledge. It suffices to explain and justify simple basic concepts and axioms. Mathematical knowledge is not the result of a passive contemplation of data given by intuition but a result of the activity of the mind which has a dynamic and cumulative character. Data provided by the intuition can be developed by a deeper study of mathematical objects and this can lead to the adoption of new statements as axioms.

In the eighties a naturalized version of Gödel's ideas has been developed by Penelope Maddy (cf. Maddy 1980, 1990a, 1990b). Gödel thought we can intuit abstract sets, Maddy claims that we can see sets of concrete objects whose members are before our eyes. We perceive sets of concrete physical objects by perceiving their elements (physical objects). Sets are located in the space-time real world. In this way we can know "simple" sets, i.e., hereditarily finite sets. More complicated sets are treated by Maddy as theoretical objects in physics – they and their properties can be known by metatheoretical considerations. One sees that in this conception Gödel's mathematical intuition has been replaced by the usual sensual perception. The advantage of Maddy's approach is that it unifies the advantages of Gödel's platonism enabling us to explain the evidence of certain mathematical facts with Quine's realism taking into account the rôle that mathematics plays in scientific theories.

We must mention also Ludwig Wittgenstein (1889–1951). His ideas concerning mathematics can be reconstructed from his remarks made at various periods – hence they are not uniform, moreover they are even inconsistent (cf. Wittgenstein 1953, 1956). They grew out from his philosophy of language as a game. He was against logicism, and especially against Russell's attempts to reduce mathematics to logic. He claimed that by such reductions the creative character of a mathematical proof disappears. A mathematical proof cannot be reduced to axioms and rules of inference, because it is in fact a rule of constructing a new concept. Logic does not play so fundamental rôle in mathematics as logicism claims – its rôle is rather auxiliary. Mathematical knowledge is independent and specific in comparison with logic. Mathematical truths are a priori, synthetic and constructive. Mathematicians are not discovering mathematical objects and their properties but creating them. Hence mathematical knowledge is of a necessary character. One can easily see here the connections of Wittgenstein's philosophy of mathematics and Kant's ideas as well as the ideas of intuitionists.

In the sixties there appeared in the philosophy of mathematics a new antifoundational tendency. It was the reaction to the limitations and one-sidedness of the classical views which are giving one-dimensional static picture of mathematics as a science and are trying to provide indubitable and infallible foundations for mathematics. They treat mathematics as a science in which one automatically and continuously collects true proved propositions. Hence they provide only one-sided reconstructions of the real mathematics in which neither the development of mathematics as a science nor the development of mathematical knowledge of a particular mathematician would be taken into account. New conceptions challenge the dogma of foundations and try to reexamine the actual research practices of mathematicians and those using mathematics and to avoid the reduction of mathematics to one dimension or aspect only. They want to consider mathematics not only in the context of justification but to take into account also the context of discovery.

One of the first attempts in this direction was the conception of Imre Lakatos (1922–1974). He attempted to apply some of Popper's ideas about the methods of natural science to episodes from the history of mathematics.[19]

Lakatos claims that mathematics is not an indubitable and infallible science – on the contrary, it is fallible. It is being developed by criticising and correcting former theories which never are free of vagueness and ambiguity. One tries to solve a problem by looking simultaneously for a proof and for a counterexample. New proofs explain old counterexamples, new counterexamples undermine old proofs. By proofs Lakatos means here usual non-formalized proofs of actual mathematics. In such a proof one uses explanations, justifications, elaborations which make the conjecture more plausible, more convincing. Lakatos does not analyze the idealized formal mathematics but the informal one actually developed by "normal" mathematicians, hence mathematics in process of growth and discovery. His main work *Proofs and Refutations* (1963a, 1963b, 1963c, 1964; see also 1976) is in fact a critical examination of dogmatic theories in the philosophy of mathematics, in particular of logicism and formalism. Main objection raised by Lakatos is that they are not applicable to actual mathematics. Lakatos claims that mathematics is a science in Popper's sense, that it is developed by successive criticism and improvement of theories and by establishing new and rival theories.

---

19 The reference to the history of mathematics is one of the characteristic features of new trends in the philosophy of mathematics.

The rôle of "basic statements" and "potential falsifiers" is played in the case of formalized mathematical theories by informal theories (cf. Lakatos 1967).

Another attempt to overcome the limitations of the classical theories of philosophy of mathematics is the conception of Raymond L. Wilder (1896–1982). His main thesis says that mathematics is a cultural system.[20] Mathematics can be seen as a subculture, mathematical knowledge belongs to the cultural tradition of a society, mathematical research practice has a social character. Thanks to such an approach the development of mathematics can be better understood and the general laws of changes in a given culture can be applied in historical and philosophical investigations of mathematics. It also enables us to see the interrelations and influences of various elements of the culture and to study their influence on the evolution of mathematics. It makes also possible to discover the mechanisms of the development and evolution of mathematics. Wilder's conception is therefore sometimes called evolutionary epistemology. He has proposed to study mathematics not only from the point of view of logic but also using methods of anthropology, sociology and history. Wilder maintains that mathematical concepts should be located in the Popper's "third world". Mathematics investigates no timeless and spaceless entities. It cannot be understood properly without regarding the culture in the framework of which it is being developed. In this sense mathematics shares many common features with ideology, religion and art. A difference between them is that mathematics is science in which one justifies theorems by providing logical proofs and not on the basis of, say, general acceptance.

Those new anti-foundational trends in the philosophy of mathematics should not be treated as competitive with respect to old theories. They should be rather seen as complements of logicism, intuitionism and formalism. One is looking here not for indubitable, unquestionable and irrefutable foundations of mathematics, one tries not to demonstrate that the actual mathematics can be reconstructed as an infallible and consistent system but one attempts to describe the actual process of building and constructing mathematics (both in individual and historical aspect).

Considering new conceptions in the philosophy of mathematics one must also mention structuralism. It can be characterized as a doctrine claiming that mathematics studies structures and that mathematical objects are featureless positions in these structures. As forerunners of such views one can see R. Dedekind, D. Hilbert, P. Bernays and N. Bourbaki. The latter is in fact a pseudonym of a group of (mainly French) mathematicians who undertook in the thirties the task of a systematization of the whole of mathematics (the result of their investigations was the series of books under the common title *Éléments de mathématique*). Their work refered to Russell's idea of reconstructing mathematics as one system

---

20 Wilder presented his ideas in many papers and lectures. A complete version of them can be found in two of his books: *Evolution of Mathematical Concepts. An Elementary Study* (1968) and *Mathematics as a Cultural System* (1981).

developed on a firm (logical) basis. For bourbakists the mathematical world is the world of structures. The very notion of a structure was explained by them in terms of set theory. They distinguished three principal types of mathematical structures: algebraic, order and topological structures.

The idea of treating mathematics as a science about structures is being developed nowadays by Michael Resnik, Stewart Shapiro and Geoffrey Hellman. Resnik claims (cf. Resnik 1981, 1982) that mathematics can be viewed as a science of patterns with its objects being positions in patterns. The identity of mathematical objects is determined by their relationships to other positions in the given structure to which they belong. He does not postulate a special mental faculty used to acquire knowledge of patterns (they are not seen through a mind's eyes). We go through a series of stages during which we conceptualize our experience in successively more abstract terms. This process do not necessarily yield necessary truths or a priori knowledge. Important is here our tendency to perceive things as structured. The transition from experience to abstract structures depends upon the culture in which it takes place. Add that the transition from simple patterns to more complicated ones and the development of pure theories of patterns rely upon deductive evidence.

S. Shapiro claims (cf. Shapiro 1989, 1991) that there is a strict connection between objects of mathematics and objects of natural sciences. An explanation of it can be provided in his opinion just by structuralism according to which mathematics studies not objects *per se* but structures of objects. Hence objects of mathematics are only "places in a structure". The advantage of such an approach is that it enables us to explain the phenomenon of applicability of abstract mathematical theories in natural sciences as well as the interrelations between various domains of mathematics itself. It enables also a holistic approach to mathematics and science.

G. Hellman argues (cf. Hellman 1989) that one can interpret mathematics (in particular arithmetic and analysis) as nominalistic theories concerned with certain logically possible ways of structuring concrete objects. He uses by such interpretations second-order logic and modal operators (hence his approach is sometimes called modal-structural).

4. Conclusions

The above presentation of conceptions in the epistemology of mathematics indicates that there were various proposals and attempts to answer the basic questions concerning the epistemological status and the methods of mathematics as a science. There is no unique answer accepted by all philosophers of mathematics (but the rôle of philosophy is not to give definite answers but rather to indicate problems and show the complexity of considered issues). On the other hand mathematics is dynamic and is being developed rather independently of philosophical settlements of questions. But of course this independence is not complete. The rôle of the philosophy of mathematics towards mathematics itself is not only descriptive but also normative, i.e., some philosophical conceptions and solutions fix certain norms and rules according to which mathematical knowledge should be (and in fact is) developed and presented.

Those norms and rules are being changed and transformed of course. For example the ideal basic method to develop mathematics since the Greek antiquity was considered to be the axiomatic-deductive method and the basic method to justify a statement was to give a proof. But the very concept of a proof has been changed very deeply from the intuitive one in which a reference to drawings and "self-evident" facts were allowed to the precise notion of a formal proof in a formalized axiomatic system. Also the idea of what is the nature of an axiom has changed very much. On the other hand methods of mathematical logic enabled us to discover several limitations of the axiomatic method and simultaneously to make precise various philosophical concepts and ideas (as for example to distinguish in mathematics between truth and provability).

The development of mathematics leads not only to the formulation of new problems and questions in the philosophy of mathematics but also to the necessity of revising the former conceptions. As an example can serve here the construction of non-Euclidean geometries in the nineteenth century. Nowadays the most spectacular examples are connected with computers and their applications.

Computers are used in mathematics not only to perform complicated (and tedious) numerical calculations but also in automated theorem proving. Studies on the mechanization and automatization of (mathematical) reasonings can be traced back to the seventeenth century, to Leibniz and his idea of the *characteristica universalis*. They received an important impulse from the mathematical logic on the turn of the nineteenth century. Recently the possibility of realization of those method on computers brings new contexts. The main question one should ask in the connection with this is: what are the reasons for accepting mathematical results obtained by using a computer. One of the ways to verify such results is to perform the given computer program several times on various machines and to check whether the results are identical. But note that this procedure is similar to the procedure of verifying experimental data in physics and is in fact quasi-empirical. So can it be used in mathematics?

Recently there appeared some results in mathematics in which computers were essentially applied. The most spectacular and most discussed example is the four-color theorem being a solution of an old problem concerning the coloring of a map. In the proof of this theorem computer calculations are heavily used. What more, computer was applied here not only to perform some computations but some important tricks and ideas used in the proof were improved by certain computer experiments, by "dialogues" with a computer. The validity of the computer program cannot be checked without a computer. On the other hand no traditional proof (i.e., a proof not referring to computers) of the four-color theorem has been given (there are doubts whether such a proof can be given). Hence the considered theorem is the first example of a mathematical theorem of a new type. Its proof is convincing and can be formalized but is not surveyable, so it has not one of the important features mathematical proofs should (traditionally) have. Consequently one can ask whether the four-color theorem has been proved and whether it can be considered as a mathematical theorem and whether it belongs to the mathematical knowledge? Certainly it is not an a priori statement. There are two possibilities: either extend the scope of methods accepted in mathematics and to allow the usage of computers (hence a type of experiments) or to admit that the four-color theorem has not been proved yet and does not belong to mathematical knowledge. The former possibility implies in particular that mathematics becomes a quasi-empirical (and not an a priori) knowledge. Such solution is represented by Ph. J. Davis, R. Hersh, Ph. Kitcher and E. R. Swart who claim that mathematics always admitted empirical elements and had in fact an empirical character. On the other hand one attempts to defend the a priori character of mathematics by arguing that proofs using computers can be transformed into traditional proofs by adding new axioms or that a computer is in fact a mathematician and it knows the result proved deductively or that procedures similar to applying computer programs have been used in mathematics for a long time, hence the applications used in the proof of the four-color theorem are in fact nothing essentially new.

Discussing here the problem of the influence of computers on the philosophy of mathematics one should mention also questions connected with the old mind-body problem, in particular with the problem whether machines can act in an intelligent way and the whole scope of problems formulated in the domain called artificial intelligence. They are not directly connected with the philosophy of mathematics – therefore we will not discuss them here. Note only that Gödel's incompleteness theorems are also used in the study of them. In particular it has been argued (cf. Chaitin 1974, 1982) that Gödel's theorems (when interpreted from the point of view of the information theory) show that if one wants to obtain more complex mathematical theorems (i.e., theorems containing more information) then one will have to continually introduce new axioms and new methods. Neither the admissible methods and rules can be fixed and codified nor the concept of a correct mathematical proof can be defined once and for ever. Hence progress in mathematics seems to be much like the progress in the natural sciences than hitherto expected. All such claims provide new arguments for the quasi-empiricism claiming that mathematics is in fact much like natural sciences.

On the Power and Weaknesses of the Axiomatic Method

It seems to be a common conviction that mathematics is a pattern of certainty and precision. This conviction is based on the fact that in mathematics only precise notions are used and that every proclaimed thesis must be justified. Admissible methods of justification are – at least to certain extent – precisely codified. In fact they are based on the so called axiomatic-deductive method.

The very axiomatic-deductive method comes from the ancient Greeks – Plato was probably the first who developed it and proclaimed to be a proper method for mathematics. It was connected with another of his claims, namely with the thesis that objects of mathematics belong to the realm of ideas and that consequently mathematics describing them and their interrelations is a science about certain (i.e., arithmetical and geometrical) ideas. This implies that the mathematical knowledge is based on the reason and the proper method of mathematics is just the axiomatic method. The latter consists of accepting certain propositions without proof and then on deducing from them all other theorems.

Plato's ideas has been developed by Aristotle in his *Posterior Analytics* where he discussed the structure of scientific theories. According to him at the base of any knowledge lie the general notions that do not need to be defined and certain general propositions that have and need no proof. All other notions should be defined and all other theses must be deduced from the basic ones. He distinguished two types of initial theses accepted without proof: axioms and postulates. Axioms are propositions of a general character describing basic properties of any entities and being consequently common to all theories (they correspond to logical axioms and axioms of identity in nowadays terminology of formal logic). Postulates on the other hand are special principles saying about specific properties of considered objects (nowadays they are called non-logical axioms). Any theory should contain beside axioms and postulates also definitions and existential hypotheses stating the existence of defined objects – existence independent of our thought and perception.

A perfect example and implementation in mathematics of the methodology described here are *Elements* – the work of Euclid, the hellenistic mathematician from Alexandria, written about 300 B.C. They were an attempt to present almost all of known then mathematics in the framework of an axiomatic-deductive system – hence in the methodological framework established by Plato and Aristotle. In particular Euclid gave in *Elements* axioms and postulates for a system of geometry and showed how one can deduce from them – using the introduced definitions – properties of the considered objects (i.e., of geometrical figures and solid bodies). It is worth noting here that one finds in *Elements* no axioms and postulates for the arithmetic (of natural numbers). It is not clear why Euclid did see no need of basing this fundamental domain of mathematics on a system of axioms and postulates.

The pattern of developing (and presenting) mathematics established by Plato, Aristotle and Euclid has been commonly accepted, Euclid's *Elements* established a paradigm called today Euclidean one. Its characteristic feature is to base all knowledge on a system of axioms and postulates. It should be noted that no attention was paid to the language in which axioms and postulates were formulated – in practice it was simply the unprecise common language. Also rules of inference with the help of which the reasonings should be performed were not precisely described and fixed – one recalled to the intuition and to the "natural" commonly accepted methods.

During long centuries *Elements* played the rôle of a pattern of any scientific discourse. Its method was admired and followed. They marked an ideal that should be realized. Descartes claimed that only mathematicians can find proofs and in this way deliver the unquestionable knowledge. Consequently he proposed to establish a universal analytical and mathematical discipline (called by him *mathesis universalis*) comprising all the knowledge about the world. It should consider – as mathematics does – only quantitative properties. He proposed to call it universal mathematics, "because it contains all thanks to which other disciplines are called mathematical" – as he wrote in *Regulae ad directionem ingenii* [Rules for the Direction of the Mind]. The certainty of arithmetic and geometry (hence of mathematics) follows from the fact that "only they consider objects so pure and simple that they assume nothing which experience can make uncertain but they depend entirely on the deduction of consequences" (*ibidem*). Describing in *Discours de la méthode* [Discourse on the Method] rules according to which the mathematical analytical method should be based on he suggested "to conduct [...] thoughts in such an order that, by commencing with objects the simplest and easiest to know, I might ascend by little and little, and, as it were, step by step, to the knowledge of the more complex" (Chapter 2). Descartes thought here about the deduction, i.e., he recommended to move from simple truths (axioms) to complex propositions that can be deduced from them.

The idea of Descartes was revived by Leibniz in his idea of developing a universal logical calculus that should enable us to reduce all scientific reasonings to a mechanical transformations of appropriate symbols of a universal symbolic language (called *characteristica universalis*) according to certain rules of a formal character fixed ahead, i.e., referring only to the shape of expressions and not to their sense or meaning (this calculus on symbols was called *calculus ratiocinator*). Hence Leibniz – recalling to ideas of Plato, Aristotle and Euclid concerning the axiomatic method – proposed even certain mechanization and automatization of reasonings (hence he can be treated in a certain sense as a precursor of artificial intelligence!).

The Euclidean paradigm was the commonly accepted and applied pattern of developing science, the axiomatic-deductive method became more and more popular. One should recall Spinoza's attempts to present with the help of it principles of ethic (cf. his *Ethica Ordine Geometrico Demonstrata* [The Ethics], 1677), Leibniz's attempts to present politics (cf. *Pattern of Political Proofs*, 1659) or Newton's attempts to present optics (cf. *Opticks*, 1704). Also in the mathematics itself (old and new) theories were provided with adequate – as one hoped – systems of axioms. In particular at the end of the 19th century the axioms for the arithmetic of natural numbers were proposed – this was done by the Italian mathematician Giuseppe Peano in the work *Arithmetices principia nova methodo exposita* (1889). It is worth adding here that Peano did a lot to popularize the axiomatic method.[1] On the other hand David Hilbert proposed in the work *Grundlagen der Geometrie* (1899) a new system of axioms and postulates for geometry completing the system of Euclid and removing in this way some gaps in it.

The intensive development of mathematical logic (being in fact a realization of Leibniz's idea of establishing logic after the example of mathematics) at the turn of 19th and 20th centuries connected mainly with works of G. Boole, A. De Morgan, G. Frege, E. Schröder, Ch. S. Peirce, B. Russell contributed also to the improvement of the axiomatic-deductive method. It has been done by introducing the principle that an artificial precise symbolic language should be used and by the codification of the inference rules. In consequence a new paradigm called logico-settheoretical paradigm has been established.

In all those studies it was implicitly assumed that any mathematical knowledge can be founded on a set of axioms and postulates that are sufficient for the deduction of all theorems that are true in a given domain and that are solutions of considered problems of this domain. Moreover, one was convinced that for any domain of mathematics an adequate set of axioms and postulates can be chosen in such a way that they would form a sufficient base for solving every problem concerning this domain, i.e. – as modern mathematical logic says – one can find a complete set of axioms and postulates.

Parallel to those works and to the process of securing the rôle of the axiomatic-deductive method in the methodology of mathematics there appeared in the 19th century some qualitatively new results that seemed to destroy and to shake the edifice of mathematics being just constructed. One should mention here the solution – by methods of algebra, more exactly by methods of the theory of algebraic equations – of three classical problems of the ancient Greek mathematics, namely of the problem of squaring the circle, the problem of trisecting the angle and the problem of doubling the cube (known also as the Delian problem). It was shown that those problems cannot be solved by so called Platonian constructions, i.e., by using only a finite number of steps with compass and straightedge. Hence

---
1   Cf. Murawski (1985) and (2010).

it was a result stating the impossibility of certain constructions – so it was a result of a qualitatively new character.

Similar character had also the result on the independence of Euclid's fifth postulate on parallels. This was shown by establishing new systems of geometry in which the negation of the fifth postulate was assumed and by proving the consistency of those systems called non-Euclidean geometries (C. F. Gauss, N. I. Lobachevsky, J. Bolyai).

Gradually it became clear that the essence of mathematics consists of deducing theorems from the postulated assumptions (axioms and postulates) and not of deciding the truth of those assumptions. Mathematics became more and more an abstract and formal science. It has been shortly expressed by Russell who said in (1901) that pure mathematics is a domain in which one does not know what we are talking about and whether that what is said is in fact true. This emphasized the rôle and meaning of the axiomatic method which in fact was (and still is) the most important method of mathematics determining the essence of this science.[2]

At the beginning of the thirties of the 20th century it has appeared that the situation is not so comfortable and satisfying as it seemed to be. In 1931 a paper written by a young Austrian mathematician Kurt Gödel was published – it showed in fact certain weakness (not to say certain defect) of the axiomatic method. We mean here the paper "Über formal unentscheidbare Sätze der 'Principia Mathematica' und verwandter Systeme. I". It was to become one of the most important and influential works in logic and the foundations of mathematics having simultaneously also many consequences for the philosophy and methodology of mathematics. Gödel showed in it that the arithmetic of natural numbers (as an axiomatic-deductive system) and all richer deductive systems are essentially incomplete if they are consistent. This means that there are sentences formulated in the language of those theories that are undecidable on the basis of their axioms, i.e., there are sentences φ such that neither φ nor $\neg\varphi$ (the negation of φ) are theorems of the given theory. Moreover, one can show (using stronger methods) which sentence of the pair φ and $\neg\varphi$ is true (in the standard model of the theory, i.e., in the structure for the description of which the considered theory has been constructed). This incompleteness is essential, that is it cannot be removed by augmenting the theory by undecidable sentences as new axioms – in fact in such a new richer theory another undecidable sentences will appear. This result is called today Gödel's first incompleteness theorem.

---

2   As an example of the meaning credited sometimes to this method let us quote a sentence from Kleinert (2002) who says that "every strong application of this method [i.e., of the axiomatic-deductive method – my remark, R.M.], independently of the fact about which objects we are talking, is already mathematics" ("jede strenge Anwendung dieser Methode, auf welchen Gegenstand auch immer, ist schon Mathematik"). It means that, according to Kleinert, the essence and specific of mathematics is determined by its method (and not by its objects).

Gödel showed even more. At the end of the paper quoted above Gödel formulated another theorem called today Gödel's second incompleteness theorem. It stated that one cannot prove in a given consistent formalized theory containing arithmetic of natural numbers its own consistency, i.e., to prove the consistency of such a theory one must use methods much stronger that those available in the theory.[3]

Gödels incompleteness results shook in a certain sense the reliance in the axiomatic-deductive method and introduced certain worry to the methodology of mathematics and generally to the opinions concerning mathematics as a science. At the beginning one did not understand fully their meaning and thought that Gödel simply discovered a new paradox. Another reaction was the disregard of them – one claimed that they do not in fact concern the essence of mathematics, i.e., that mathematics is complete with respect to sentences that are mathematically interesting (whatever it may mean) and that examples of undecidable sentences given by Gödel are artificial from the point of view of real mathematics. So they were taken with certain disbelief and specific reserve.[4]

But it was impossible (and irrational) to ignore at long range Gödel's results and their consequences. At the turn of the seventies and eighties of the 20th century examples of new undecidable sentences of the (directly) mathematical contents were given by J. Paris, L. Harrington and L. Kirby.[5] All those forced some changes and corrections in the thinking about mathematics as a science and its methods.

Gödel's theorems showed that the axiomatic-deductive method – though the best method of presenting (mathematical) knowledge developed by the mankind – has in fact some weaknesses. In particular it does not give complete knowledge, more exactly, one cannot comprise the whole of mathematics in a consistent formalized axiomatic-deductive system based on the predicate logic of the first order – moreover, one cannot comprise in such a system even the whole truth about such basic and "natural" domain as natural numbers. So one sees here the inexhaustibility of the domain of mathematical truth, even of the arithmetical truth concerning natural numbers. In fact there will always exist undecidable sentences, independently of the fact how strong axioms one assumes (providing they form a recursive set, i.e., that one can effectively decide what is an axiom and what is not – this is a natural requirement!). Moreover, there will be undecidable sentences of the form "for all x, $\varphi(x)$" such that every particular instance of φ, i.e., sentences of the form $\varphi(0)$, $\varphi(1)$, $\varphi(2)$, etc. are theorems of the considered theory. On the other hand by using methods beyond those available in the considered theory one can show whether the undecidable

---

3 Technical details of Gödel's incompleteness theorems can be found for example in Murawski (1999b).
4 On the reception of Gödel's incompleteness results see Murawski (2003).
5 Cf. for example Murawski (1994).

sentences are true or false (in a given structure). Hence they can be decided! Consequently it became necessary to apply infinitistic methods by proving theorems about finite objects like natural numbers. In this way one comes to the conclusion that formalized systems and inference rules admissible in them are not (and cannot be) fully adequate with respect to usual research practice of mathematicians. In fact, no mathematician does limit ahead in his researches the admissible methods that can be used in solving a given problem. Just the opposite, any correct methods are admissible. But what does it mean "correct method"? We must rely on intuition and on the common consent of specialists – in this way we leave the strict ground of logic and move towards pragmatics! A mathematician trying to solve a problem does not work in a fixed formalized system and does not strictly follow requirements of the axiomatic-deductive method. The latter helps rather to order the knowledge and to communicate it to others. It makes also possible the study of mathematics (and its theories) *as a science* and in this way to develop strict and precise *meta*mathematics. Using methods of logic and considering axiomatic theories one can study the deductive power of a given set of axioms and given inference rules and to compare various theories from this point of view. Hence it provides us with strict and precise methodological tools.

Gödel's incompleteness theorems showed also that provability in a given axiomatic-deductive system is not fully identical with the truth – in fact, every theorem of the formalized arithmetic is true but there are true sentences of arithmetic that are not theorems (undecidable sentences). Consequently the concept of the semantical truth cannot be adequately expressed in terms of the syntactical truth (i.e., of provability).

Since formalized axiomatic systems are usually incomplete and as such they are inadequate with respect to really developed mathematics, hence one cannot *a priori* set any bounds to methods applied in mathematics, one cannot limit the creative invention of mathematicians. Human mind does not work in such a way as that presented in axiomatic-deductive systems, hence it does not work like (Turing) machine. Consequently the axiomatic-deductive method does not give us real information on how the mind of a mathematician works, does not give us an adequate model of mind. The mind of a mathematician works probably along other principles. Hence Gödel's theorems are also a contribution to the problems of artificial intelligence. They indicate among others that the research work of a mathematician cannot be completely and fully mechanized.

Incompleteness theorems indicate some boundaries of cognitive abilities of the axiomatic-deductive method as well as of the possibilities of (Turing) machines. Do they indicate also limits of cognitive abilities of human mind? Since we can decide (with the help of richer methods) undecidable sentences, hence it seems that the answer is negative. But are we able to answer any formally undecidable question? We come to this problem later.

Second incompleteness theorem states that one cannot prove in a given theory (satisfying appropriate conditions) its own consistency. Hence to prove that a theory T is consistent one needs means stronger (so less certain and more questionable) than those available in T. Consequently there exist no absolute consistency proofs! On the other hand consistency is one of the most important conditions in mathematics (it has also ontological consequences because sometimes it is assumed that consistency implies existence – this was claimed for example by Hilbert). In such a way developing mathematics and having no absolute consistency proofs we must rely on the belief that our theories are consistent. This belief can be based only on the fact that so far the inconsistency of mathematics has not been shown and always when any inconsistencies appeared one was able to remove them. On the other hand without this belief that mathematics is consistent it would be impossible to develop mathematics – indeed, by Duns Scotus' law $p \to (\neg p \to q)$ from a pair of inconsistent sentences one can deduce any sentence! Consequently if mathematics appeared to be inconsistent then one could prove in it for example that $2+2=4$ and that $2+2=5$ (sic!). And this would make of it an absolutely useless theory.

The existence of undecidable sentences in any richer formalized axiomatic-deductive system puts also another light on the problem of analyticity of mathematical, in particular arithmetical, truths. Since not all such truths follow from axioms and definitions, hence are they still analytic?

Beside Gödel's incompleteness theorems one proved also in the 20th century mathematical logic other theorems indicating some limits of the axiomatic-deductive method. They are all (together with Gödel's theorems) called limitation theorems. We mean here Skolem–Löwenheim theorem and Tarski's theorem.

Let us start with Tarski's theorem. It states the undefinability of the concept of truth. More exactly it says that for a given theory T (satisfying certain natural conditions) there exists no formula Φ in the language of this theory that would be a definition of truth, i.e., such that for any sentence φ of the language of T the equivalence $\varphi \equiv \Phi(\varphi)$ would be a theorem of T. Consequently in T one cannot define the concept of truth for formulas of T. In particular one cannot define in the arithmetic of natural numbers the concept of the arithmetical truth – one must use here stronger means. And what to say about the general concept of mathematical truth? And about the concept of truth as such?

Skolem–Löwenheim theorem says (among others) that every formalized axiomatic-deductive system for the arithmetic of natural numbers has so called nonstandard models, i.e., if such a system is consistent (and consequently has models) then it has also nonstandard models, hence models different from the model consisting of numbers 0, 1, 2, ... together with the operations of addition $+$ and multiplication $\cdot$ – recall that one develops axiomatic arithmetic just to describe this model and its properties. Moreover, any such system (providing it is consistent) has arbitrarily large models (models of any cardinality – as mathematicians used to say), hence models completely different than the distinguished standard model. So we see that even natural numbers and natural operations on them cannot be uniquely characterized by formalized axiomatic-deductive systems. There exists no adequate axiomatic description of the domain of natural numbers – any such description will be in fact inadequate, i.e., it will admit (semantic) interpretations different from the intended standard one. And this is not a consequence of accepted set of axioms but it follows from the very nature of the axiomatic method.

Results described above, though formulated for the simplicity only for the arithmetic of natural numbers, hold also in the case of other mathematical theories. They are interesting especially in the case of set theory, i.e., a theory treated as a foundation of all of mathematics.

Set theory founded by Georg Cantor at the end of the 19th century became the foundation of mathematics in this sense that all mathematical concepts are defined in terms of set-theoretical notions, moreover it seems that the whole mathematics can be reduced to set theory and logic (this is the main thesis of logicism, one of the main tendency in the modern philosophy of mathematics). Cantor understood the notion of a set in an intuitive way. This led to certain paradoxes in set theory. A solution was the axiomatization of the concept of a set and the acceptance of an appropriate set of axioms. Such an axiomatic system has been proposed in 1908 by Ernst Zermelo. After some changes and corrections a system called today Zermelo–Fraenkel set theory ZF was established – it is accepted today as a standard axiomatic system of set theory.[6]

It turns out that the theory ZF (and similarly also other systems of set theory) is not a sufficient base for solving questions concerning sets. On the one hand one can apply to ZF general incompleteness results of Gödel described above and stating the existence of undecidable sentences and on the other there are (at least) two undecidable sentences formulated not by logicians in their studies of set theory as a theory but by (real) mathematicians in their normal mathematical research, namely the axiom of choice and the continuum hypothesis. We cannot describe here (rather complicated) technical details of them. So let us say that the axiom of choice is a sentence stating that for any nonempty family of nonempty disjoint sets there exists a set containing exactly one representative of every set of this family. This rather innocent sentence plays very important rôle in mathematics – it forms essential component of many very important theorems in various fields of mathematics, i.e., those theorems could not be proved without accepting and using the axiom of choice. On the other hand this axiom has also some objectionable consequences. The most known is Banach–Tarski's theorem on the paradoxical decomposition of the ball (known also as Banach–Tarski paradox). It is a theorem in set theoretic geometry stating that a solid ball in

---

6   There are also other axiomatic systems of set theory – cf. for example Chapter 4 of Bedürftig––Murawski (2010) or Appendix I to Murawski (1995).

3-dimensional space can be split into a finite number of non-overlapping pieces, which can then be put back together in a different way to yield two identical copies of the original ball. In the proof of the theorem the axiom of choice is essentially used. The theorem can be paraphrased by saying that a mathematician having at his disposal the axiom of choice would be like God having the possibility of creating new objects – the only difference would be the fact that God can create *ex nihilo* (from nothing) and a mathematician must have something given at the very beginning (e.g., the initial ball).

The following problem arises: should the axiom of choice be accepted or rejected in mathematics? On the one hand it is useful and needed in mathematics, on the other it leads to undesired (inconsistent with the intuition) consequences. Crucial could be here the answer to the question whether it follows from other (not so controversial) axioms of set theory or not (i.e., whether other properties of sets expressed in the axioms imply it or not). Unfortunately researches undertaken on this have shown that this will not lead to a solution. In fact it turns out that the axiom of choice is on the one hand consistent with other axioms of set theory (this was proved by Gödel in 1939) and that it is independent of other axioms (as shown by P. J. Cohen in 1963). It means that other axioms of set theory neither prove the negation of the axiom of choice (hence there is no reason to reject it) nor the axiom itself (hence there is no reason to accept it). So what can be done? In fact one has full (rather uncomfortable) freedom: one can accept the axiom of choice or reject it (and possibly accept its negation). In each case one obtains a consistent system of set theory. But since set theory is the foundation of the whole mathematics and the axiom of choice interferes in many parts of it hence we come to the conclusion that mathematics can be developed on the base of various systems of set theory and that one will get in this way various systems of mathematics (they will differ by theorems concerning the same problems[7])!

Similar is the situation with the second indicated proposition of set theory, namely with the continuum hypothesis. It is a statement connected with the continuum problem considered already by Cantor who was unable to solve it on the base of his intuitive concept of a set. Continuum hypothesis is a sentence stating that every infinite set of reals is either countable or has the power of the continuum (hence is equipollent with the set of all reals). It turns out that the continuum hypothesis is, similarly to the axiom of choice, undecidable in Zermelo–Fraenkel set theory ZF, i.e., neither it nor its negation can be deduced from other axioms of ZF, what means that properties of sets "encoded" in accepted

---

7 For example in mathematics based on set theory with the axiom of choice one will have a theorem stating that Cauchy's definition of continuity of a function is equivalent to the definition of Heine and consequently there is only one notion of continuity of a function whereas in mathematics based on set theory without the axiom of choice (or with its negation) one will not have this equivalence and consequently there will be two non-equivalent notions of continuity

axioms do not suffice to decide the problem of continuum.[8] Hence two consistent (though mutually inconsistent) systems of set theory, namely ZF + "continuum hypothesis" and ZF + "the negation of continuum hypothesis" are possible. Since the continuum hypothesis is applied in various domains of mathematics, hence two different systems of mathematics are possible according to the fact whether one develops mathematics on the base of one or another system of set theory.

So we see that the system of axioms of Zermelo–Fraenkel set theory does not exhaust the concept of a set, does not fully and uniquely characterize it. Similar phenomenon holds for other axiomatic systems of set theory. All this indicates certain weakness of the axiomatic-deductive method!

Since there is no decisive argument which system of set theory should be chosen as the foundation of mathematics, different systems of mathematics are possible. So there is an insurmountable uncertainty concerning the foundations of mathematics and the latter seems to be rather (or better: only) a formal play, a study of what can be deduced from what and what will happen when one changes this or another axiom – it would have nothing to do with the search how the things really are, how in fact it is in ... Well, where? In Plato's realm of ideal mathematical objects? But how can it be cognitively reached? The axiomatic method does not allow to describe it comprehensively! Hence we depend on the eternal uncertainty in this most certain of scientific disciplines and we must content ourselves with the strict knowledge – obtained by the usage of the axiomatic method! – of our ignorance?!

---

8   It was proved by Gödel in 1939 and Cohen in 1963 who shown, resp., that the continuum hypothesis is consistent with the axioms of ZF (i.e., its negation is not a theorem of ZF) and that it is independent of ZF (i.e., it cannot be proved in ZF).

Remarks on the Mathematical Universe

1. Introduction

Let us begin with the following question: did Leibniz and Newton discover or create the calculus. The question only on the surface deals with the linguistical problem. In fact it concerns one of the most fundamental problems of the philosophy of mathematics: is the reality of mathematical objects given to us or is the realm of mathematics a product (a creation) of human mind? This ontological aspect of the question has also an epistemological dimension: if one answers that the mathematical reality exists independently of space, time and the human mind (such answer is usually called realism or platonism) then there arises a next question: how are we able to get the knowledge about such a reality? Similarly when the answer to the starting question is that mathematical objects are constructions and creations of the human mind (in this case one speaks about conceptualism in the broad sense) then one can ask how does this process of creating or constructing the world of mathematics look like, where are its sources and which criteria should it satisfy? In both cases the question of truth and of the decidability of mathematical statements can be given different answers.

The aim of this essay is to present in a historical perspective various answers given to the above questions, to consider how the answers were justified and what are their consequences.

2. Realism

As a father of the conception that the realm of mathematical objects is given to us and that consequently a mathematician is a discoverer (and not a creator) one should consider Plato (427–347 B.C.). This explains also the origin of the name 'platonism'.[1]

Plato's philosophy of mathematics grew out of his theory of ideas. He claimed that the subject of mathematics are mathematical (arithmetical and geometrical) ideas (or forms).[2] They are real entities conceived as being independent of perception and being apprehended, as being capable of absolutely precise definition and as being absolutely permanent, that is to say timeless or eternal. Hence a mathematician does not create mathematical objects and their properties

---

1   The very term 'platonism' was introduced by Paul Bernays in (1935a).
2   According to Aristotle, Plato distinguished between the arithmetical and geometrical ideas (forms) and the so-called mathematicals, each of which is an instance of some unique form – each form having many such instances.

but does discover them. Consequently the mathematical knowledge is based on the reason and the proper method of mathematics is the axiomatic method – Plato was probably the first who introduced it. Mathematics is very close to Plato's ideal of knowledge because it abstracts from changeable phenomena and concentrates on unchangeable, timeless, mind-independent and definite objects and relations between them. Plato admitted that a mathematician uses in his research practice observations and drawings or performing constructions but they serve only the process of remembering the proper mathematical objects (ideas) and not the creation of them (Plato refers here to his theory of *anamnesis*). Hence mathematics is a science whose aim is the description of timeless, mind-independent and definite mathematical objects (ideas) and their mutual relations. Consequently all mathematical propositions are necessarily true. Their necessity is independent of their being apprehended by a mathematician, independent of any formulation and thus of any rules governing a natural or artificial language. Mathematical theorems can be applied to the description of the objects of sense-experience because the latter are to a certain degree similar to, or better, approximate the ideas (Plato says here that, for example, one apple participates in the arithmetical idea One).

An adherent of Plato's ideas was in particular Euclid (365 (?)–300 (?) B.C.) whose *Elements* established a paradigm in mathematics prevailing up until the end of the nineteenth century and called today Euclidean paradigm. He demonstrated in a mastery way how the Plato's axiomatic method can be applied to develop mathematics. In his approach to it one can easily recognize also other traces of Plato's influence, in particular in Euclid's static-Eleatic approach to geometry, in attempts to catch in definitions this what is constant and eternal, in negative attitude towards applying geometrical theories in the measuring practice as well as in his rejection of any approximative solutions.

Another follower of the doctrine of Plato who explicitly formulated his theses was Georg Cantor (1845–1918), the founder of set theory. He was convinced that concepts of mathematics posses the immanent reality (in the mind of a mathematician) as well as the extra-subjective reality. He was convinced that a mathematician does not create or produce mathematical objects but discovers them. He expressed his beliefs in the third thesis of his *Habilitationsschrift* from 1869 when he wrote (cf. 1932):

> *Numeros integros simili modo atque corpora coelestia totum quoddam legibus et relationibus compositum efficere.*

This was confirmed in his fundamental (composed of two parts) work *Beiträge zur transfiniten Mengenlehre* (1895) which began by the following three mottos:

> *Hipotheses non fingo.*
>
> *Neque enim leges intellectui aut rebus damus ad arbitrium nostrum, sed tanquam scribae fideles ab ipsius naturae voce latas et prolatas excipimus et describimus.*

Veniet tempus, quo ista quae nunc latent, in lucem dies extrahat et longioris aevi diligentia.

In a letter to G. Mittag-Leffler from 1884 Cantor wrote: "what concerns other things [namely things other than style and the way of presenting – R.M.], all those is not my merit; with respect to the contents of my work I am only a reporter and a servant *(nur Berichterstatter und Beamter)*".

Note that Cantor ascribed to set-theoretical notions (and consequently to all notions of the whole of mathematics) real existence not only in the world of ideas but also in the physical world. Hence he was convinced of the real existence in this world of sets of cardinality aleph zero or continuum. He did not treat consistency as the unique and sufficient criterion of existence in mathematics.

On the other hand it should be added that Cantor stressed that the intrasubjective and immanent reality of mathematical concepts is a source and warrant of the pure and free mathematics.

In the twentieth century platonism found expression in the philosophy of mathematics by Kurt Gödel (1906–1978). He claimed that mathematical objects exist in the reality independently of time, space and the knowing subject. He stressed the analogy between logic and mathematics on the one hand and natural sciences on the other. Mathematical objects are transcendental with respect to their representation in mathematical theories. The basic source of mathematical knowledge is intuition though it should not be understood as giving us the immediate knowledge. It suffices to explain and justify simple basic concepts and axioms. Mathematical knowledge is not the result of a passive contemplation of data given by intuition but a result of the activity of the mind which has a dynamic and cumulative character. Data provided by the intuition can be developed by a deeper study of mathematical objects and this can lead to the adoption of new statements as axioms.

3. Conceptualism

Consider now the second group of ideas concerning the title question, namely the conceptualism claiming that a mathematician is a creator (and not a discoverer).

As an example of such conception one can mention Nicolaus Cusanus (1401–1464), a mathematician, theologian, cardinal and the last representative of scholasticism. He claimed that numbers as well as geometrical objects are creations of human mind. Their source are numbers and geometrical objects existing in the mind of God. Consequently he distinguished numbers being objects of mathematics – they are created by mathematicians, their source is in the human mind, and numbers coming from God and having their source in His mind. The first are reflections, pictures (*ymago*) of the latter. And similarly for geometrical objects. How does a man create mathematical objects in his mind? Nicolaus refers to the human ability called ability to assimilate (cf. *Liber de mente*) as well as abstracting (cf. *De docta ignorantia*). Add that a form of an empiricism is also referred to when in *Liber de mente* Nicolaus writes: "[...] there is nothing in the mind that earlier was not in the senses".

Still another approach one finds by Immanuel Kant (1724–1804). According to him mathematical knowledge is not based on experience, it is a pure product of human mind. On the other hand it is synthetic. The source of concepts and objects of mathematics are time and space as *a priori* forms of our intuition (*Anschauung*), in particular time is the foundation of arithmetic and space – of geometry.

Kant did not claim that passive contemplation of the structure of time and space are enough to develop mathematics. On the contrary – the activity of the human mind is necessary. Human mind constructs mathematical concepts in the sense that going out of their verbal definitions it provides appropriate objects *a priori*.

It is necessary to stress here the distinction which Kant made between the thought of a mathematical concept and its construction. The former requires merely internal consistency while the latter requires that perceptual space should have a certain structure. Consequently one can postulate the existence of, for example, 5-dimensional sphere but one cannot construct it. Hence Kant did not deny the possibility of consistent geometries other than Euclidean one. So the development (construction or discovery?) of non-Euclidean geometries in the nineteenth century did not refute Kant's philosophy of mathematics.

Mathematical objects, in particular numbers, were treated as free product of the human mind also by Richard Dedekind (1831–1916). He was true to opinions of his master Carl Friedrich Gauss. Dedekind did not develop a complete conception of the philosophy of mathematics and did not make explicite claims concerning his philosophical ideas (accompanying or making the base of his strictly mathematical theories) but the idea mentioned above can be recognized and identified in his works at least by the terminology he was using. For example in the *Stetigkeit und irrationale Zahlen* (1872) where Dedekind developed the correct theory of real numbers (defining irrational numbers by cuts), he wrote:

> Always when a cut (A_1, A_2) is given that is not determined by a rational number, we *create* (*erschaffen wir*) [emphasis is mine – R.M.] a new number, irrational number α which we will treat as completely determined by this cut (A_1, A_2); it will be said that the number α corresponds to this cut or that it determines this cut.[3]

It is worth noting here that the very term "create" (*erschaffen*) is used in the Bible, in Genesis in the description of the creation of the world by God!

---

3 „Jedesmal nun, wenn ein Schnitt (A_1, A_2) vorliegt, welcher durch keine rationale Zahl hervorgebracht wird, so erschaffen wir eine neue, eine irrationale Zahl, welche wir als durch diesen Schnitt (A_1, A_2) vollständig definirt ansehen; wir werden sagen, daß die Zahl diesem Schnitt entspricht, oder daß sie diesen Schnitt hervorbringt."

Conceptions concerning the status of mathematical objects were explicitly formulated also by Henri Poincaré (1854–1912) who was considered as one of the predecessors of intuitionism. According to him the fundamental rôle in the mathematical knowledge is played by creative activity of the mind and by its ability to construct concepts. Objects of mathematics are constructed by a human being, there exists no domain of mathematical knowledge that would be independent of knowing subject. The creative rôle of the human mind is being demonstrated in various ways. One of them is intuition. In its scope there is – according to Poincaré – the possibility (preexisting in the human mind) to construct a concept of a group as a pure, and not sensorial form of knowledge. This conception played an important rôle in particular in Poincaré's philosophy of geometry in which he claimed conventionalism. By Poincaré all axioms of geometry are simply conventions. We are choosing these or those axioms on the base of experimental facts but the choice is free and is limited only by the necessity of avoiding inconsistencies (cf. Poincaré 1902). Hence the main criterion of the choice are not truth or falsity but the convenience of the formulation.

The idea that mathematical objects are creations of human mind one finds with the full clarity by intuitionists, in particular by the founder of this doctrine in the philosophy of mathematics Luitzen Egbertus Jan Brouwer (1881–1966). His main fundamental thesis is the rejection of platonism in the philosophy of mathematics. According to Brouwer mathematics is a function of the human intellect and a free activity of the human mind, it is a creation of the mind and not a theory or a system of rules and theorems. Mathematical objects are mental constructions of an (idealized) mathematician. As a consequence one should reject the axiomatic-deductive method as a method of developing and founding mathematics. It is not sufficient to postulate only the existence of mathematical objects (as it is done in the axiomatic method) but one must first construct them. One must also reject the actual infinity. An infinite set can be understood only as a law or a rule of forming more and more of its elements, but they will never exist as forming an actual totality. Hence there are no uncountable sets and no cardinal numbers other than $\aleph_0$.

The conceptualistic thesis of intuitionism implies also the rejection of any nonconstructive proofs of the existential theses, i.e., of proofs giving no constructions of the postulated objects. In fact in intuitionism "to be" equals "to be constructed".

According to Brouwer and the intuitionism mathematics is based on the fundamental intuition of time. One sees here the connection with the philosophy of Kant. Indeed the intuitionism accepts Kant's thesis about the *a priori* time and rejects his thesis about the *a priori* space. The intuition of time is a basis for the mental construction of natural numbers. Moreover this construction is the basic mathematical activity from which all other mathematical activities spring.

A mathematical theorem is a declaration that a certain mental construction has been completed. All mathematical constructions are independent of any language. Hence there is no language (formal or informal) which would be safe for mathematics and would protect it from inconsistencies. It is a mistake to analyze the language of mathematics instead of analyzing the mathematical thinking. Mathematics should be justified not "on paper" but "in the human mind".

As an intermediate (in a certain sense) position between realism (platonism) and conceptualism one can treat the conception of Aristotle (384–322 B. C.). He developed his philosophy of mathematics partly in opposition to that of Plato and partly independently of it. He rejected Plato's theory of ideas claiming that mathematical objects are forms of things, are idealizations obtained by the process of abstraction. Hence they do not exist timelessly and independently of things but are in a sense in things.

Hence a mathematician constructs and creates in a certain sense mathematical objects but there is no absolute freedom – on the contrary, he is limited and determined by the physical reality.

Another example of a conception that mathematical objects are created by human mind in the process of abstracting from the physical reality is the empiristic conception of John Stuart Mill (1806–1873). He developed the methodological version of the empiricism and attempted to justify it using logic. He applied it also to mathematics and attempted to argue that mathematics is in fact an empirical science. In particular he claimed that the source of mathematics is the reality perceived by senses. Mathematical concepts have been simply abstracted from the reality by omitting some properties of real objects and by generalizing and idealizing other properties. Hence mathematical propositions are not necessarily true. Their necessity can be reduced only to the fact that they are logical consequences of assumptions. But the very assumptions are far from being necessary and certain, on the contrary, they are only hypotheses and in fact one can adopt any propositions as assumptions.

Similar ideas will be proclaimed later by materialistic philosophers, in particular by marxists who claimed that all mathematical concepts are obtained in the process of abstraction and idealization from the material reality. A mathematician discovers only some dependencies and relations in the corporeal nature and expresses it using appropriate notions. Hence the subject of mathematics is in a certain sense similar to the subject of natural sciences.

4. Conclusions

So far we have presented various answers to the main question: does a mathematician discover mathematical objects and their properties or are they the product of human mind, are they human creations? The adoption of any of the positions, either realistic (platonistic) or conceptualistic or an intermediate has various consequences.

Note that the adoption of realism (platonism) implies that the problem of truth and being true in mathematics becomes very simple and easy. In fact, "true" means simply "being in accordance with (to correspond to) the mathematical reality". Hence one can speak about objective truth in mathematics. Any yes-or-no question has an objectively existing answer and solution. So there are no objectively undecidable problems. Consequently both the fifth postulate of Euclid and the continuum hypothesis and the axiom of choice (that are undecidable on the ground of axiomatic systems accepted today) have uniquely determined truth value, i.e., are true or false (in the mathematical world). The task of a mathematician consists only of finding appropriate axioms that can be accepted by specialists and would enable to decide a given topic. Hence Gödel was convinced of the necessity of looking for new axioms of set theory that would better describe the world of sets and its properties and would enable us to decide the continuum hypothesis and the axiom of choice. He tried to find them among infinity axioms postulating the existence of large cardinals. He hoped that such axioms would help not only to answer set-theoretical questions but also to solve various problems concerning for example the arithmetic of natural numbers.

And similarly for other unsolved (so far) mathematical hypotheses, for example with Riemann's hypothesis of Goldbach's hypothesis. They are objectively true or false, the problem is only to decide which of the both possibilities in fact does hold.

Here arises the next problem – the problem of adopting new axioms. Where should they come from? The platonistic answer is: they should simply describe true properties of objects under consideration. But how can we obtain any knowledge about that objective and given to us abstract reality? Various answers can be found: Plato spoke about *anamnesis*, Gödel about intuition. Add that according to Gödel new axioms might be justified also from outside, i.e., by their consequences, hence by the fact that they enable us to solve problems which were unsolvable so far. He meant here both consequences in mathematics and outside mathematics, e.g., in physics.

The realistic (platonic) position meets some difficulties. In fact, how can one explain (without referring to formalism) the existence of various mutually inconsistent axiomatic systems of geometry and set theory? Only one of them can be the true description of mathematical reality. Which one? And what do the others describe, what are they talking about?

By Gödel's completeness theorem, any consistent (first order) theory has a (countable) model. On the other hand by Gödel's incompleteness theorem there are (in the arithmetic of natural numbers and in any richer theory) undecidable sentences. Hence theories PA + φ_G and PA + $\neg\varphi_G$ (where PA is Peano arithmetic and φ_G denotes Gödel's undecidable sentence) are consistent provided PA itself is consistent. Hence they have models. What are those models? And what do they describe? What are their relations towards the platonistic world of mathematical objects? Where and in what sense do they exist?

If one adopts the conception that a mathematician creates and constructs mathematical objects then the question about truth and falsity has no sense any more. Hence the question whether the axiom of choice or the continuum hypothesis are true or the question which of the geometries: Euclidean or non-Euclidean is true, are senseless. One can speak only about provability or unprovability of certain statements in a given theory. And from here there is only one step to strict formalism claiming that mathematics is nothing more than a game of symbols (*Glassperlenspiel*).

Another consequence of adopting conceptualism is the necessity of rejecting actual infinity and the possibility of accepting only the countable potential one what is in fact a big restriction.

Some forms of conceptualism (e.g., intuitionism) force also the rejection of certain methods of proof, for example nonconstructive proofs of existence statements based on the application of the laws of excluded middle and/or of the double negation. This implies limitations of acceptable results and acceptable methods (hence a limitation in the methodological sense).

The realistic attitude allows to use in mathematics nonpredicative definitions whereas the conceptualism sees in them the source of all antinomies and inconsistencies. A platonist can assume the existence of everything that is consistent but a conceptualist is allowed to use in the case of existence statements only certain accepted methods and should construct (whatever this means) a postulated object.

There arises another problem: on which base, with reference to what can mathematical objects be constructed? One speaks here about abstracting from the physical/corporeal/sensory reality, about intuition of *a priori* time and space. This suggests that a mathematician is in a certain sense determined and restricted. On the other hand in the framework of conceptualism, various ways of grasping the ontological status of mathematical objects are possible. In particular one can distinguish:

– objectivism claiming that mathematical objects are objective products of constructing processes and do exist independently of the constructing subject,

– intentionalism according to which mathematical objects do exist intentionally (in such a way as cultural elements, in particular musical or literary works do),

– mentalism which treats mathematical objects as products of mental acts and claims that they exist only in those acts,

– an attitude claiming that mathematical objects are concrete objects existing in space and time.

One should stress that just thanks to his platonistic attitude Gödel was able to obtain incompleteness theorems (he admitted this himself!) – in fact without platonism their deep sense disappears! On the other hand the intellectual atmosphere of his time – in particular the attitude of Vienna Circle as well as the ideas of Hilbert's formalism and finitism, hence the stressing of the rôle of syntax, language and of the finitistic effective methods – forced him to avoid in his writings any notice of his platonism and even to avoid the terms "true" and "truth" themselves.

* * *

So there are two main positions: realism/platonism according to which even if there were no mathematician at all there would be the world of numbers, geometrical figures and other mathematical objects as well as their mutual relations though they would be not described in any language and there would be no mathematics as a collection of definitions and theorems. An eternal, independent and constant reality of mathematical objects is given to us and the task of a mathematician is to describe it. Consequently a mathematician *discovers* the objective truth and *describes* it.

On the other hand one has conceptualism according to which a mathematician is a creator or constructor. There disappears an element of absoluteness and a certain form of relativism appears. If one takes into account the influence of culture in the frame of which mathematics is being developed then the picture becomes more complicated. A mathematician cannot say that he/she discovers the objective truth – on the contrary he/she does only *construct* and *develop* systems and theories. The latter can prove to be more or less useful in certain domains and situations, more or less elegant, more or less constructive, but they are only products of a human being.

Add at the end that still one position is possible – one can claim that there are in fact no mathematical objects at all. All we call mathematical objects are only useful fictions which enable us to express and formulate certain dependencies and relations (for example in natural sciences) in a simple and well-phrased way but in fact they are superfluous, hence in particular one can develop physics without mathematical concepts. Such an attitude is called nominalism and it has its sources in the medieval controversy concerning the existence and nature of universals. Its adherents were among others B. Russell or Polish logicians such as A. Tarski, T. Kotarbiński, S. Leśniewski.[4]

What is the answer to our starting question? Did Leibniz and Newton discover or create the differential and integral calculus? The answer is: we do not know. There are various theories and conceptions and each of them has its arguments. Fortunately this does not disturb the development of mathematics! Mathematics is being developed despite of the fact that the philosophical questions concerning it are still unsolved. On the other hand the observation shows that most of mathematicians behave in their research work like platonists (even if they declare to be formalists or conceptualists).

---

[4] It should be stressed that this attitude had in fact no influence on their logical and (meta)mathematical studies.

Structuralism and Category Theory in the Contemporary Philosophy of Mathematics

Co-authored by Izabela Bondecka-Krzykowska

1. One of the most popular trends of the contemporary philosophy of mathematics is structuralism usually connected with the slogan: mathematics is the science of structures. Mathematical structuralism can be characterized as a view that the subject of any branch of mathematics is a structure or structures. For example, we can define a natural number system to be a countably infinite collection of objects with one distinguished initial object and the successor relation that satisfy the principle of mathematical induction. Therefore the natural number is nothing more then a place in the structure of natural numbers. According to the structuralism, arithmetic is a science about the form or structure common to natural number systems.

Structuralism is consonant with current mathematical practice at least in two points: (1) objects considered by mathematicians are determined up to isomorphism, (2) at least some features of mathematical objects, some mathematical facts about them, depend solely on their structure. But what is a mathematical structure? What do we mean by "having the same structure"?

In the contemporary philosophy of mathematics various structuralistic conceptions have been formulated. They differ with respect to the way of defining structures and their existence. One can divide them into two groups corresponding to two main perspectives for the structural mathematics: foundational (set-theoretical) and categorical one.

Let us begin with the foundational (set-theoretical) perspective in defining structures due to Bourbaki. Bourbaki structure is a domain of objects with some relations and functions defined on it. In this case we use such terms as set, function or relation, which are terms of set theory. Thus we can apply methods of model theory to investigate it.1 Such descriptions of mathematical objects leads us to a useful structural perspective but in many cases methods of model theory itself do not suffice to describe mathematical structures well enough.

Generally among set-theoretic structuralistic conceptions one can distinguish two main attitudes towards ontological problems:

(a) *in re* structuralism (called also eliminative structuralism) which states in particular that all statements about numbers are only generalizations. The *in re* structuralism claims that the natural number structure is nothing more than systems which are its instantiations. If such particular systems were destroyed then there would be also no structure of natural numbers.

(b) *ante rem* structuralism which claims that structures do exist apart from the existence of their particular examples. It is often said that *ante rem* structures have ontological priority with respect to their instantiations.

The main thesis of the eliminative structuralism is: statements about some kind of objects should be treated as universal statements about specific kind of structures. Thus number theory examines properties of all structures of order type. In case of arithmetic every sentence A expresses a property of all natural number systems, and can be understood as an implication: *For every system S, if S is an instance of natural number system, then A(S)*.

This treatment of the natural numbers rests on two claims: the claim that simply infinite systems do exist and on the categoricity theorem. It is necessary to prove the existence of a natural number system, otherwise the above implication is always true, because every implication with a false predecessor is always true. So eliminative structuralism needs a basic ontology, a domain of considerations whose objects could take up places in structures *in re*. Such an ontology should be rich enough and we are not interested in the very nature of objects but rather in their quantity. The ontology of the *in re* structuralism requires an infinite base. One of the methods proposed to eliminate this problem is to apply modalities. Hellman introduced, using second order modal logic, a theory containing the axiom stating the possibility of the existence of an infinite system.

In any case, in taking structures to be objects, we either run into the problem of having to assume a foundational background (eliminative structuralism) or of 'reification of structures' (*ante rem* structuralism) or we make mathematics dependent of the logic of possibility (modal structuralism).

2. The object of modern mathematical studies is rarely a specific set with relations or functions defined on it. As said above mathematicians investigate mostly objects determined up to isomorphism, relations between such objects bearing the same structure, relations between different kinds of structures on such objects and so on. So there is a need for a language and methods well suited to problems involving different kinds of structures. In response to this need category theory arose.

Category theory is an algebraic theory, which is a general mathematical theory of structures and of systems of structures. It is still evolving. At minimum, it is a powerful language, or conceptual framework, allowing us to see the common parts of a family of structures of a given kind as well as how structures of different kinds are interrelated.

The central rôle in this theory is played by the notion of category, which consists of objects $A, B, C, \ldots$ and morphism $f, g, h, \ldots$ such that:

(i) every f has a unique domain A and a unique codomain B, written $f: A \to B$;

(ii) given any $g: B \to C$ there is a unique composite $g \circ f: A \to C$, with composition being associative;

(iii) each B has an identity $1_B: B \to B$, which is a unit for composition, i.e., $1_B \circ f = f$ and $g \circ 1_B = g$ for any f and g as stated. A category is anything satisfying these axioms.

The objects need not to have elements; nor need the morphism be functions (for example category associated with any formal system of logic is a category, the objects of which are formulas and morphisms of which are deductions from premises).

Consider now whether a categorical perspective in structuralism is better, at least in some points, then the foundational (set-theoretical) one.

First of all the categorical notion of a structure is "syntax invariant", it does not depend on particular choice among the different possible set theoretic descriptions of a given kind of mathematical structures. For example spaces may be defined in several ways, the objects of the category Top (i.e., the category of all topological spaces) are described by various different Bourbaki structures.

The categorical notion of an isomorphism may serve as a definition of "having the same structure of a given type". Category theory provides a uniform notion of a structure: given any category, one automatically knows the right notion of having the same structure. Two objects may be said to bear the same structure if they are structurally indistinguishable, i.e., if any structural property enjoyed by one is also enjoyed by the other.

According to structuralism objects of mathematics (such as numbers, functions or points) are only places in structures, they do not have any properties which are not structural. Structuralists claim that mathematical objects have no important features outside the structure and all of their features have to and can be explained in terms of structural relations. For example the number 2 is nothing more then the successor of 1 and the predecessor of 3, so the essence of a natural number (for example 2) is determined by relations to other natural numbers (1 and 3). (Thus arithmetic is the science about relations between places of any system similar to the structure of natural numbers.) Category theory allows us to express structural properties of objects in a convenient way. Any mathematical property or construction given in terms of structure preserving mappings (in a given category) will necessarily respect isomorphism in that category and thus will be structural. Since all categorical properties are structural, the only properties which a given object in a given category may have, qua object in that category, are structural ones. As Awodey states in (1996, pp. 214–215):

> Thus doing mathematics 'arrow-theoretically' automatically provides a structural approach, and this has proven quite effective in attacking certain kinds of mathematical problems having to do with mathematical structure.

Furthermore many useful categories describe some structures which are not structures in the sense of Bourbaki. For example category whose objects are the open sets of a particular space and whose morphisms are inclusion maps between these is a kind of a mathematical structure on objects which is not a model of a Bourbaki structure in any conventional sense.

A further and very important advantage of the categorical approach to mathematical structure is that representing different kinds of structures as different categories provides uniform notion of a structure. For example from a categorical point of view, a Cartesian product in set theory, a direct product of groups (Abelian or otherwise), a product of topological spaces, and a conjunction of propositions in a deductive system are all instances of a categorical product characterized by a universal property.

Formally, a *product* of two objects X and Y in a category **C** is an object Z of **C** together with two morphisms, called the projections, $p\colon Z \to X$ and $q\colon Z \to Y$ such that – and this is the universal property – for all objects W with morphisms $f\colon W \to X$ and $g\colon W \to Y$, there is a unique morphism $h\colon W \to Z$ such that $p \circ h = f$ and $q \circ h = g$.

It is totally consonant with mathematical structuralism. Note that we have defined a product for X and Y and not the product for X and Y. Indeed, products and other objects with a universal property are defined only up to a (unique) isomorphism.

In category theory, the nature of elements constituting a certain construction is irrelevant. What matters is the way in which an object is related to other objects of the category, that is, the morphisms going in and the morphisms going out, or, in other words, how certain structures can be mapped into a given object and how a given object can map its structure into other structures of the same kind.

Category theory reveals how different kinds of structures are related to one another (it is not so easy in the case of set-theoretic approach to structuralism). For instance, in algebraic topology, topological spaces are related to groups (and modules, rings, etc.) in various ways (such as homology, cohomology, homotopy, K-theory). Groups with group homomorphisms constitute a category. Eilenberg and Mac Lane invented category theory precisely in order to clarify and compare these connections. What matters are the morphisms between categories, given by functors. Homology, cohomology, homotopy, K-theory are all examples of functors. Informally, functors are structure-preserving maps between categories. Given two categories **C** and **D**, a functor F from **C** to **D** sends objects of **C** to objects of **D**, and morphisms of **C** to morphisms of **D**, in such a way that composition of morphisms in **C** is preserved, i.e., $F(g \circ f) = F(g) \circ F(f)$, and identity morphisms are preserved, i.e., $F(id_X) = id_{FX}$. It immediately follows that a functor preserves commutativity of diagrams between categories.

Following Awodey (1996) we can characterize categorical structuralism in the following way (p. 235):

> The structural perspective on mathematics codified by categorical methods might be summarized in the slogan: The subject matter of pure mathematics is invariant form, not a universe of mathematical objects consisting of logical atoms.

3. Some philosophers claim that category theory is an alternative to set theory as a foundation for mathematics and that methods of category theory will suffice for many present-day mathematical purposes. But there are some problems and objections connected with this claim.

The first one is a problem of the autonomy of category theory from set theory. Is category theory really independent from set theory? If we agree that category theory uses set-theoretic notions such as domain, codomain and function, then structuralism framed by category theory falls under set-theoretic variety of structuralism. Moreover category theory can not be treated as an alternative for set theory in any reasonable sense of 'alternative'.

Another important problem announced by Hellman in (2003) is a problem of mathematical existence. "This problem as it confronts category theory can be put very simply: the question just does not seem to be addressed! (*We might dub this the problem of the 'home address': where do categories come from and where do they live?*)" (p. 136)

Axioms defining categories include existence claims, but if we want to read this axioms 'structurally' (á la algebra), they are only defining conditions, not absolute assertions of truths based on established meanings of primitive terms (the axioms of set theory, as usually read, are not 'structural' in this sense).

To sum up, Hellman (2003) claims that category theory is defective as a framework for structuralism in at least two major interrelated ways: (1) it is not independent from set theory and (2) it lacks substantive axioms of mathematical existence.

As Awodey noticed in (2004) the questions asked about mathematical existence such as: "Where do categories come from and where do they exist" are reasonable only from the foundational perspective. He proposed to use category theory to avoid the whole business of 'foundations'. The idea of 'doing mathematics categorically' involves a point of view different from the foundational one, which is based on the idea of specifying for a given theorem or theory only the required or relevant degree of information or structure, the important features of a given situation, without assuming some knowledge or specification of the 'objects' involved. He writes in (2004, p. 56):

> The laws, rules and axioms involved in a particular piece of reasoning, or a field of mathematics, may vary from one to the next, or even from one mathematician or epoch to another.

Mathematical theorems are schematic, they do not involve the specific nature of structures or their components in an absolute sense. It does not matter what structures are supposed to be or to 'consists of'. In mathematical statements particular nature of the entities involved plays no rôle. Rather their relations, operations, etc. are important and crucial. In this sense mathematical statements (theorems, proofs, even definitions) are about connections, operations, relations, properties of connections, operations on relations, connections between those operations and so on.

Thus according to this view there is no absolute universe of all mathematical objects, there is no unique context that provides us with conditions for the actual or possible existence of structure or structured systems. In a categorical framework the context, systematized by the category-theoretic axioms, varies, so mathematical concepts has to be thought of in a context that can be varied in a systematic fashion. Categorical framework provides us with the conditions a context has to satisfy in order to talk about or to do mathematics. So we cannot say what the natural numbers are, but in which contexts we can talk about them.

Category theory describes conditions under which we can talk about the same type of systems. Category should be treated not as a system of statements about objects (i.e., neither about "structures" nor about possible types of systems possessing a structure), but rather as a context describing conditions, which have to be fulfilled to talk about particular type of objects. Axioms of a given category provide context in which one can talk about the common structure of systems in terms of morphism between them, without necessity of appealing to the theory of sets, theory of structures or modal logic.

Supporter of such structuralism does not have to determine what is a structure or what is a category, in ontological or modal sense of the word "is". Everything what has to be done is to provide a proper context, in which one can talk about a common structure of systems.

An advantage of such an approach to structuralism is that it does not provide "constructive basis" for mathematics, but rather provides "descriptive basis" for the structuralistic claim that mathematics is a science of structures (it is interpreted as a claim that mathematics is a science of systems possessing structures).

So what is the difference between set-theoretical structuralism and the categorical one? Hale named categorical structuralism the *pure* structuralism and described it as algebraic structuralism *in re* from the *top-down* perspective.

Now the natural question appears: what is the difference between *top-down* structuralism and the *bottom-up* (set-theoretical) one? Structuralism from the *bottom-up* perspective should have, as said above, a basic ontology. The notion of structure is built from the objects of this ontology in the process of abstraction. "The direction" of this abstraction is clear: from details to the whole, so *bottom-up*. For all versions of set-theoretical structuralism the same conditions, actual or modal, for the existence or possible existence of systems possessing structure, have to be assumed.

In the case of *top-down* structuralism this demand can be omitted by introducing a basic theory in Hilbert's sense. Instead of asking what structures are, there appears the question: what does it mean that two systems have the same structure. *Top-down* structuralism is called pure because axioms of a category provide a framework for talking about particular structural systems without considering what those systems are built from.

In the *top-down* perspective one starts from the concept of an abstract system, in the algebraic sense, understood as a language for description of the common structure of systems: it allows to talk about systems possessing the same structure as examples of the same type of structure without the necessity of considering from what those systems are built. So in this perspective instead of asking what the structure is one asks what does it mean "to have the same structure".

Therefore category theory provides a framework for *top-down*, *in-re* interpretations of mathematical structuralism, because category provides context, in which one can talk about "common structure" of systems, regardless of what this systems are built from. Such *top-down* algebraic structuralism, expressed in the language of the category theory, does not require neither treating structures as "objects" (actual or possible) nor understanding axioms as truths or assertions. In contrast to Shapiro, categorical structuralist does not have to claim that categories exist as objects independently of abstract systems, which are examples of them: he/she does not even claim that categories exist in the sense of "objects" in some system. Categorical structuralism can be summarized by words of Awodey (1996, p. 235):

> The subject matter of pure mathematics is an invariant form and not a universe of mathematical objects consisting of logical atoms.

To sum up, we must distinguish the claim that category theory can be the language of mathematical structuralism from the claim that it can be an alternative for set theory as a basic theory for mathematics. Category theory is a more convenient tool for exploring mathematical structuralism than set theory but one should be careful to claim that category theory can serve as a foundation for the whole of mathematics. Indeed, it is not clear if category theory is really independent from set theory, moreover we do not know enough about the ontological and epistemological status of categories. Category theory is the useful language for talking about mathematical structuralism but it is not a tool for "doing" mathematics structurally.

Part II
Hilbert's Program vs. Incompleteness Phenomenon

Hilbert's Program:
Incompleteness Theorems vs. Partial Realizations

1. Hilbert's Program

Mathematics on the turn of the 19th century was characterized by the intense development on the one hand and by the appearance of some difficulties in its foundations on the other. Main controversy centered around the problem of the legitimacy of abstract objects. The works of K. Weierstrass have contributed to the clarification of the rôle of the infinite in calculus. Set theory founded and developed by G. Cantor promised to mathematics new heights of generality, clarity and rigor. Unfortunately paradoxes appeared. Some of them were known already to Cantor (e.g., the paradox of the set of all ordinals and the paradox of the set of all sets[1]) and they could be removed by appropriate modifications of set theory (cf. Cantor's distinction between *absolut unendliche* or *inkonsistente Vielheiten* and *konsistente Vielheiten*, i.e., between classes and sets[2]). Frege's attempt to realize the idea of the reduction of mathematics to logic (which was in fact a continuation of the idea of the arithmetization of analysis developed among others by Weierstrass) led to a really embarrassing contradiction discovered in Frege's system by B. Russel and known today as Russell's antinomy or as the antinomy of nonreflexive classes. This meant a crisis of the foundations of mathematics (called the second crisis the first being the crisis caused by the discovery of incommensurable segments in the ancient Greek mathematics).

Various ways of overcoming those difficulties and of securing the edifice of mathematics were proposed. Great mathematicians, e.g., L. Kronecker, H. Poincaré, L. E. J. Brouwer, H. Weyl challenged the validity of all infinitistic reasonings and proposed to restrict methods of mathematics to secure finite ones. L. Kronecker rejected any infinite objects restricting mathematics to integers only (*"Die ganzen Zahlen hat der liebe Gott gemacht, alles andere ist Menschenwerk"* as he formulated his scientific and methodological credo during a meeting in Berlin in 1886). H. Poincaré saw the source of antinomies in impredicativity of mathematics and demanded a restriction to predicative methods only. The radical proposal of L. E. J. Brouwer, known today as intuitionism, was based on the idea that mathematics should be founded on the primitive intuition of natural number. He claimed that mathematics is a free activity of human mind, it can (and should) be developed independently of any language, one should restrict only to constructive methods,

---

[1] The first paradox was communicated by G. Cantor in a letter to R. Dedekind from 1896, the second in a letter to R. Dedekind from 31st August 1899 (cf. Cantor 1932, p. 448; see also Murawski 1984b).

[2] Cf. G. Cantor's letters to R. Dedekind from 28th July 1899 (Cantor 1932, pp. 443–447) and from 31st August 1899 (Cantor 1932, p. 448; see also Murawski 1984b).

hence in particular any nonconstructive proofs of existential sentences should be rejected (Brouwer claimed that proofs of that type were the source of all antinomies). He accepted only countable infinity but rejected any uncountable one.

All those proposals meant in fact a restriction of mathematics and a rejection of a great part of it, especially that dealing with infinite objects. D. Hilbert was definitely against it. He wrote (cf. Reid 1970, p. 155):

> What Weyl and Brouwer do come to the same thing as to follow in the footsteps of Kronecker! They seek to save mathematics by throwing overboard all that which is troublesome. [...] They should chop up and mangle the science. If we would follow such a reform as the one they suggest, we would run the risk of losing a great part of our most valuable treasures!

And added (cf. Reid 1970, p. 157):

> I believe that as little as Kronecker was able to abolish the irrational numbers [...] just as little will Weyl and Brouwer today be able to succeed. Brouwer is not, as Weyl believes him to be, the Revolution only the repetition of an attempted Putch.

And he stressed firmly that (cf. Hilbert 1926):

> *Aus dem Paradis, das Cantor uns geschaffen hat, soll uns niemand vertreiben können.*

Hilbert proposed a method of justification of (infinite) mathematics known today as Hilbert's program. It was the core of a new doctrine in the philosophy of mathematics called formalism (which became one of the main trends of the modern philosophy of mathematics beside Frege's and Russell's logicism and Brouwer's intuitionism).

Hilbert was first of all a mathematician and – as Smoryński writes – "had little patience with philosophy, his own philosophy of mathematics being perhaps best described as naïve optimism – a faith in the mathematician's ability to solve any problem he might set for himself" (cf. Smoryński 1988). His aim was to save the integrity of classical mathematics (dealing with actual infinity) by showing that it is secure.[3] This problem was first stated by him in his lecture at the Second International Congress of Mathematicians held in Paris in 1900 (cf. Hilbert 1901), then repeated in a number of articles in the twenties where he proposed a method of solving it (a good account of the development of Hilbert's views can be found in Smoryński (1988); see also Peckhaus (1990) where detailed analysis of Hilbert's scientific activity in the field of foundations of mathematics in the

---

3 Detlefsen writes that "Hilbert did want to preserve classical mathematics, but this was not for him an end in itself. What he valued in classical mathematics was its efficiency (including its psychological naturalness) as a mean of locating the truths of real or finitary mathematics. Hence, any alternative to classical mathematics having the same benefits of efficiency would presumably have been equally welcome to Hilbert" (cf. Detlefsen 1990, p. 374).

period 1899–1917 can be found). Hilbert saw the supramathematical significance of this issue. He wrote in (1926):

> The definitive clarification of the nature of the infinite has become necessary, not merely for the special interests of the individual sciences but for the honor of human understanding itself.

Hilbert's program of clarification and justification of mathematics was kantian in character. Following Kant he claimed that the mathematician's infinity does not correspond to anything in the physical world, that it is "an idea of pure reason" as Kant used to say. On the other hand (cf. (Hilbert 1926):

> Kant taught and it is an integral part of his doctrine that mathematics treats a subject matter which is given independently of logic. Mathematics, therefore can never be grounded solely on logic. Consequently, Frege's and Dedekind's attempts to so ground it were doomed to failure.
>
> As a further precondition for using logical deduction and carrying out logical operations, something must be given in conception, viz., certain extralogical concrete objects which are intuited as directly experienced prior to all thinking. For logical deduction to be certain, we must be able to see every aspect of these objects, and their properties, differences, sequences, and contiguities must be given, together with the objects themselves, as something which cannot be reduced to something else and which requires no reduction. This is the basic philosophy which I find necessary not just for mathematics, but for all scientific thinking, understanding and communicating. The subject matter of mathematics is, in accordance with this theory, the concrete symbols themselves whose structure is immediately clear and recognizable.

According to this Hilbert distinguished between the unproblematic, finitistic part of mathematics and the infinitistic part which needed justification. Finitistic mathematics deals with so called real sentences, which are completely meaningful because they refer only to given concrete objects. Infinitistic mathematics on the other hand deals with so called ideal sentences that contain reference to infinite totalities. Hilbert believed that every true finitary proposition had a finitary proof. Infinitistic objects and methods played only an auxiliary rôle. They enabled us to give easier, shorter and more elegant proofs but every such proof could be replaced by a finitary one. He was convinced that consistency implies existence and that every proof of existence not giving a construction of postulated objects is in fact a presage of such a construction. (Compare in connection with this Hilbert's solution in 1888 to Gordan's problem in the theory of invariants in which he proved, without construction, the existence of a finite base for any ideal in the polynomial ring $K[X_0,\ldots,X_{n-1}]$ over a field K.)

Unfortunately Hilbert did not give a precise definition of finitism – one finds by him only some hints how to understand it. Hence various interpretations are possible. Usually it is assumed that a finitist reasoning is essentially a primitive recursive reasoning in the sense of Skolem (cf. Tait 1981, Resnik 1974). But there are also other interpretations, cf., e.g., Detlefsen (1979) where it is suggested that even some variants of ω-rule can be regarded as finitistic or Smoryński (1988) where instead of a dichotomy real/ideal a trichotomy real/finitary general/ideal is proposed (cf. also the criticism of this proposal in Detlefsen 1990). Prawitz argues in (1981) that real sentences are the decidable ones (i.e., numerical equations and truth functional compositions of them) as well as formulas of the form $\forall x A(x)$ where each instance $A(t)$ is decidable. The rest are considered to be ideal. This emphasizes the rôle of Π_1^0 sentences in Hilbert's program (cf. Kitcher 1976; Tait 1981).

The infinitistic mathematics can be justified only by finitistic methods because only they can give it security (*Sicherheit*). Hilbert's proposal was to base mathematics on finitistic mathematics via proof theory. Its main goal was to show that proofs which use ideal elements in order to prove results in the real part of mathematics always yield correct results. One can distinguish here two aspects: consistency problem and conservation problem. In some of Hilbert's publications (cf., e.g., Hilbert 1926 and 1927) both aspects are stressed but usually (cf., e.g., Hilbert's last publication on this subject, namely the first volume of Hilbert and Bernays 1934, 1939) the onesided emphasis is put on the consistency problem only (cf. also Kreisel 1968; 1976) where the author calls for a proper formulation taking into account both aspects). The consistency problem consists in showing (by finitistic methods, of course) that the infinitistic mathematics is consistent; the conservation problem consists in showing by finitistic methods that any real sentence which can be proved in the infinitistic part of mathematics can be proved also in the finitistic part, i.e., that infinitistic mathematics is conservative over finitistic mathematics with respect to real sentences and, even more, that there is a finitistic method of translating infinitistic proofs of real sentences into finitistic ones. Both those aspects are interconnected (what was indicated by Kreisel – we shall discuss this problem later).

Hilbert's proposal to carry out this program consisted of two steps. The first step was to formalize mathematics, i.e., to reconstitute infinitistic mathematics as a big, elaborate formal system (containing classical logic, infinite set theory, arithmetic of natural numbers, analysis). An artificial symbolic language and rules of building well-formed formulas should be fixed. Next axioms and rules of inference (referring only to the form, to the shape of formulas and not to their sense or meaning) ought to be introduced. In such a way theorems of mathematics become those formulas of the formal language which have a formal proof based on a given set of axioms and given rules of inference. There was one condition put on the set of axioms (and rules of inference): they ought to be chosen in such a way that they suffice to solve any problem formulated in the language of the considered theory as a real sentence, i.e., they ought to form a complete set of axioms with respect to real sentences.

The second step of Hilbert s program was to give a proof of the consistency and conservativeness of mathematics. Such a proof should be carried out by finitistic methods. This was possible since the formulas of the system of formalized mathematics are strings of symbols and proofs are strings of formulas. i.e., strings of strings of symbols. As such they can be manipulated finitistically. To prove the consistency it suffices to show that there are not two sequences of formulas (two formal proofs) such that one of them has as its end element a formula φ and the other $\neg\varphi$ (the negation of the formula φ). To show conservativeness it should be proved that any proof of a real sentence can be transformed into a proof not referring to ideal objects.

One should note here that formalization was for Hilbert only an instrument used to prove the correctness of (infinitistic) mathematics. Hence the objections raised to him by Brouwer are mistaken. As indicated in Kreisel (1964) the real opposition between Brouwer's and Hilbert's approach to mathematics was between: (i) the conception of what constitutes a foundation and (ii) two informal ways of reasoning, namely finitist and intuitionist. Recall that Brouwer ignored nonconstructive mathematics altogether.

Note also that if one identifies real sentences with Π_1^0 sentences (see above) then – as shown by Kreisel – a solution to the consistency problem yields a solution to the conservation problem (Kreisel's results are presented, e.g., in Smoryński 1977, pp. 858–860).

2. Incompleteness Results

Hilbert and his school had scored some successes in realization of the program of justifying infinite mathematics. In particular Hilbert's student W. Ackermann showed by finitistic methods the consistency of a fragment of arithmetic of natural numbers (cf. Ackermann 1924/1925 and 1940). But soon something was to happen that undermined Hilbert's program.

In September 1930 a Conference on Epistemology of Exact Sciences (organized by Gesellschaft für empirische Philosophie) was held in Königsberg. On 7th September (the last day of the conference) a young Austrian mathematician Kurt Gödel presented a short announcement in which he reported on his recent result on incompleteness of arithmetic of natural numbers. The result known today as Gödel's first incompleteness theorem was published in January 1931 in a paper "Über formal unentscheidbare Sätze der Principia Mathematica und verwandter Systeme. I" (cf. Gödel 1931a). It states, roughly speaking, that arithmetic of natural numbers and all systems containing it are essentially incomplete provided they are consistent. It means that there are sentences which are undecidable in them, i.e., sentences φ such that neither φ nor $\neg\varphi$ are theorems. What more one knows which sentence of the pair φ, $\neg\varphi$ is true (in the basic model of the theory, i.e., in the model to the description of which the theory was formulated). This incompleteness is essential, i.e., it cannot be removed by adding the undecidable sentences as new axioms because new undecidable sentences will appear (undecidable in the new stronger theory).

Hence Gödel's theorem indicated certain cognitive limitations of the deductive method. It showed that one cannot include the whole mathematics in a consistent formalized system based on the first-order predicate calculus, moreover, one cannot even include in such a system all truths about natural numbers. There will be always, as Gödel proved, undecidable sentences of the form $\forall x \varphi(x)$ such that all instances of φ, i.e., sentences $\varphi(0)$, $\varphi(1)$, $\varphi(2)$,... are theorems. Gödel's undecidable sentence constructed by the diagonal method stated its own unprovability ("I am not a theorem") (the construction of such a sentence was possible thanks to Gödel's idea of arithmetization of syntax). On the other hand one can prove, using some infinitistic methods (model-theoretical or set-theoretical ones) that this sentence is true. Hence we have an example of a (real) sentence (referring to natural numbers only) which can be proved by some infinitistic methods but which has no arithmetical proof.

Gödel's paper contained also at the end an announcement (with a promise to give a proof in the second part of the paper which in fact was never written[4]) of another theorem, called today Gödel's second incompleteness theorem and stating that no formal theory containing arithmetic of natural numbers can prove its own consistency.

Note that Gödel's remark (in the paper from 1931) that one can prove the second incompleteness theorem by formalizing the proof of the first incompleteness theorem was not correct. The first full proof of the unprovability of consistency was given in Hilbert and Bernays (1934, 1939). It has turned out that the way of formalizing the metamathematical sentence "the theory T is consistent" in the formal language of T is significant. Hilbert and Bernays formulated certain so called derivability conditions for formulas representing in T the metamathematical notion of provability (in fact those conditions require certain internal properties of provability to be formally derivable in T). If those conditions are fulfilled then the second incompleteness theorem holds. M. H. Löb gave in 1954 another, more elegant form of derivability conditions (cf. Löb 1955). It was proved also that there exist formal translations of the sentence "T is consistent" which are provable in T (hence for them Gödel's second theorem fails). An example of such a formula was given in Rosser (1936). A detailed analysis of this problem can be found in Feferman (1960).

Gödel's results struck Hilbert's program. Did they reject it? This question cannot be answered definitely for the simple reason that Hilbert's program was not formulated precisely enough, it used vague terms as finitistic, real, ideal which

---

4   Gödel explained it by the fact that in the meantime the theorem became well known, hence no such proof was needed any longer.

Hilbert's Program: Incompleteness Theorems vs. Partial Realizations 89

were never precisely defined. In this situation various opinions are formulated and defended.

On the one hand it is claimed that Gödel's incompleteness results showed the failure of Hilbert's program – compare, for example, Smoryński (1977; 1985; 1988).

On the other hand it is argued that it is not the case. Several reasons are given here. For example Detlefsen (1979) says that the second Gödel's theorem does not imply the rejection of Hilbert's proposal because the unprovable formal sentence stating the consistency of the theory "does not really 'express' consistency" in the sense meant by Hilbert. In Detlefsen (1990) it is argued that Gödel's first incompleteness theorem does not refute the program because Hilbert did not demand the conservation property but only a weak conservation, i.e., conservation with respect to real sentences which can be finitistically decided and he did not claim that every real sentence may be decided in such a way.[5]

As indicated above, Gödel's second incompleteness theorem requires certain assumptions about the formal translation of some metamathematical notions (such as proof, provability, etc.). Detlefsen observes in (1990) that they are not satisfied by various theories or 'theory-like' arrangements of proofs and theorems which nonetheless do satisfy the conditions required by Gödel's first theorem.[6] He considers so called 'consistency-minded' theories which incorporate consistency constraints into the very conditions on proof, provability, etc. Two types of such theories are distinguished: Rosser systems (studied in Rosser 1936; Kreisel and Takeuti 1974; Guaspari and Solovay 1979; Visser 1989; Arai 1990) and Feferman systems (introduced in Feferman 1960 and studied in Jaroslov 1975 and Visser 1989). Since some of those theories constitute plausible ways in which the Hilbertian might go about constructing his ideal theories, Detlefsen formulates the following open question:[7] "whether G2 [i.e., Gödel's incompleteness theorem – R.M.]

---

5   Let us remark here that Detlefsen's criticism of Smoryński's thesis that Gödel's first incompleteness theorem refutes Hilbert's program is mistaken. The argument of Smoryński does not need in fact the assumption that an ideal theory T is complete with respect to *all* real sentences. It is enough to know that a particular sentence (i.e., Gödel's undecidable sentence) is true and this can be shown for example by set-theoretical methods. Hence we get a sentence which can be proved in an ideal theory but which is undecidable (hence cannot be proved nor disproved) in a real, finitistic part of mathematics.

6   The first incompleteness theorem requires only that the set of theorems is representable, i.e., the formal system considered may be "identified" with its set of theorems. Consequently, all we need to know about it is its set of theorems. Recall that the second incompleteness theorem requires certain internal properties of the provability to be formally derivable.

7   Detlefsen's argument is based on the fact that for example $PA \vdash Con_{PA}^R$ where PA denotes Peano arithmetic and Con_{PA}^R is the Rosser translation of the metamathematical sentence "PA is consistent" (using Rosser's provability notion $\vdash_R$). So we have here two different notions of provability: Rosser's one $\vdash_R$ and the usual one $\vdash$ based on the classical predicate calculus. To be consequent one should use also in metamathematics the Rosser's provability. But then it should be proved that $PA \vdash_R Con_{PA}^R$.

applies to Hilbert's Program *per se*, or only to those versions of it which needlessly restrict themselves to theory construction of the usual static variety" (p. 345).

Resnik argues in (1974) that the incompleteness theorem "has less bearing upon the [Hilbert's] program than is often credited to it!" This thesis is based on the claim that: "every formal system to which Gödel's theorems apply is complete with respect to its real sentences. Thus the undecidable sentences are ideal sentences."

Note also that the failure of Hilbert's program for a certain formalized system of arithmetic need not be a failure of Hilbert's program for elementary number theory in the informal sense. In fact one cannot exclude the possibility that the latter can be formalized in a system which can be justified on finitistic grounds.

And what were the reactions and opinions of the main heros of the whole story, that is of Hilbert and Gödel? Hilbert, though taking part in the conference in Königsberg, did not learn about Gödel's result. He did it only by the early part of 1931. As Smoryński (1988) writes: "He was angry at first, but was soon trying to find a way around it". He proposed to add to the rules of inference a simple form of the ω-rule (which allows the derivation of all true arithmetic sentences). In Preface to the first volume of Hilbert and Bernays (1934, 1939) Hilbert wrote:

> [...] the occasionally held opinion, that from the results of Gödel follows the nonexecutability of my Proof Theory, is shown to be erroneous. This result shows indeed only that for more advanced consistency proofs one must use the finite standpoint in a deeper way than is necessary for the consideration of elementary formalisms.

K. Gödel in his 1931 paper wrote (English translation taken from Heijenoort 1967, p. 615):

> I wish to note expressly that Theorem XI (and the corresponding results for M and A) do not contradict Hilbert's formalistic viewpoint. For this viewpoint presupposes only the existence of a consistency proof in which nothing but finitary means of proof is used, and it is conceivable that there exist finitary proofs that cannot be expressed in the formalism of P (or M or A).[8]

At the Vienna Circle meeting on 15th January 1931 Gödel argued that it is doubtful, "whether all intuitionistically correct proofs can be captured in a *single* formal system. That is the weak spot in Neumann's argumentation" (quotation taken from Sieg 1988). In (1946) he explicitly called for an effort to use progressively more powerful transfinite theories to derive new arithmetical theorems.

---

8   Theorem XI states that if P is consistent then its consistency is not provable in P, P being Peano arithmetic extended by simple type theory; M is set theory; A is classical analysis.

3. Generalizations and Strengthenings of Gödel's Theorems

Beside all the discussions and doubts to connections between Gödel's incompleteness theorems and Hilbert's program described above one question more should be raised.

Gödel in his first incompleteness theorem indicated a true arithmetic sentence which is undecidable in the given formal system of arithmetic. This sentence has not a mathematical but in fact a metamathematical contents (it states: "I am not a theorem"). This diminished the meaning and significance of Gödel's theorem. Also other undecidable sentences constructed later (e.g., sentences of Rosser, Kreisel and Levy, Kent, Mostowski, Shepherdson – see Smoryński 1981) have this failure. Though interesting for logicians they are rather artificial from the mathematical point of view. Hence an open question arose: is it possible to indicate examples of undecidable sentences of mathematical, in particular number-theoretical, contents? This question was even more interesting because it was still possible to cherish hopes that all sentences which are interesting and reasonable from the mathematical point of view (whatever it means) are decidable.

This problem was solved in 1977 – it was done by J. Paris (cf. Paris 1978). Working on nonstandard models of Peano arithmetic PA he has invented a new method of constructing sentences which are independent of PA but true (in the standard model $\mathfrak{N}_0$). The sentences of Paris were simplified by L. Harrington and at the end a new elegant undecidable sentence of a combinatorial contents was obtained (cf. Paris and Harrington 1977).

To describe Paris–Harrington sentence we need to fix some notation. By Peano arithmetic PA (which is now a standard formal system used in studies of the foundations of mathematics) we mean a first-order theory based on the classical predicate calculus and on the following nonlogical axioms:

$$Sx = Sy \to x = y,$$
$$Sx \neq 0,$$
$$x + 0 = x,$$
$$x + S(y) = S(x+y),$$
$$x \cdot 0 = 0,$$
$$x \cdot S(y) = x \cdot y + x,$$
$$\varphi(0) \wedge \forall x[\varphi(x) \to \varphi(S(x))] \longrightarrow \forall x \varphi(x)$$

where $\varphi(x)$ is any formula of the language of PA with the free variable x (and possibly some other free variables treated as parameters). The standard model $\mathfrak{N}_0$ of PA is the structure $\langle \mathbb{N}, S, +, \cdot, 0 \rangle$ where $\mathbb{N}$ is the set of natural numbers, S is the successor function, and $+$ and $\cdot$ are addition and multiplication, resp., and 0 denotes the number zero.

If X is a set of natural numbers then $[X]^n$ denotes the family of all n-element subsets of X. A function $C: [X]^n \to c$ (c being a natural number which we identify with the set of its predecessors, i.e., $c = \{0, 1, \ldots, c\}$) is said to be a *colouring function*. It may be interpreted as a colouring of n-element subsets of X by colours $0, 1, 2, \ldots, c$. In 1929 the English mathematician F. P. Ramsey proved that if C is a function colouring $[X]^n$ and X is big with respect to c and n then there exists a big set Y such that all its n-element subsets are coloured by one colour. Such a set $Y \subseteq X$ we call *homogeneous* with respect to C. In fact Ramsey proved the following two theorems:

Theorem 1 (Infinite Ramsey Theorem). *Let n and c be positive natural numbers. For any colouring function $C: [\mathbb{N}]^n \to c$ there is an infinite $Y \subseteq \mathbb{N}$ such that Y is homogeneous with respect to C, i.e., $C|[Y]^n$ is constant.*

Theorem 2 (Finite Ramsey Theorem). *Let s, n and c be positive natural numbers such that $s \geqslant n+1$. Then there is a number $R(s,n,c)$ such that for every $r > R(s,n,c)$, for any set X having r elements and any colouring function $C: [X]^n \to c$ there exists a set homogeneous with respect to C having s elements.*

These theorems are not intuitively obvious and need proofs. They can be treated as generalizations of Dirichlet's *Schubfachprinzip*. For $n = 1$ Theorem 1 says that if one divides an infinite set into a finite number of disjoint parts then one of these parts must be infinite. Theorem 2 for $n = 1$, $s = 2$ and $R(2, 1, c) = c + 1$ is exactly Dirichlet's principle: if one divides a set containing $c + 1$ (or more) elements into c parts then one of them must contain at least 2 elements.

It turns out that Finite Ramsey Theorem can be proved in PA. Harrington observed that modifying it a bit we obtain a sentence independent of PA. Call a set $X \subseteq \mathbb{N}$ *relatively large* if and only if $card(X) > min(X)$. Then for example the set $\{2, 3, 89, 92\}$ is relatively large but the set $\{10, 13, 7, 9\}$ is not relatively large. The Paris–Harrington sentence φ_0 says now: *for any natural numbers s, n, c there exists a natural number $H(s,n,c)$ such that for any $h > H(s,n,c)$, any set X of cardinality h, any $C: [X]^n \to c$ there is a set Y homogeneous with respect to the function C and such that $card(Y) \geqslant s$ and Y is relatively large.* It can be proved (for example by set-theoretical methods) that $\mathfrak{N}_0 \models \varphi_0$ (in fact φ_0 is a consequence of Infinite Ramsey Theorem) but PA non $\vdash \varphi_0$. Hence φ_0 is an undecidable sentence of a combinatorial contents.

Paris–Harrington sentence is still not a fully satisfying solution to the problem stated above: it has not a purely arithmetical (number-theoretical) contents. Such a sentence was constructed in 1982 by L. Kirby and J. Paris (cf. Kirby and Paris 1982). The construction uses some ideas of Goodstein (1944). To describe this sentence let m, n be natural numbers and define a *representation of m by the basis n* in the following way: we write m as a sum of powers of n (e.g., if $m = 266$, $n = 2$ then $266 = 2^8 + 2^3 + 2^1$). We do this same with all exponents and at the

end we get:
$$266 = 2^{2^{2+1}} + 2^{2+1} + 2^1.$$

We define now a number $G_n(m)$ as follows:
if $m = 0$ then $G_n(m) = 0$,
if $m \neq 0$ then $G_n(m)$ is a number obtained by substituting everywhere in the representation of m (by the basis n) the number n by $n+1$ and subtracting 1.

For example: $G_2(266) = 3^{3^{3+1}} + 3^{3+1} + 2 \approx 10^{38}$. *The Goodstein sequence for m is defined in the following way:*

$$m_0 = m,$$
$$m_1 = G_2(m_0),$$
$$m_2 = G_3(m_1),$$
$$\vdots$$

For example:

$$m_0 = 266_0 = 2^{2^{2+1}} + 2^{2+1} + 2^1,$$
$$m_1 = 266_1 = G(m_0) = 3^{3^{3+1}} + 3^{3+1} + 2 \approx 10^{38},$$
$$m_2 = 266_2 = G_3(m_1) = 4^{4^{4+1}} + 4^{4+1} + 1 \approx 10^{616},$$
$$m_3 = 266_3 = G_4(m_2) = 5^{5^{5+1}} + 5^{5+1} \approx 10^{10000}, \quad \text{etc.}$$

Observe that this procedure of constructing the sequence m_k can be described in the language L(PA) of Peano arithmetic. Consider now the following sentence φ_1 of L(PA): $\forall m \exists k (m_k = 0)$. It can be proved that $\mathfrak{N}_0 \models \varphi_1$. The unprovability of φ_1 has its source, roughly speaking, in the fact that $m_k = 0$ only for very big k, e.g., for $m = 4$, $m_k = 0$ for $k = 3 \cdot 2^{402653211} - 3 \approx 10^{121000000}$ (observe that the whole number of atoms in the Universe is estimated as 10^{80}).

Proofs of Paris–Harington and Kirby–Paris results are technical and cannot be presented here. They use sophisticated machinery of the model theory of arithmetic and of the indicator theory. For some remarks on the proofs as well as for the discussion of sources of undecidability of the considered sentences cf. Murawski (1987).

The results of Paris–Harrington and Kirby–Paris being strengthenings of Gödel's incompleteness theorem demonstrated that the hopes described at the beginning of this section cannot be realized.

4. Generalized Hilbert's Program

Incompleteness results of Gödel indicated various obstacles in carrying out the validation and justification of classical mathematics on finitistic grounds postulated by Hilbert. The natural consequence of it was the idea of extending the admissible methods and allowing general constructive methods. It seems that P. Bernays was among the first to recognize this need. He wrote (cf. Bernays 1967, p. 502):

> It thus became apparent that the *"finite Standpunkt"* is not the only alternative to classical ways of reasoning and is not necessary implied by the idea of proof theory. An enlarging of the methods of proof theory was therefore suggested: instead of a restriction to finitist methods of reasoning, it was required only that the arguments be of a constructive character, allowing us to deal with more general forms of inference.

One of the motivations of this shift from the original Hilbert's program can be sought in Gödel's reduction (independently found also by G. Gentzen) of Peano arithmetic PA to the intuitionistic system HA of Heyting's arithmetic – it was shown that PA is consistent relative to HA (cf. Gödel 1933a; Gentzen 1969).[9] Another reason may be Gentzen's proof of consistency of PA by transfinite induction up to ε_0 (cf. Gentzen 1936; 1938)[10] which was apparently accepted by Hilbert and Bernays in the second volume of *Grundlagen der Mathematik*.

There arises, of course, a problem: what is meant here by constructivity – this concept is in general much less clear than that of finitism. Nevertheless the broadening of original Hilbert's proof theory postulated by Bernays has become an accepted paradigm. Investigations were carried out in this direction and several results were obtained. Studies following the idea of Gentzen of using a transfinite induction on a certain recursive ordering for some ordinal form a part of them (cf. Schütte 1960; Takeuti 1975). Another program of reductionism was elaborated by Feferman, namely the program of predicative reductionism (cf. Feferman 1964; 1968; see also Simpson 1985a). Gödel has proposed an "extension" of the *"finite Standpunkt"* by way of primitive recursive functionals of higher type (cf. Gödel 1958).

A further refinement of the original Hilbert's program was suggested in Kreisel (1958) and elaborated in Kreisel (1968) where a call for a hierarchy of Hilbert's programs can be found. Beside reduction to finitary and constructive

---

9   Gentzen's paper "Über das Verhältniss zwischen intuitionistische und klassische Arithmetik" was submitted and accepted by *Mathematische Annalen* in 1933 but withdrawn on account of Gödel's publication. An English translation was published in 1969 – cf. Gentzen (1969).
10 The first version of Gentzen's proof was submitted in 1935 but was withdrawn after criticism directed against the means used in the proof which were considered to be too strong. This version became publicly known because of a paper Bernays (1970) and was recently published in the name of Gentzen, cf. Gentzen (1974).

conceptions, Kreisel considers also the nonconstructive predicative conception and within constructivity itself he analyzes which specific principles would be needed for various pieces of reductive work.

It is impossible to report here on particular results obtained along those lines (it would require a lot of technical notions and would change completely the character of this paper). One can find a survey of them in Feferman (1988). Let us note only that the researches have led to two surprising insights: (i) classical analysis can be formally developed in conservative extensions of elementary number theory and (ii) strong impredicative subsystems of analysis can be reduced to constructively meaningful theories, i.e. relative consistency proofs can be given by constructive means for impredicative parts of second order arithmetic.

We want to stress at the end of this section that all the generalizations discussed above are very different from the original Hilbert's program. Hilbert's postulate was the validation and justification of classical mathematics by a reduction to finitistic mathematics. The latter was important here for philosophical and, say, ideological reasons: finitistic objects and reasonings have clear physical meaning and are indispensable for all scientific thought. None of the proposed generalizations can be viewed as finitistic (whatever it means). Hence they have another value and meaning from the methodological and generally philosophical point of view. They are not contributing directly to Hilbert's program but on the other hand they are in our opinion compatible with Hilbert's reductionist philosophy.

5. Relativized Hilbert's Program and Reverse Mathematics

Another consequence of Gödel's incompleteness results (beside those described above) is the so called relativized Hilbert's program. If the entire infinitistic mathematics cannot be reduced to and justified by finitistic mathematics then one can ask for which part of it is that possible? In other words: how much of infinitistic mathematics can be developed within formal systems which are conservative over finitistic mathematics with respect to real sentences? This constitutes the relativized version of the program of Hilbert. In what follows we would like to show how the so called reverse mathematics of Friedman and Simpson contributes to it providing us with a partial realization of Hilbert's original program.

To be able to consider the problem one should specify what is meant by finitistic mathematics and by real sentences. We follow here Tait (1981) where it is claimed that Hilbert's finitism is captured by the formal system PRA of primitive recursive arithmetic (called also Skolem arithmetic). Its language contains the constant 0, successor function S and a function symbol for each primitive recursive function. Its nonlogical axioms are: some trivial axioms concerning the constant terms 0, S0, SS0 etc. and the successor function, the defining equations of the primitive recursive functions and induction on quantifier-free formulas. The theory PRA is certainly finitistic and "logic-free". On the other hand it is powerful enough to accommodate all elementary reasonings about natural numbers and manipulations of finite strings of symbols. By real sentences we shall understand Π_1^0 sentences, i.e., sentences of the language of Peano arithmetic of the form $\forall x \varphi(x,\ldots)$ where φ contains only atomic formulas, connectives and bounded quantifiers.

It turns out that one can formalize classical mathematics not only in set theory but most of its parts (such as geometry, number theory, analysis, differential equations, complex analysis etc.) can be formalized in a weaker system called second order arithmetic A_2^- (denoted also sometimes as Z). This is a system formalized in a language with two sorts of variables: number variables $x, y, z, \ldots$ and set variables $X, Y, Z, \ldots$ Its nonlogical constants are those of Peano arithmetic, i.e., $0, S, +, \cdot$ and the membership relation $\in$. Nonlogical axioms of A_2^- are the following:

(1) axioms of PA without the axiom scheme of induction,
(2) (extensionality) $\forall x(x \in X \equiv x \in Y) \to (X = Y)$,
(3) (induction axiom)

$$0 \in X \wedge \forall x(x \in X \to Sx \in X) \to \forall x(x \in X),$$

(4) (axiom scheme of comprehension) $\exists X \forall x[x \in X \equiv \varphi(x,\ldots)]$,

where $\varphi(x,\ldots)$ is any formula of the language of A_2^- (possibly with free number- or set-variables) in which X does not occur free. Possible models of A_2^- are structures of the form $\langle \mathfrak{M}, \mathscr{X} \rangle$ where $\mathfrak{M}$ a model of PA and $\mathscr{X}$ is a family of subsets of M (the universe of $\mathfrak{M}$). Observe that for the standard model $\mathfrak{N}_0$ of PA the structure $\langle \mathfrak{N}_0, \mathscr{P}(\mathbb{N}) \rangle$ is a model of A_2^- but this is not the case for a nonstandard model $\mathfrak{M}$ of PA, i.e., $\langle \mathfrak{M}, \mathscr{P}(M) \rangle$ is never a model of A_2^- (for standard models of A_2^- see, e.g., Apt and Marek 1974 and for nonstandard models of A_2^- see Murawski 1976a, 1976b, 1977 and 1984a where a bibliography can also be found).

Second order arithmetic is a nice system because one avoids here troubles connected with set theory (for example paradoxical consequences of the axiom of choice or philosophical problems connected with the adoption of these or those axioms) and on the other hand it is strong enough to prove many important theorems of classical mathematics. There is only a problem of impredicativity of A_2^-: a formula φ in the comprehension axiom may be *any* formula of the language of A_2^-, so it may be of the form $\forall Y \psi(Y, x, \ldots)$, i.e., defining one particular set it may refer to the family of all sets (recall Poincaré's objections against such definitions mentioned above). But it turns out that in many cases certain fragments of A_2^- suffice, i.e., only particular special forms of the comprehension axiom are needed.

At the Congress of Mathematicians in Montreal in 1974 H. Friedman formulated a program of foundations of mathematics called today reverse mathematics (cf. Friedman 1975). Its aim is to study the rôle of set existence axioms, i.e., comprehension axioms in ordinary mathematics. The main problem is: Given a specific theorem τ of ordinary mathematics, which set existence axioms are needed in order to prove τ? This research program turned out to be very fruitful and led to many interesting results[11] and ... showed that Hilbert's program can be partially realized! There is a rich literature connected with the reverse mathematics. The most comprehensive report on it is the book Simpson (1998), one can consult also Simpson (1985b; 1987; 1988a).

The procedure used in the reverse mathematics (it reveals the inspiration for its name) is the following: assume we know that a given theorem τ can be proved in a particular fragment $S(\tau)$ of A_2^-. Is $S(\tau)$ the weakest fragment with this property? To answer this question positively one shows that the principal set existence axiom of $S(\tau)$ is equivalent to τ, the equivalence being provable in some weaker system in which τ itself is not provable.

Some specific systems – fragments of A_2^- – arose in this context; the most important are: RCA_0, WKL_0, ACA_0, ATR_0 and $\Pi_1^1\text{-}CA0$. We shall describe only the first three of them. To do this we need a hierarchy of formulas of the language of A_2^-. By an arithmetical formula we mean a formula containing no quantifiers bounding set-variables. Bounded formulas are arithmetical formulas with only bounded quantifiers, i.e., quantifiers of o the form: $\forall x < y$, $\exists x < y$. The class of all such formulas is denoted by Δ_0^0. Formulas of the class Σ_1^0 are formulas of the form $\exists x_1 \ldots \exists x_n \psi(x_1, \ldots, x_n, y_1, \ldots, y_k)$ where $\psi \in \Delta_0^0$ and formulas of the class Π_1^0 are negations of Σ_1^0 formulas.

The system RCA_0 is a theory in the language of A_2^- based on the following axioms: (i) PA^- (i.e., axioms of Peano arithmetic PA without the axiom scheme of induction), (ii) scheme of induction for Σ_1^0 formulas, i.e.:

$$\varphi(0) \wedge \forall x[\varphi(x) \to \varphi(Sx))] \to \forall x \varphi(x)$$

where φ is a Σ_1^0 formula, (iii) (recursive comprehension axiom)

$$\forall x[\varphi(x) \equiv \psi(x)] \to \exists X \forall x[x \in X \equiv \varphi(x)],$$

where φ is Σ_1^0 and ψ is Π_1^0. [Axiom (iii) explains the name RCA_0 of the theory.] It can be shown that $\langle \mathfrak{N}_0, \text{Rec} \rangle$ where Rec is the family of all recursive sets is a model of RCA_0.

The theory WKL_0 consists of RCA_0 plus a further axiom known as weak König's lemma (therefore the name WKL_0) which states that every infinite binary

---

[11] It is even claimed that the implications of the results of reverse mathematics "make much of what was written in the past on the philosophy of mathematics, obsolete" (cf. Drake 1987).

tree has an infinite path (this can be formulated in the language of A_2^- using coding). It is stronger than RCA_0 what follows for example from the fact that $\langle \mathfrak{N}_0, \text{Rec} \rangle$ is not a model of WKL_0 (this is a consequence of Gödel's theorem on essential undecidability of Peano arithmetic).

The theory ACA_0 is PA^- plus induction axiom plus arithmetical comprehension, i.e., comprehension scheme for any arithmetical formula (possibly containing set parameters). This theory is not weaker than WKL_0 because it proves weak König's lemma. It is in fact stronger than WKL_0 what follows from the fact that there are models of WKL_0 consisting of sets definable in $\mathfrak{N}_0$ by formulas of a given class[12] whereas for any model $\langle \mathfrak{N}_0, \mathscr{X} \rangle$ of ACA_0 the family $\mathscr{X}$ must be closed with respect to arithmetical definability.

The specified subsystems of A_2^- are appropriate for particular parts of classical mathematics. In RCA_0 one can construct reals, define notions of the convergence of a sequence, of a continuous function, of Riemann's integrability etc. and prove positive results of recursive analysis and recursive algebra. For example one can prove in RCA_0 that every countable field has an algebraic closure, that every countable ordered field has an extension to a real closed field as well as the intermediate value theorem.

The theory WKL_0 turns out to be a quite strong theory, in particular one can prove in it the following theorems:

– the Heine–Borel covering theorem: every covering of $[0,1]$ by a countable sequence of open intervals has a finite subcovering,
– every continuous function on $[0,1]$ is uniformly continuous,
– every continuous function on $[0,1]$ is bounded,
– every continuous function on $[0,1]$ has a supremum,
– every uniformly continuous function on $[0,1]$, which has a supremum, attains it,
– every continuous function on $[0,1]$ attains a maximum value,
– the Hahn–Banach theorem,
– the Cauchy–Peano theorem on the existence of solutions of ordinary differential equations,
– every countable commutative ring has a prime ideal,
– every countable formally real field can be ordered,
– every countable formally real field has a real closure,
– Gödel's completeness theorem for the predicate calculus.

Even more: if S is one of the above stated theorems then $RCA_0 + S$ is equivalent to WKL_0.

To indicate the strength of ACA_0 let us mention that the following theorems can be proved in it:

---

[12] A family of Δ_2^0 sets is an example of a model of WKL_0 where Δ_2^0 sets are sets definable in $\mathfrak{N}_0$ simultaneously by Σ_2^0 and Π_2^0 formulas; recall that Σ_2^0 formulas are formulas of the form $\exists x_1 \ldots x_n \psi(x_1,\ldots,x_n,y_1,\ldots,y_k)$, $\psi \in \Pi_1^0$ and Π_2^0 formulas are negations of Σ_2^0 formulas.

– the Bolzano–Weierstrass theorem (every bounded sequence of real numbers has a convergent subsequence),
– every Cauchy sequence of reals is convergent,
– every bounded sequence of reals has a supremum,
– every bounded increasing sequence of real numbers is convergent,
– the Arzela–Ascoli lemma (any bounded equicontinuous sequence of functions on [0,1] has a uniformly convergent subsequence),
– every countable vector space has a basis,
– every countable commutative ring has a maximal ideal,
– every countable Abelian group has a unique divisible closure.

And again, if S is any of those theorems then $RCA_0 + S$ is equivalent to ACA_0.

Those results indicate how much of A_2^- do we need in fact to prove various particular theorems of classical mathematics. It also appears that proofs (in the formalized subsystems of A_2^-) of uniqueness are usually more difficult and more complicated than proofs of the existence (in mathematical practice the former are usually simple consequences of the latter). There is also no direct connection between the complexity of a classical proof of a theorem and the level in the hierarchy of subsystems of A_2^- in which a formalized version of it can be proved (as an example can serve here the theorem that every Abelian group has a torsion subgroup which is trivial in classical algebra but RCA_0 plus this theorem is equivalent to ACA_0, hence it is not a theorem of, say, WKL_0). Results of the reverse mathematics have also interesting mathematical, not only logical, applications. For example: since Cauchy–Peano theorem on the existence of solutions of ordinary differential equations is equivalent to WKL_0 and $\langle \mathfrak{N}_0, \text{Rec} \rangle$ is not a model of this theory, it follows that there exists a differential equation with a recursive continuous function on the right hand side which has no recursive solution.

Observe that not every mathematical theorem can be classified in Friedman's hierarchy of subsystems of A_2^-. Hilbert's basis theorem can serve here as an example (cf. Simpson 1988b) – it is provable in ACA_0 but RCA_0 plus 'Hilbert's basis theorem' is not equivalent to any of the considered systems. There are also sentences unprovable in A_2^- but provable in Zermelo–Fraenkel set theory (cf. Friedman 1981).

The impatient reader is certainly asking already: well, but what are the connections of reverse mathematics with Hilbert's program? We are just coming to that problem.

In early eighties L. Harrington and Z. Ratajczyk proved a theorem on conservativeness of WKL_0 (none of them published it, the proof can be found in Simpson 1998). It says that if $\langle \mathfrak{M}, \mathscr{X} \rangle$ is a countable model of RCA_0 and $A \in \mathscr{X}$ then there exists a family $\mathscr{Y} \subseteq \mathscr{P}(M)$ such that $A \in \mathscr{Y}$ and $\langle \mathfrak{M}, \mathscr{Y} \rangle$ is a model of WKL_0. To indicate the syntactical context of the theorem recall that Π_1^1 formulas are formulas of the language of A_2^- of the form $\forall X \psi$ where ψ is an arithmetical formula. From Harrington–Ratajczyk theorem follows that the theory

WKL_0 is conservative over RCA_0 with respect to Π_1^1 sentences, i.e., that every Π_1^1 sentence provable in WKL_0 can be proved in RCA_0. Friedman showed by model-theoretical methods that WKL_0 is a conservative extension of PRA with respect to Π_2^0 sentences (cf. Friedman 1977; this result can be also found in Kirby and Paris 1977). W. Sieg improved this theorem giving an alternative proof (which uses Gentzen-style method) and exhibiting a primitive recursive proof transformation. Thus the reducibility of WKL_0 to PRA is itself provable in PRA.

Combining those results together with the fact that WKL_0 is a strong theory (as indicated above) we come to the conclusion that a large and significant part of classical mathematics is finitistically reducible. This means in fact a partial realization of Hilbert's program!

All those facts have various "practical" consequences. First observe that the class of Π_2^0 sentences is rather broad – many theorems of number theory can be formulated as sentences belonging to that class. Since one can formalize within WKL_0 the technique of contour integration, any Π_2^0 number-theoretic theorem which is provable with the help of it can also be proved "elementarily", i.e., within PRA and, even more, one can effectively (at least theoretically) find such an "elementary" proof. To give one more example consider Artin's theorem (being a solution to Hilbert's seventeenth problem – cf. Hilbert 1901). It can be written as a Π_2^0 sentence. Since all results of the theory of real closed fields needed in the proof of Artin's theorem are provable in WKL_0, it follows by Friedman's and Sieg's theorems that Artin's theorem can be proved in PRA, i.e., in an elementary way.

6. Conclusions

In this way we came to the end of our story. Conclusions of it are in our opinion rather optimistic. Though the original Hilbert's program of validation and justification of classical mathematics cannot be realized it proved to be fruitful, both philosophically and mathematically by stimulating various investigations which contributed to the clarification of the nature of mathematics. We learned that not all classical mathematics can be reduced to finitary one, that for a (rather broad) part of it this is still possible, we learned also the limitations of the axiomatic method. Maybe Hilbert was right claiming that "There is no *ignorabimus* in mathematics" (cf. Hilbert 1926) and finishing his speech over the local radio station in Königsberg (in September 1930, just after the conference during which Gödel announced his incompleteness result) with the motto:

Wie müssen wissen. Wir werden wissen.
(We must know. We shall know.)

On the Distinction Proof-Truth in Mathematics

The essay is devoted to some historical, philosophical and logical considerations connected with the distinction between proof and truth in mathematics. The crucial rôle of Gödel's incompleteness theorems as well as of the undefinability of truth vs. definability of provability and the rôle of finitary vs. infinitary methods are stressed. The problem of the need of extending the available methods by new rules of inference and new axioms is also considered.

Concepts of proof and truth are (even in mathematics) ambiguous. It is commonly accepted that proof is the ultimate warrant for a mathematical proposition, that proof is a source of truth in mathematics. One can say that a proposition A is true if it holds in a considered structure or if we can prove it. But what is a proof? And what is truth?

The axiomatic method was considered (since Plato, Aristotle and Euclid) to be the best method to justify and to organize mathematical knowledge. The first mature and most representative example of its usage in mathematics were *Elements* of Euclid. They established a pattern of a scientific theory and a paradigm in mathematics. Since Euclid till the end of the nineteenth century mathematics was developed as an axiomatic (in fact rather a quasi-axiomatic) theory based on axioms and postulates. Proofs of theorems contained several gaps – in fact the lists of axioms and postulates were not complete, one freely used in proofs various "obvious" truths or referred to the intuition. Proofs were informal and intuitive, they were rather demonstrations and the very concept of a proof was of a psychological (and not of a logical) nature. Note that almost no attention was paid to the making precise and to the specification of the language of theories – in fact the language of theories was simply the unprecise colloquial language. One should also note here that in fact till the end of the nineteenth century mathematicians were convinced that axioms and postulates should be true statements. It seems to be connected with Aristotle's view that a proposition is demonstrated (proved to be true) by showing that it is a logical consequence of propositions already known to be true. Demonstration was conceived here of as a deduction whose premises are known to be true and a deduction was conceived of as a chaining of immediate inferences.

Basic concepts underlying the Euclidean paradigm have been clarified on the turn of the nineteenth century. In particular the intuitive (and rather psychological in nature) concept of an informal proof (demonstration) was replaced by a precise notion of a formal proof and of a consequence. This was the result of the development of mathematical logic and of a crisis of the foundations of mathematics on the turn of the nineteenth century which stimulated foundational investigations.

One of the directions of those foundational investigations was the program of David Hilbert and his *Beweistheorie*. Note at the very beginning that "this program was never intended as a comprehensive philosophy of mathematics; its purpose was instead to legitimate the entire corpus of mathematical knowledge" (cf. Rowe 1989, p. 200). Note also that Hilbert's views were changing over the years, but always took a formalist direction.

Hilbert sought to justify mathematical theories by means of formal systems, i.e., using the axiomatic method. He viewed the latter as holding the key to a systematic organization of any sufficiently developed subject. In "Axiomatisches Denken" (1918, p. 405) Hilbert wrote:

> When we put together the facts of a given more or less comprehensive field of our knowledge, then we notice soon that those facts can be ordered. This ordering is always introduced with the help of a certain *network of concepts* (*Fachwerk von Begriffen*) in such a way that to every object of the given field corresponds a concept of this network and to every fact within this field corresponds a logical relation between concepts. The network of concepts is nothing else than the *theory* of the field of knowledge.

By Hilbert the formal frames were contentually motivated. First-order theories were viewed by him together with suitable non-empty domains, *Bereiche*, which indicated the range of the individual variables of the theory and the interpretations of the nonlogical vocabulary. Hilbert, as a mathematician, was not interested in establishing precisely the ontological status of mathematical objects. Moreover, one can say that his program was calling on people to turn their mathematical and philosophical attention away from the problem of the object of mathematical theories and turn it toward a critical examination of the methods and assertions of theories. On the other hand he was aware that once a formal theory has been constructed, it can admit various interpretations. Recall here his famous sentence from a letter to G. Frege of 29th December 1899 (cf. Frege 1976, p. 67):

> Yes, it is evident that one can treat any such theory only as a network or schema of concepts besides their necessary interrelations, and to think of basic elements as being any objects. If I think of my points as being any system of objects, for example the system: love, law, chimney-sweep [...], and I treat my axioms as [expressing] interconnections between those objects, then my theorems, e.g., the theorem of Pythagoras, hold also for those things. In other words: any such theory can always be applied to infinitely many systems of basic elements.

The essence of the axiomatic study of mathematical truths consisted for him in the clarification of the position of a given theorem (truth) within the given axiomatic system and of the logical interconnections between propositions.

Hilbert sought to secure the validity of mathematical knowledge by syntactical considerations without appeal to semantic ones. The basis of his approach was the distinction between the unproblematic 'finitistic' part of mathematics and the 'infinitistic' part that needed justification. As is well known Hilbert proposed to base mathematics on finitistic mathematics via proof theory (*Beweistheorie*). The latter was planned as a new mathematical discipline in which mathematical proofs are studied by mathematical methods. Its main goal was to show that proofs which use ideal elements (in particular actual infinity) in order to prove results in the real part of mathematics always yield correct results. One can distinguish here two aspects: consistency problem and conservation problem. The consistency problem consists in showing (by finitistic methods, of course) that the infinitistic mathematics is consistent; the conservation problem consists in showing by finitistic methods that any real sentence which can be proved in the infinitistic part of mathematics can be proved also in the finitistic part. One should stress here the emphasis on consistency (instead of correctness).

To realize this program one should formalize mathematical theories (even the whole of mathematics) and then study them as systems of symbols governed by specified and fixed combinatorial rules.

The formal axiomatic system should satisfy three conditions: it should be complete, consistent and based on independent axioms. The consistency of a given system was the criterion for mathematical truth and for the very existence of mathematical objects. It was also presumed that any consistent theory would be categorical, that is, would (up to isomorphism) characterize a unique domain of objects. This demand was connected with the completeness.

The meaning and understanding of completeness by Hilbert plays a crucial rôle from the point of view of our subject. Note at the beginning that in the *Grundlagen der Geometrie* completeness was postulated as one of the axioms (the axiom was not present in the first edition, but was included first in the French translation and then in the second edition of 1903). In fact the axiom V(2) stated that it is not possible to extend the system of points, lines and planes by adding new entities so that the other axioms are still satisfied. In Hilbert's lecture at the Congress at Heidelberg in 1904 (cf. 1905b) one finds such an axiom system for the real numbers. Later there appears completeness as a property of a system. In lectures "Logische Principien des mathematischen Denkens" (1905a, p. 13) Hilbert explains the demand of the completeness as the demand that the axioms suffice to prove all "facts" of the theory in question. He says: "We will have to demand that all other facts (*Thatsachen*) of the given field are consequences of the axioms". On the other hand one can say that Hilbert's early conviction as to the solvability of every mathematical problem – expressed for example in his 1900 Paris lecture (cf. Hilbert 1901) and repeated in his opening address "Naturerkennen und Logik" (cf. Hilbert 1930) before the Society of German Scientists and Physicians in Königsberg in September 1930 – can be treated as informal reflection of a belief in completeness.

In his 1900 Paris lecture Hilbert spoke about completeness in the following words (see the second problem): "When we are engaged in investigating the foundations of a science, we must set up a system of axioms which contains an exact and complete description of the relations subsisting between the elementary ideas of that science."

One can take the "exact and complete description" to be complete enough to decide the truth or falsity of every statement. Semantically such completeness follows from categoricity, i.e., from the fact that any two models of a given axiomatic system are isomorphic; syntactically it means that every sentence or its negation is derivable from the given axioms. Hilbert's own axiomatizations were complete in the sense of being categorical. But notice that they were not first-order, indeed his axiomatization of geometry from *Grundlagen* as well as his axiomatization of arithmetic published in 1900 were second-order.

The demand discussed here would imply that a complete (in this sense) system of axioms is possible only for sufficiently advanced theories. On the other hand Hilbert called for complete systems of axioms also for theories being developed. One should also add here that Hilbert admitted the possibility that a mathematical problem may have a negative solution, i.e., that one can show the impossibility of a positive solution on the basis of a considered axiom system (cf. Hilbert 1901).

In Hilbert's lectures from 1917–1918 (cf. Hilbert 1917/1918) one finds completeness in the sense of maximal consistency, i.e., a system is complete if and only if for any non-derivable sentence, if it is added to the system then the system becomes inconsistent. In his lecture at the International Congress of Mathematicians in Bologna in 1928 Hilbert stated two problems of completeness: one for the first-order predicate calculus (completeness with respect to validity in all interpretations, hence the semantic completeness) and the second for a system of elementary number theory (formal completeness, in the sense of maximal consistency, i.e., Post-completeness, hence the syntactical completeness) (cf. Hilbert 1929).

Hilbert's emphasis on the finitary and syntactical methods together with the demand of (and belief in) the completeness of formal systems seem to be the source and reason of the fact that, as Gödel put it (cf. Wang 1974, p. 9), "[...] formalists considered formal demonstrability to be an *analysis* of the concept of mathematical truth and, therefore were of course not in a position to *distinguish* the two". Indeed, the informal concept of truth was not commonly accepted as a definite mathematical notion at that time. As Gödel wrote in a crossed-out passage of a draft of his reply to a letter of the student Yossef Balas: "[...] a concept of objective mathematical truth as opposed to demonstrability was viewed with greatest suspicion and widely rejected as meaningless" (cf. Wang 1987, pp. 84–85). Therefore Hilbert preferred to deal in his metamathematics solely with the forms of the formulas, using only finitary reasonings which were considered to be save – contrary to semantical reasonings which were non-finitary and consequently not save. Non-finitary reasonings in mathematics were considered to be meaningful only to the extent to which they could be interpreted or justified in terms of finitary metamathematics.[1]

On the other hand there was no clear distinction between syntax and semantics at that time. Recall for example that, as indicated earlier, the axiom systems came by Hilbert often with a built-in interpretation. Add also that the very notions necessary to formulate properly the difference syntax-semantics were not available to Hilbert.

The problem of the completeness of the first-order logic, i.e., the fourth problem of Hilbert's Bologna lecture, was also posed as a question in the book by Hilbert and Ackermann *Grundzüge der theoretischen Logik* (1928). It was solved by Kurt Gödel in his doctoral dissertation (1929; cf. also 1930a) where he showed that the first-order logic is complete, i.e., every true statement can be derived from the axioms. Moreover he proved that, in the first-order logic, every consistent axiom system has a model. More exactly Gödel wrote that by completeness he meant that "every valid formula expressible in the restricted functional calculus [...] can be derived from the axioms by means of a finite sequence of formal inferences". And added that this is equivalent to the assertion that "Every consistent axiom system [formalized within that restricted calculus] [...] has a realization" and to the statement that "Every logical expression is either satisfiable or refutable" (this is the form in which he actually proved the result). The importance of this result is, according to Gödel, that it justifies the "usual method of proving consistency". One should notice here that the notion of truth in a structure, central to the very definition of satisfiability or validity, was nowhere analyzed in either Gödel's dissertation or his published revision of it. There was in fact a long tradition of use of the informal notion of satisfiability (compare the work of Löwenheim, Skolem and others).

Some months later, in 1930, Gödel solved three other problems posed by Hilbert in Bologna by showing that arithmetic of natural numbers and all richer theories are essentially incomplete (provided they are consistent) (cf. Gödel 1931a). It is interesting to see how Gödel arrived at this result.

Gödel himself wrote on his discovery in a draft reply to a letter dated 27th May 1970 from Yossef Balas, then a student at the University of Northern Iowa (cf. Wang 1987, pp. 84–85). Gödel indicated there that it was precisely his recognition of the contrast between the formal definability of provability and the formal undefinability of truth that led him to his discovery of incompleteness. One finds also there the following statement:

> [...] long before, I had found the *correct* solution of the semantic paradoxes in the fact that truth in a language cannot be defined in itself.

---

[1] Cf. Gödel's letter to Hao Wang dated 7th December 1967 – see Wang (1974, p. 8).

On the base of this quotation one can argue that Gödel obtained the result on the undefinability of truth independently of A. Tarski (cf. Tarski 1933).[2]

Note also that Gödel was convinced of the objectivity of the concept of mathematical truth. In a latter to Hao Wang (cf. Wang 1974, p. 9) he wrote:

> [...] it should be noted that the heuristic principle of my construction of undecidable number theoretical propositions in the formal systems of mathematics is the highly transfinite concept of 'objective mathematical truth' as *opposed* to that of 'demonstrability'.

In this situation one should ask why Gödel did not mention the undefinability of truth in his writings. In fact, Gödel even avoided the terms "true" and "truth" as well as the very concept of being true (he used the term "richtige Formel" and not the term "wahre Formel"). In the paper 'Über formal unentscheidbare Sätze ...' (1931a) the concept of a true formula occurs only at the end of Section 1 where Gödel explains the main idea of the proof of the first incompleteness theorem (but again the term "inhaltlich richtige Formel" and not the term "wahre Formel" appears here). Indeed, talking about the construction of a formula which should express its own unprovability invokes the interpretation of the formal system.

On the other hand the term "truth" occurred in Gödel's lectures on the incompleteness theorems at the Institute for Advanced Study in Princeton in the spring of 1934. He discussed there, among other things, the relation between the existence of undecidable propositions and the possibility of defining the concept "true (false) sentence" of a given language in the language itself. Considering the relation of his arguments to the paradoxes, in particular to the paradox of "The Liar", Gödel indicates that the paradox disappears when one notes that the notion "false statement in a language B" cannot be expressed in B. Even more, 'the paradox can be considered as a proof that "false statement in B" cannot be expressed in B.'

What were the reasons of avoiding the concept of truth by Gödel? An answer can be found in a crossed-out passage of a draft of Gödel's reply to the letter of the student Yossef Balas (mentioned already above). Gödel wrote there:

> However in consequence of the philosophical prejudices of our times 1. nobody was looking for a relative consistency proof because [it] was considered axiomatic that a consistency proof must be finitary in order to make sense, 2. a concept of objective mathematical truth as opposed to demonstrability was viewed with greatest suspicion and widely rejected as meaningless.

Hence it leads us to the conclusion formulated by S. Feferman in (1984) in the following way:

---

[2] For the problem of the priority of proving the undefinability of truth see Woleński (1991) and Murawski (1998).

[...] Gödel feared that work assuming such a concept [i.e., the concept of mathematical truth – R.M.] would be rejected by the foundational establishment, dominated as it was by Hilbert's ideas. Thus he sought to extract results from it which would make perfectly good sense even to those who eschewed all non-finitary methods in mathematics.

Though Gödel tried to avoid concepts not accepted by the foundational establishment, his own philosophy of mathematics was in fact platonist. He was convinced that (cf. Wang 1996, p. 83):

> It was the anti-Platonic prejudice which prevented people from getting my results. This fact is a clear proof that the prejudice is a mistake.

Gödel's theorem on the completeness of first-order logic and his discovery of the incompleteness phenomenon together with the undefinability of truth vs. definability of formal demonstrability showed that formal provability cannot be treated as an analysis of truth, that the former is in fact weaker than the latter. It was also shown in this way that Hilbert's dreams to justify classical mathematics by means of finitistic methods cannot be fully realized. Those results together with Tarski's definition of truth (in the structure) and Carnap's work on the syntax of a language led also to the establishing of syntax and semantics in the 1930s.

On the other hand it should be added that Gödel shared Hilbert's "rationalistic optimism" (to use Hao Wang's term) insofar as informal proofs were concerned. In fact Gödel retained the idea of mathematics as a system of truth, which is complete in the sense that "every precisely formulated yes-or-no question in mathematics must have a clear-cut answer" (cf. Gödel 1961). He rejected however – in the light of his incompleteness theorem – the idea that the basis of these truths is their derivability from axioms. In his Gibbs lecture of 1951 Gödel distinguishes between the system of all true mathematical propositions from that of all demonstrable mathematical propositions, calling them, respectively, mathematics in the objective and subjective sense. He claimed also that it is objective mathematics that no axiom system can fully comprise.

Gödel's incompleteness theorems and in particular his recognition of the undefinability of the concept of truth indicated a certain gap in Hilbert's program and showed in particular, roughly speaking, that (full) truth cannot be comprised by provability and, generally, by syntactic means. The former can be only approximated by the latter. Hence there arose a problem: how should Hilbert's finitistic point of view be extended?

Hilbert in his lecture in Hamburg in December 1930 (cf. Hilbert 1931) proposed to admit a new rule of inference. This rule was similar to the ω-rule, but it had rather informal character (a system obtain by admitting it would be semi-formal). In fact Hilbert proposed that whenever $A(z)$ is a quantifier-free formula for which it can be shown (finitarily) that $A(z)$ is a correct (*richtig*) numerical formula for each particular numerical instance z, then its universal generalization $\forall x A(x)$ may be taken as a new premise (*Ausgangsformel*) in all further proofs.

Gödel pointed in many places that new axioms are needed to settle both undecidable arithmetical and set-theoretic propositions. In (1931?, p. 35) he stated that "[...] there are number-theoretic problems that cannot be solved with number-theoretic, but only with analytic or, respectively, set-theoretic methods". And in (1933b, p. 48) he wrote: "there are arithmetic propositions which cannot be proved even by analysis but only by methods involving extremely large infinite cardinals and similar things". In (1961) Gödel proposed "cultivating (deepening) knowledge of the abstract concepts themselves which lead to the setting up of these mechanical systems". In (1972) (this paper was a revised and expanded English version of (1958)) Gödel claimed that concrete finitary methods are insufficient to prove the consistency of elementary number theory and some abstract concepts must be used in addition. In the paper (1946) Gödel explicitly called for an effort to use progressively more powerful transfinite theories to derive new arithmetical theorems.

Also Zermelo proposed to allow infinitary methods to overcome restrictions revealed by Gödel. According to Zermelo the existence of undecidable propositions was a consequence of the restriction of the notion of proof to finitistic methods (he said here about "finitistic prejudice"). This situation could be changed if one used a more general "scheme" of proof. Zermelo had here in mind an infinitary logic, in which there were infinitely long sentences and rules of inference with infinitely many premises. In such a logic, he insisted, "*all* propositions are decidable!". He thought of quantifiers as infinitary conjunctions or disjunctions of unrestricted cardinality and conceived of proofs not as formal deductions from given axioms but as metamathematical determinations of the truth or falsity of a proposition. Thus syntactic considerations played no rôle in his thinking.

To give a rough account of how those suggestions and proposals to extend the finitistic point of view do in fact work let us quote some technical results. We restrict ourselves to the case of the arithmetic of natural numbers, more exactly to Peano arithmetic PA.

Generally speaking one can obtain completions of PA by:
– admitting the ω-rule,
– adding new axioms (in particular reflection principles),
– adding (partial) notion(s) of truth.

Let us start by considering the case of the ω-rule, i.e., of the following rule:

$$\frac{\varphi(0), \varphi(\overline{1}), \varphi(\overline{2}), \ldots, \varphi(\overline{n}), \ldots \quad (n \in \mathbb{N})}{\forall x \varphi(x)}.$$

Denote by $(PA)_\omega$ Peano arithmetic PA with the ω-rule. One can easily see that $(PA)_\omega$ is complete – it follows from the fact that its unique model up to isomorphism is the standard model $\mathfrak{N}_0 = \langle \mathbb{N}, S, 0, +, \cdot \rangle$. Hence $(PA)_\omega = \text{Th}(\mathfrak{N}_0)$.

One can ask: how many times must the ω-rule be applied to obtain a complete extension of PA? To give an answer let us define the following hierarchy of theories where T is any first-order theory in the language L(PA) of Peano arithmetic:

$T^0 = T$,
$T^{\alpha+\frac{1}{2}} = T^\alpha \cup \{\varphi : \varphi \text{ is of the form } \forall x \psi(x) \text{ and } \psi(\bar{n}) \in T^\alpha \text{ for every } n \in \mathbb{N}\}$,
$T^{\alpha+1} = $ the smallest set of formulas containing $T^{\alpha+\frac{1}{2}}$ and closed under the rules of inference of PA,
$T^\lambda = \bigcup_{\alpha < \lambda} T^\alpha$ for λ limit.

One can now prove that

Theorem 1. $\text{Th}(\mathfrak{N}_0) = (PA)_\omega = PA^\omega$.

Recall the hierarchy of formulas of the language L(PA). Let $\Sigma_0^0 = \Pi_0^0 = \Delta_0^0$ be the set of all quantifier free formulas and all formulas with bounded quantifiers. Define Σ_{n+1}^0 to be the set of all formulas of the form $\exists x \psi$ for $\psi \in \Pi_n^0$ and Π_{n+1}^0 to be the set of all formulas of the form $\forall x \psi$ for $\psi \in \Sigma_n^0$. We also define Δ_n^0 as the set of all formulas equivalent (in PA) to a Σ_n^0 formula and to a Π_n^0 formula. One can prove that

Theorem 2. *For every $n \in \mathbb{N}$ the theory PA^n is complete with respect to Σ_{2n+1}^0 sentences.*

In PA one can define partial notions of truth, i.e., one can define satisfaction and truth for formulas of a given class of the arithmetical hierarchy. Denote by $Sat_{\Sigma_n^0}$ the definition of satisfaction for Σ_n^0 formulas; similarly let $Sat_{\Pi_n^0}$ denote the definition of satisfaction for Π_n^0 formulas. Note that $Sat_{\Sigma_0^0}$ and $Sat_{\Pi_0^0}$ are Σ_1^0 formulas and that $Sat_{\Sigma_n^0}$ and $Sat_{\Pi_n^0}$ (for $n \geq 1$) are Σ_n^0 and Π_n^0, respectively. Let further $Tr_{\Sigma_n^0}$ and $Tr_{\Pi_n^0}$ denote truth predicates for Σ_n^0 and Π_n^0 sentences.[3] In the sequel we shall identify formulas defining satisfaction and truth and their extensions in the standard model $\mathfrak{N}_0$.

The previous theorem can now be formulated as:

$$PA^n \supseteq PA + Tr_{\Sigma_{2n+1}^0}.$$

In the definition of the hierarchy T^α no restriction was put on formulas to which the ω-rule was applied. Consider now a hierarchy in which such a restriction

---

[3] Construction of $Sat_{\Sigma_n^0}$ and $Sat_{\Pi_n^0}$ can be found in Kaye (1991) and Murawski (1999b).

is put. So let T be any theory in the language L(PA). Define the following hierarchy of theories (cf. Niebergall 1996):

$$T^{(0)} = T,$$
$$T^{(\alpha+\frac{1}{2})} = T^{(\alpha)} \cup \{\varphi : \varphi \text{ is of the form } \forall x \psi(x) \text{ and } \psi(x) \in \Sigma^0_{2\alpha+1}$$
$$\text{and } \psi(\bar{n}) \in T^{(\alpha)} \text{ for every } n \in \mathbb{N}\},$$
$$T^{(\alpha+1)} = \text{the smallest set of formulas containing } T^{(\alpha+\frac{1}{2})} \text{ and closed under}$$
$$\text{the rules of inference of PA,}$$
$$T^{(\lambda)} = \bigcup_{\alpha<\lambda} T^{(\alpha)} \text{ for } \lambda \text{ limit.}$$

Hence the ω-rule is now applied at stage n to Σ^0_{2n+1} formulas only.

One has the following

Theorem 3 (Niebergall 1996). *For any $n \in \mathbb{N}$,*

$$PA^{(n)} = PA + Tr_{\Sigma^0_{2n+1}}.$$

The above theorems[4] indicate interconnections between Peano arithmetic augmented with the ω-rule and the partial truths. Other connections between them can be formulated in the language of interpretability. So let $S \preccurlyeq T$ denote that a theory S is relatively interpretable in the theory T (in the sense of (Tarski 1953)). We have now the following facts (cf. Niebergall 1996):

Theorem 4. *Let Con_S (for an appropriate theory S) denote a statement of L(PA) stating that S is consistent. Then*

$$PA^n \preccurlyeq PA + Tr_{\Sigma^0_{2n+1}} + \mathrm{Con}_{PA^n},$$
$$PA^n \preccurlyeq PA + Tr_{\Sigma^0_{2n+2}},$$
$$PA + \mathrm{Con}_{PA+Tr_{\Sigma^0_n}} \preccurlyeq PA + Tr_{\Sigma^0_2},$$
$$PA + \mathrm{Con}_{PA^n} \preccurlyeq PA + Tr_{\Sigma^0_2}.$$

The above theorems indicate that the arithmetical truth, i.e., the set $\mathrm{Th}(\mathfrak{N}_0)$ of all arithmetical sentences true in the standard model $\mathfrak{N}_0$, can be approximated by syntactical methods, i.e., by demonstrability – though not by finitary means (one uses here the ω-rule).

---

[4] Note that many of those theorems hold not only for Peano arithmetic PA but for a broad class of theories – cf. Niebergall (1996).

So far we have considered Peano arithmetic and partial truths. Ask now: what about PA and the full truth? Gödel's and Tarski's theorem shows that the truth predicate for L(PA) cannot be defined in PA. But one can extend the language L(PA) by adding a new binary predicate S called satisfaction class and characterizing it axiomatically by adding to Peano arithmetic PA axioms being an appropriate modification of Tarski's definition of satisfaction (cf. Krajewski 1976, where this notion was introduced, or Murawski 1997b). Note that since those axioms form a finite set of axioms one can write them as a single formula of the language L(PA) $\cup$ S. Denote this theory as PA + "S is a satisfaction class". One can extend this theory by adding new axioms stating special properties of S. In particular one can demand that S is full, i.e., S decides any formula of L(PA) on any valuation or that S is Γ-inductive for Γ being a given class of formulas of the language L(PA) $\cup$ S, i.e., that the induction axiom holds for all formulas of the class Γ (if Γ is the class of all formulas of L(PA) $\cup$ S then one says that the satisfaction class S is inductive).

Since theories T of the indicated type are extensions of PA one can ask what about natural numbers can be proved in T, i.e., one can consider theories of the type

$$\text{PA}^\text{T} = \{\varphi \in \text{L(PA)} : \text{T} \vdash \varphi\}.$$

Theorems of PA$^\text{T}$ are those sentences of the language L(PA) of Peano arithmetic (hence sentences about natural numbers) which can be proved in the stronger theory T. A natural problem of finding an axiomatics of the theory PA$^\text{T}$ arises.

One can easily see that the following theories are conservative extensions of PA:

(a) PA + "S is a satisfaction class",
(b) PA + "S is a full satisfaction class",
(c) PA + "S is an inductive satisfaction class".

This means that one can prove in those theories exactly the same theorems about natural numbers (i.e., formulas of the language L(PA)) as in Peano arithmetic PA. Hence the addition of a new notion, i.e., of a notion of a satisfaction class (and consequently a notion of truth), with properties indicated in (a)–(c) does not increase the proof-theoretical power of a theory with respect to sentences of the language L(PA). On the other hand the assumption that a satisfaction class is full and Δ_0^0-inductive gives a nonconservative extension of PA. In fact one can prove in this theory the consistency of PA.

The theories PA$^\text{T}$ for T being PA + "S is a full (Σ_m^0-)inductive satisfaction class" can be characterized by transfinite induction or the consistency of appropriate ω-logics. Denote by $\Gamma - \text{PA}(S)$ the theory PA + "S is a full Γ-inductive satisfaction class" and by PA(S) the theory PA + "S is a full inductive satisfaction class".

Consider the following sequence of formulas of the language L(PA) (one uses here arithmetization):

$\Gamma_0(\varphi) =$ "PA $\vdash \varphi$",
$\Gamma_{n+\frac{1}{2}}(\varphi) =$ "φ is of the form $\eta \vee \forall z\, \psi(z)$ and $\forall z\, \Gamma_n(\eta \vee \psi(S^z 0))$",
$\Gamma_{n+1}(\varphi) =$ "there exists a proof of the formula φ based on
$\quad$ PA$\cup \{\psi : \Gamma_{n+\frac{1}{2}}(\psi)\}$".

Observe that in this system of ω-logic only the application of the ω-rule increases the degree of complexity of a proof.

Theorem 5 (Kotlarski 1986).

$$\text{PA}^{\Delta_0^0 - \text{PA}(S)} = \text{PA} \cup \{\neg \Gamma_n(0=1) : n \in \mathbb{N}\}.$$

It can also be proved (cf. Kotlarski 1986) that the theory $\Delta_0^0 - \text{PA}(S)$ is equal to the theory

$\quad$ PA + S is a full satisfaction class $+ \forall \varphi[(\text{PA} \vdash \varphi) \rightarrow S(\varphi)]$.

The last sentence can be read as: "S makes all theorems of PA true". It is equivalent to the Δ_0^0-inductiveness of the satisfaction class S.

The system of ω-logic described above can be iterated in the transfinite and one can axiomatize theories PA$^{\Sigma_m^0 - \text{PA}(S)}$ ($m \in \mathbb{N}$) and PA$^{\text{PA}(S)}$ by consistency statements of appropriate systems of this logic (cf. Kotlarski and Ratajczyk 1990a).

Define for an ordinal α a sequence $\omega_m(\alpha)$ in the following way: $\omega_0(\alpha) = \alpha$, $\omega_{m+1}(\alpha) = \omega^{\omega_m(\alpha)}$ and put $\omega_n = \omega_n(\omega)$. Let now $TI(\rho)$, where ρ is an ordinal, denote the scheme of transfinite induction up to ρ. Then the following theorem holds.

Theorem 6 (Kotlarski and Ratajczyk 1990b). *Let m be a natural number. Then*

$$\text{PA}^{\Sigma_m^0 - \text{PA}(S)} = \text{PA} \cup \{TI(\varepsilon_{\omega_m(k)}) : k \in \mathbb{N}\}.$$
$$\text{PA}^{\text{PA}(S)} = \text{PA} \cup \{TI(\varepsilon_{\omega_k}) : k \in \mathbb{N}\}.$$

The above theorems show how strong is Peano arithmetic augmented with an appropriate notion of satisfaction (and truth). One can see that only by assuming that the added notion of satisfaction (truth) is full and at least Δ_0^0-inductive one obtains a proper extension of PA. It is interesting that such extensions are equivalent to PA extended by appropriate forms of transfinite induction or by the statements of the consistency of appropriate systems of ω-logic. In other words, the above theorems show in particular that what can be proved about natural numbers using Peano axioms and the notion of satisfaction (truth) that is assumed to be full and Σ_m^0-inductive is exactly the same as what can be proved in PA plus transfinite induction for ordinals $\varepsilon_{\omega_m(k)}$ (for all $k \in \mathbb{N}$) or in PA plus appropriate consistency statements. Similarly for PA plus full inductive satisfaction (truth) on the one hand and PA plus transfinite induction for ordinals ε_{ω_k} (for all $k \in \mathbb{N}$) or PA plus appropriate consistency statements on the other. They show also that by adding to PA the notion of satisfaction (truth) and assuming that it is full and makes all theorems of PA true one obtains a theory with exactly the same theorems about natural numbers as by taking PA augmented with a concept of a full and Δ_0^0-inductive satisfaction (truth) or PA plus appropriate consistency statements.

In the above considerations we restricted ourselves to formal proofs and to the semantical notion of truth *in* mathematics. We tried to show how was developing the awareness of differences between them (from the hopes that formal proofs provide sufficient means to exhaust the mathematical truth to the discovery of various limitations of them). Let us finish with some general remarks.

Concepts of proof and truth are (even in mathematics) ambiguous. One should distinguish between working proofs of everyday mathematics and idealized formal proofs used by logicians. On the other hand a proof in mathematics has various aspects and can be studied from various points of view. One can distinguish psychological, social, cultural and logical aspects of proofs. A proof can be studied as a mathematical or as an epistemological object. The former is precisely defined on the basis of mathematical logic, the latter is a vague concept. The former is an idealization of proofs occurring in a research practice of mathematicians, is a reconstruction of them. Recently one can observe in the philosophy of mathematics a tendency to concentrate on the actual research practice of mathematicians rather than on idealized foundational reconstructions of it and consequently to study the methods actually used by mathematicians.

Similar distinctions can be made with respect to the concept of truth. The semantical concept of truth precisely defined by Tarski is in fact a mathematical notion. It provides a definition of truth *in* mathematics,[5] it is the concept of truth for a model in a formal language (its essential feature is to define truth in terms of reference or satisfaction on the basis of a particular kind of syntactico-semantical analysis of the language). But one can also speak about epistemic truth – cf. for example Isaacson (1987) and (1992) where it is argued that Peano arithmetic is complete with respect to an epistemic notion of arithmetical truth.

The distinction between proof and truth in mathematics presupposes of course some philosophical assumptions. In fact for pure formalists and for intuitionists there exists no truth/proof problem. For them a mathematical statement is true just in case it is provable, and proofs are syntactic or mental constructions of our

---
5   One should distinguish the truth *in* mathematics and the truth *of* mathematics.

own making. In the case of a platonist (realist) philosophy of mathematics the situation is different. One can say that platonist approach to mathematics enabled Gödel to state the problem and to be able to distinguish between proof and truth, between syntax and semantics.[6]

---

[6] Note that, as indicated above, Hilbert was not interested in philosophical questions and did not consider them.

Reactions to the Discovery of the Incompleteness Phenomenon

1. Gödel's incompleteness theorems belong to the most important results of logic and the foundations of mathematics. They indicated the phenomenon of incompleteness of first order systems and in this way struck Hilbert's program of clarification and justification of the classical (infinite) mathematics by finitistic methods. They showed that this program cannot be fully realized in the original form by indicating some limitations of the axiomatic-deductive method.[1]

Gödel learned about Hilbert's program through Hans Hahn, the founder of the Vienna Circle.[2] Hahn took part in the Congress of Mathematicians in Bologna in 1928 during which Hilbert gave his lecture "Probleme der Grundlegung der Mathematik" (cf. Hilbert 1929). After his return to Vienna he communicated Hilbert's views to the Vienna Circle and encouraged his student Kurt Gödel to set to work on problems formulated by Hilbert. By mid 1929 Gödel had solved the fourth problem of Hilbert's Bologna address.[3] Taking first-order logic as presented in Hilbert and Ackermann (1928), he showed that the first-order logic is complete, i.e., every true statement can be derived from the axioms. Moreover he proved that, in the first-order logic, every consistent axiom system has a model. Those results were included in Gödel's doctoral dissertation (1929) (the results were published in the paper (1930a) – some remarks included in (1929) were omitted now).

Some months later, in 1930, Gödel solved three other problems posed by Hilbert in Bologna by showing that arithmetic of natural numbers and all richer

---

[1] On the development of Hilbert's program see, e.g., Murawski (1993, 1994 and 2002b) and Mancosu (1998).

[2] It was just through his contacts with the Vienna Circle that Gödel became interested in mathematical logic and the foundations of mathematics. It is worth mentioning here that he did not accept the positivistic philosophy of the Circle. As he wrote in a letter to Grandjean of 19th August 1975 (not sent): "It is true that my interest in the foundations of mathematics was aroused by the "Vienna Circle", but the philosophical consequences of my results, as well as the heuristic principles leading to them, are anything but positivistic or empiricistic" (cf. Wang Hao 1987, p. 20). In the thirties Gödel's conntacts with the Vienna Circle were weaker. He was engaged instead in Menger's *Mathematisches Kolloquium* and helped to publish *Ergebnisse eines Mathematischen Kolloquiums*. For further information on Gödel's contacts with the Vienna Circle cf. Köhler (1991) and Sigmund (1995) as well as Wang Hao (1987).

[3] In his lecture Hilbert set out four open problems connected with the justification of classical mathematics which should be solved: (1) to give a (finitist) consistency proof of the basic parts of analysis (or second-order functional calculus), (2) to extend the proof for higher-order functional calculi, (3) to prove the completeness of the axiom systems for number theory and analysis, (4) to solve the problem of completeness of the system of logical rules (i.e., the first-order logic) in the sense that all (universally) valid sentences are provable.

theories (i.e., theories extending it) are essentially incomplete. This means that there are sentences undecidable in them, hence sentences φ such that neither φ nor $\neg \varphi$ can be proved in the given theory. This property cannot be changed by adding new axioms (provided that the set of axioms is recursive, i.e., it can be effectively recognized whether a given sentence is an axiom or not). Furthermore, Gödel also showed that no such theory can prove its own consistency. Those results are called today as, respectively, First and Second Gödel's Incompleteness Theorem.

Gödel's answers to Hilbert's questions from Bologna were the opposite of what the latter had expected. In this way Hilbert's hopes and plans to justify classical mathematics were essentially weakened or even destroyed. In fact Gödel's theorems showed that there is even no consistent and complete formal system comprising the theory of natural numbers, moreover, to prove the consistency of a theory containing the arithmetic of natural numbers one needs stronger means than those available in the considered theory.

2. Gödel's results on the incompleteness of arithmetic were announced for the first time during the Second Conference on Epistemology of Exact Sciences (organized by Die Gesellschaft für Empirische Philosophie) held in Königsberg, 5–7th September 1930. Kurt Gödel presented there a twenty-minute talk devoted to the results contained in his doctoral dissertation, i.e., the completeness theorem for first-order logic (or, as it was then called 'the restricted functional calculus'). Next day, on 7th September Gödel took part in a discussion on the foundations of mathematics and told about his recent result on the incompleteness of arithmetic. After having critized the formalistic thesis that the consistency of axioms suffices to guarantee the truth of deduced theorems, he said:[4]

> (Assuming the consistency of classical mathematics) one can even give examples of propositions (and in fact of those of the type of Goldbach or Fermat) that, while contentually true, are unprovable in the formal system of classical mathematics. Therefore, if one adjoins the negation of such a proposition to the axioms of classical mathematics, one obtains a consistent system in which a contentually false proposition is provable.

The first announcement of Gödel's results had appeared in (1930b) – it was an abstract of (1931a) and was presented to the Vienna Academy of Sciences by Hans Hahn on 23rd October 1930. The full version of the results were published in Gödel's paper "Über formal unentscheidbare Sätze der 'Principia Mathematica' und verwandter Systeme. I" in January 1931 (the manuscript was received by the editors on 17th November 1930). In this paper the theorem today called Gödel's First Incompleteness Theorem was proved and the theorem called Gödel's Second Incompleteness Theorem was announced and promised to be published soon – in the second part of the paper – with a full proof. In

---

4   Cf. Gödel (1931a, p. 203).

fact the second part was never written and Gödel never published a proof of the Second Theorem. He explained this by saying that "the prompt acceptance of his results was one of the reasons that made him change his plan [of publishing the proof of the Second Theorem]" (cf. Heijenoort 1967, footnote 68a, p. 616) and that there was then no need to publish it. Another explanation was given by Bakley Rosser in (1939). In footnote 1 he wrote: "Due to ill health, Gödel has never written this second half". One should also add that the hint to the proof given in Gödel (1931a) has turned out to be incorrect (cf. Feferman 1960). The first correct proof of the Second Incompleteness Theorem was published in the monograph (1934, 1939) by D. Hilbert and P. Bernays.[5]

The system of arithmetic considered by Gödel in (1931a) was not the first-order Peano arithmetic but a system of the theory of types with individual constants for natural numbers and arithmetical axioms of Peano. In (1932) Gödel gave a more general presentation of his results using Peano arithmetic instead of the simple theory of types as the basic system.

On 25th June 1932 the paper "Über formal unentscheidbare Sätze ..." was presented to the University of Vienna as a *Habilitationsschrift*. On 1st December 1932 Gödel was granted the *Habilitation* and on 11th March 1933 the *venia legendi* and became *Privatdozent*. One of the referees in this procedure was H. Hahn. He evaluated Gödel's work as: "... a scientific achievement of the first order ... it can be safely predicted to earn a place in the history of mathematics ... Herr Gödel is already acknowledged as the foremost authority on symbolic logic and on the foundations of mathematics."

Gödel's incompleteness results were presented to a popular audience for the first time in the spring of 1932 by Karl Menger in his lecture "Die neue Logik" (the lecture was published by F. Deuticke in 1933 as one of "Fünf Wiener Vorträge" in the booklet *Krise und Neuaufbau in der exakten Wissenschaften* – for an English translation cf. Menger 1978).

3. It seems that the participants of the conference in Königsberg in 1930 were unaware of the meaning and importance of those results. There was no discussion after Gödel's pronouncement. In the proceedings of the conference Gödel's name does not occur (cf. Dawson 1984).

At least two persons among the participants of the conference in Königsberg should have had foreknowledge of Gödel's incompleteness result: Hans Hahn, the

---

5 J. W. Dawson writes in his book (1997, p. 109) that, as Bernays later told Kreisel, during his second voyage to America in 1935 aboard Cunard liner *Georgic*, Gödel met Bernays and "there was an exchange of considerable importance" between them. In fact "during that voyage and the few weeks following their arrival in Princeton (...) Gödel explained to him [i.e. Bernays – R.M.] the details of the proof of the second incompleteness theorem, as they were subsequently presented in Hilbert and Bernay's text."

supervisor of Gödel's doctoral dissertation, and Rudolf Carnap. Indeed, Carnap wrote in his diary:

> August 26, 1930: 6 to half past 8, in the Café Reichsrat coffee-house with Feigl, Gödel, later Waismann. Plan of travelling to Königsberg by boat. Gödel's discovery: incompleteness of the system of *Principia Mathematica*. Difficulties with the consistency proof.

On the other hand it seems that they failed to understand it. Carnap wrote in his diary:

> February 7, 1931: At 4 Gödel here. What concerns his work [i.e., Gödel's paper (1931a) – R.M.], I must say that it is difficult to understand.

On the other hand analysing the correspondence between Carnap, von Neumann and Reichenbach concerning the publication of the proceedings of the Königsberg meeting in 1930 one can come to the conclusion that although Carnap had difficulties with understanding of the technical details of the proof of Gödel's theorem, the importance of this result was immediately clear to him (cf. Mancosu 1999).

It seems that the only participant of the conference in Königsberg who immediately grasped the meaning of Gödel's theorem and understood it was John von Neumann. After Gödel's talk he had a long discussion with him and asked him about details of the proof. Soon after coming back from the conference to Berlin he wrote a letter to Gödel (on 20th November 1930) in which he announced that he had received a remarkable corollary from Gödel's First Theorem, namely a theorem on the unprovability of the consistency of arithmetic in arithmetic itself. In the meantime Gödel developed his Second Incompleteness Theorem and included it in his paper "Über formal unentscheidbare Sätze ...". In this situation von Neumann decided to leave the priority of the discovery to Gödel.

Note also that von Neumann was of the opinion that "Gödel's result has shown the unrealizability of Hilbert's program" and that "there is no more reason to reject intuitionism" (cf. his letter to Carnap of 6th June 1931 – see Mancosu, 1999, p. 39–41). He added in this letter:

> Therefore I consider the state of the foundational discussion in Königsberg to be outdated, for Gödel's fundamental discoveries have brought the question to a completely different level. (I know that Gödel is much more careful in the evoluation of his results, but in my opinion on this point he does not see the connections correctly).

Indeed, already in his paper from 1931 Gödel wrote:

> I wish to note expressly that Theorem XI (and the corresponding results for M and A) do not contradict Hilbert's formalistic viewpoint. For this viewpoint presupposes only

the existence of a consistency proof in which nothing but finitary means of proof is used, and it is conceivable that there exist finitary proofs that cannot be expressed in the formalism of P (or M or A).

At the Vienna Circle meeting on 15th January 1931 Gödel argued that it is doubtful, "whether all intuitionistically correct proofs can be captured in a *single* formal system. That is the weak spot in Neumann's argumentation" (quotation taken from Sieg 1988).

Gödel suggested that Hilbert's program may be continued by allowing two principles which can be treated as finitistic, namely (1) the principle of transfinite induction on certain primitive recursive well-orderings, and (2) a notion of computable functions of a finite type (i.e., of computable functionals), to which the process of primitive recursion can be extended in a natural way.

He was convinced of the power of human reasoning and was of the opinion that his incompleteness theorems are not establishing limitations of it but that "the kind of reasoning necessary in mathematics cannot be completely mechanized" (as he formulated it in a letter to David F. Plummer of 31st July 1967 – cf. Dawson 1997, p. 263) and consequently the human intellect plays essential rôle in mathematical research.

Through von Neumann about Gödel's incompleteness theorems learned (in November 1930) Jacques Herbrand. He found them to be of great interest. They also stimulated him to reflect among others on the recursive functions. In a letter to Gödel of 7th April 1931 Herbrand suggested the idea of extending the schemes for the recursive definition of functions. His remarks inspired Gödel to formulate the notion of general recursive function (in the lectures he gave at Princeton in 1934).

Gödel's paper "Über formal unentscheidbare Sätze ..." was published in January 1931. Already before its publication Paul Bernays, then the secretary and assistant of D. Hilbert took interest in it. He had learned of these results from Richard Courant (cf. Moore 1991). Around Christmas 1930 he wrote to Gödel requesting a copy of the galley proofs. Bernays and Gödel corresponded a lot with each other discussing Gödel's results, their relation to Hilbert's program and the chances and possibilities of developing this program. Mid January 1931, Bernays wrote to Gödel that the incompleteness theorems were "an important step forward in research on foundational problems", and in a letter of mid-April he called the results "surprising and significant" (quotations after Moore 1991). In a letter to Constance Reid of 3rd August 1966 Bernays wrote:

> I was doubtful already sometime before [1931] about the completeness of the formal system [for number theory], and I uttered [my doubts] to Hilbert, who was much angry [...]. Likewise he was angry at Gödel's results.[6]

---

6   Quotation according to Dawson (1998).

Hilbert took part in the conference in Königsberg but he did not attend the discussion during which Gödel announced his incompleteness results – he was busy preparing his speech for the local radio on the occassion of his honorary citizenship of Königsberg.[7] He learned about them from P. Bernays only in January 1931. And, as C. Reid writes in the biography (1970), was at the beginning "somewhat angry". The irritation and frustration passed in course of time and Hilbert tried to approach the new situation in a more constructive way. Reid writes (1970, pp. 198–199):

> Bernays found himself impressed that even now, at the very end of his career, Hilbert was able to make great changes in his program. It was not yet clear just what influence Gödel's work would ultimately have. Gödel himself felt – and expressed the thought in his paper – that his work did not contradict Hilbert's formalistic point of view.

In his lecture in Hamburg in December 1930 (cf. Hilbert 1931) Hilbert proposed to admit a new rule of inference to be able to realize his program. This rule is similar to the ω-rule, but it has rather informal character and a system obtained by admitting it would be semi-formal. In fact Hilbert proposed that whenever $A(z)$ is a quantifier-free formula for which it can be shown (finitarily) that $A(z)$ is a correct (richtig) numerical formula for each particular instance z, then its universal generalization $\forall x A(x)$ may be taken as a new premise (Ausgangsformel) in all further proofs.

In Preface to the first volume of Hilbert and Bernays' monograph *Grundlagen der Mathematik* (1934, 1939) Hilbert wrote:[8]

> [...] the occasionally held opinion, that from the results of Gödel follows the non-executability of my Proof Theory, is shown to be erroneous. This result shows indeed only that for more advanced consistency proofs one must use the finite standpoint in a deeper way than is necessary for the consideration of elementary formalism.

In fact researches showed that Hilbert's program can be successfully developed after Gödel's incompleteness theorems. One of the main problems (indicated here by Gödel's results) was the problem of means that are necessary for the con-

---

7   The honorary citizenship was presented to Hilbert at the meeting of the Society of German Scientists and Physicians. Hilbert delivered on this occasion an address "Naturerkennen und Logik" (cf. Hilbert 1930) attended by Gödel. This was the only time Gödel ever saw Hilbert. They never met or corresponded – cf. (Wang Hao 1987 p. 85) and (Dawson 1985a, footnote 4). Dawson writes: "In a letter to Constance Reid of March 22, 1966, Gödel stated that he "never met Hilbert ... nor [had] any correspondence with him". The stratification of the German academic system may have discouraged contact between the two men."
8   For remarks on some connections between those ideas of Hilbert and some ideas of Leibniz see (Murawski 2002b).

sistency proofs – for example G. Gentzen proved the consistency of the arithmetic of natural numbers using the ε_0-induction.[9]

Gödel exchanged letters (trying to explain his results) with another leading mathematician Ernst Zermelo who turned out to be one of the greatest critics of Gödel. They met at the meeting of the German Mathematical Union (Deutsche Mathematiker-Vereinigung) in Bad Elster[10] in 1931. Zermelo claimed in his lecture at this meeting (cf. Zermelo, 1932) that:

> From our standpoint every 'true' statement is at the same time *'provable'* ... There are *no* (objectively) *undecidable* statements. Mr. Gödel, on the contrary, attempts to prove the *opposite* ... But Gödel's proof succeeds only by applying the 'finitisti' restriction solely to the 'provable' statements of the system, not to *all* the statements belonging to it. The former but not the latter thus form a *countable set*, so of course in *that* sense there must be 'undecidable' statements.

The letters ZermeloIndexZermelo E. and Gödel exchanged after the meeeting indicate that the main obstacle to understand each other was deep disagreement in their philosophical views on mathematics (cf. Dawson 1985b; Grattan–Guinness 1979). In particular, Zermelo saw a contradiction in the proof of Gödel's First Incompleteness Theorem (in his letter he called it "an essential gap"), namely that a certain proposition was neither true nor false. Gödel replied that the source of the error in Zermelo's argument was the fact that the notion of truth was not expressible in the formal system used in incompleteness results.[11] Zermelo claimed that the existence of undecidable propositions was a consequence of the restriction of the notion of proof to finitistic methods (he said here about "finitistic prejudice"). This situation could be changed if one used a more general "scheme" of proof. Zermelo had here in mind an infinitary logic, in which there were infinitely long sentences and rules of inference with infinitely many premises. In such a logic, he insisted, "*all* propositions are decidable!"[12] He thought of quantifiers as infinitary conjunctions or disjunctions of unrestricted cardinality and conceived of proofs not as formal deduc-

---

9   For the development of Hilbert's program after Gödel's theorems see, e.g., (Murawski 1993 and 1994) where the generalized and relativized Hilbert's programs are described and the meaning of results of the reverse mathematics is indicated.
10 Gödel spoke there about the incompleteness results – this was the first presentation of those results outside Vienna after the publication of his paper (1931a).
11 This is one of the evidences that Gödel discovered the undefinability of truth independently of Tarski. Another evidence of Gödel's awareness of the undefinability is his letter to Bernays of 3rd May 1931 – cf. Dawson (1985a). For the discussion of this problem see Woleński (1991) and Murawski (1998).
12 Note that time was not yet ripe for such an infinitary logic. Systems of such a logic, though in a more restricted form than demanded by Zermelo, and without escaping incompleteness, were constructed in the mid-fifties in works of Henkin, Karp and Tarski (cf. Barwise 1980 and Moore 1980).

tions from given axioms but as metamathematical determinations of the truth or falsity of a proposition. Thus syntactic considerations played no rôle in his thinking.

4. Gödel was really sensitive to the fact that he was the first in discovering the incompleteness of arithmetic and reacted vividly on every attempt to deprive him of it. And there were many such attempts. For example Paul Finsler from Zurich claimed that he had obtained similar results already in 1926. In his paper (1926) he wrote:

> [...] we shall now show by means of an example that we can in fact exhibit propositions that are not formally decidable by general methods, that are therefore formally consistent, but in which we can nevertheless recognize a contradiction in another way. It follows, therefore, that the proof of the formal consistency of a system does not afford a guarantee against recognizable contradictions.

In the letter to Gödel of 11the March 1933 Finsler wrote that Gödel's results were "in principle somewhat similar" to his own from the paper (1926).

A careful inspection of Finsler's paper (cf. Heijenoort 1967, pp. 438–440) proved that the similarity was apparent and in fact Gödel was the first who proved the incompleteness. Finsler's basic notions (such as proof, formal undecidability etc.) were vague and not precisely defined. In fact he rejected formal systems, put by Gödel "at the very center of his investigations" (cf. Heijenoort 1967) as artificially restrictive. That led to incorrect corollaries. This fact was mentioned already by Gödel in his reply (of 25th March) to Finsler's letter. He wrote there:[13]

> The system [...] with which you operate is not really [well] defined, because in its definition you employ the notion of "logically unobjectionable proof," which, without being made more precise, allows arbitrariness of the widest scope.

Finsler did not give in. It seems that he did not understand Gödel's proofs.[14] On the other hand one should admit that Finsler's merit was the fact that he stated (already in the twenties) the thesis on the incompleteness of mathematics.

Another mathematician who considered the problem of incompleteness before Gödel was Emil Post. His reactions to Gödel's results were quite different than those of Finsler. At the beginning of the twenties (hence nearly ten years before Gödel!) he tried to show that there are absolutely undecidable problems in mathematics. He realized that his method could be applied to yield a statement undecidable within *Principia Mathematica* whose truth could nevertheless be established by metamathematical considerations. His results (on the decision

---

13 Quotation according to Dawson (1997, p. 89).
14 In connection with this problem see also (Ketelsen 1994, pp. 131–133) and (Ladrière 1957, p. 95).

problem for normal systems) anticipated results by Gödel and Church on the incompleteness and undecidability of systems of first-order logic. Post knew of course that his results were, as he wrote, "fragmentary". He never published them and gave up the researches. However, twenty years later, hence already after the publication of Gödel's and Church's results, he attempted to publish results of his investigations from 1920–1921. At the beginning of the fourties he wrote a paper "Absolutely unsolvable problems and relatively undecidable propositons – account of an anticipation". It was submitted in 1941 to *American Journal of Mathematics*. In a letter to H. Weyl accompanying the manuscript Post explained why he did not publish his results twenty years earlier and wants to do it now, i.e., after the publications by Gödel and Church. Among reasons he mentions problems he had with publishing his earlier papers which did not find a recognition and appreciation by mathematicians as well as the problems with the health which delayed the preparation of full detailed proofs. Though the editors appreciated the significance of Post's investigations and results, the paper has been rejected. Communicating this decision H. Weyl wrote in a letter to Post of 2nd March 1942:

> [...] I have little doubt that twenty years ago your work, partly because of its then revolutionary character, did not find its due recognition. However, we cannot turn the clock back; in the meantime Gödel, Church and others have done what they have done, and the American Journal is no place for historical accounts; ... (Personally, you may be comforted by the certainty that most of the leading logicians, at least in this country, know in a general way of your anticipation.)

Only a small part of Post's paper has been published, i.e., the part containg his Normal Form Theorem (cf. Post 1944). The full version of the paper "Absolutely unsolvable problems and relatively undecidable propositions – account of an anticipation" was published posthumously in 1965 in Davis' book *The Undecidable* (cf. Post 1965).

Post had always had great esteem of Gödel's results and expressed "the greatest admiration for them". He never sought to diminish Gödel's achievement. On the contrary, he wrote:

> The plan [i.e., his plan to prove the incompleteness of *Principia* – R.M.], however, included prior calisthenics at other mathematical and logical work, and did not count on the appearance of a Gödel! (1965, p. 418)

In a postcard to Gödel sent on 19th October 1938 Post wrote:

> I am afraid that I took advantage of you on this, I hope but our first meeting. But for fifteen years I had carried around the thought of astounding the mathematical world with my unorthodox ideas, and meeting the man chiefly responsible for the vanishing of that dream rather carried me away.

Since you seemed interested in my way of arriving at these new developments perhaps Church can show you a long letter I wrote to him about them. As for any claims I might make perhaps the best I can say is that I would have *proved* Gödel's Theorem in 1921 – had I been Gödel.

And in the letter to Gödel of 30th October 1938 he wrote:

> [...] after all it is not ideas but the execution of ideas that constitute a mark of greatness.

There also appeared some people claiming that Gödel's theorems were incorrect and false, that Gödel simply discovered a new antinomy. Such were the opinions of Charles Perelman (cf. his paper from 1936), Marcel Barzin (in a paper from 1940) or Jerzy Kuczyński (cf. his paper 1938).

Perelman in (1936) claimed that Gödel had in fact discovered an antinomy. He attempted to show there that the Gödel's method could be employed to prove two false equivalences. His argumentation turns out to be wrong. It seems that he read only the heuristic introduction of Gödel's paper (1931a) and then treated the definition of K formulated there metamathematically as a strict definition in the system P.

Gödel did not enter into the controversy with Perelman, but Kurt Grelling and Olat Helmer did (Grelling requested in a letter to Gödel his permission to react on his behalf). His answer to Perleman was the paper "Gibt es eine Gödelsche Antynomie?", Helmer's answer was his paper "Perelman versus Gödel", both published in 1937. They indicated there that the source of Perleman's error was the fact that he did not distinguish between the object language and the metalanguage. Unfortunately in their considerations they did not avoid themselves some mistakes.

Barzin and Kuczyński claimed that Gödel had discovered an antinomy. Though their argumentations were based on the formal proof of Gödel's paper (1931a) and not – as it was the case by Perelman – on the informal introductory remarks, the arguments they formulated turned out to be wrong. In fact Barzin confused formal expressions with their Gödel numbers and Kuczyński overlook the formal antecedent $Wid(\kappa)$ in Gödel's second theorem.

5. So far we have considered the reactions of logicians (hence specialists) on the Gödel's discovery of the incompleteness phenomenon. Ask now about the reactions of philosophers. We consider this problem discussing the reactions of Bertrand Russell and Ludwig Wittgenstein.

Wittgensteins' remarks on Gödel's theorems are contained in Appendix I to his posthumous *Remarks on the Foundations of Mathematics*. They seem to be an embarrassment to his work. So M. Dummett (1978), in many respects an admirer of Wittgenstein's philosophy, writes that the remarks on Gödel and on the notion of consistency are "of poor quality and contain definite errors" (p. 166). It turns out that Wittgenstein failed to understand Gödel's results and –

as Gödel himself put it in a letter to Abraham Robinson from 2nd July 1973 – "advance[d] a completely trivial and uninteresting misinterpretation" of them. On the other hand, recently some attempts were made to find a perspective from which Wittgenstein's view towards Gödel's theorems becomes more understandable (cf. Wang Hao 1987, 1991a, 1991b; Shanker 1988; see also Floyd 1995).

Russell reacted on Gödel's results in an ambiguous way. In (1959, p. 114) he seems to make the same point as Gödel (1932) made that by passing to higher types one can obtain formal systems such that the undecidable propositions constructed within each system are decidable in higher systems. Quite different are his views expressed in an unpublished letter to Leon Henkin of 1st April 1963 (now in the Russell Archive) in which he wrote:

> I realized, of course, that Gödel's work is of fundamental importance, but I was puzzled by it. It made me glad that I was no longer working at mathematical logic. If a given set of axioms leads to a contradiction, it is clear that at least one of the axioms must be false. Does this apply to school-boy's arithmetic, and, if so, can we believe anything that we were taught in youth? Are we to think that $2 + 2$ is not 4, but 4.001? Obviously this is not what is intended.

It is not clear if Russell "had recognized the futility of Hilbert's scheme for proving the consistency of arithmetic but had failed to consider the possibility of rigorously proving that futility", or "is he revealing a belief that Gödel had in fact shown arithmetic to be inconsistent" (cf. Dawson 1985a). Gödel wrote in a letter to A. Robinson of 2nd July 1973: "Russell evidently misinterprets my results; however he does so in a very interesting manner ..."

Another interpretation of Russell's attitude towards Gödel's incompleteness results suggests F.A. Rodríguez-Consuegra in (1993). He mentions Russell's very little known addendum in the 1971 edition of his Schilpp volume (pp. xviii f) where he gave a correct exposition of the essentials of the incompleteness results, insisted that this result cannot be regarded as a fatal objection against the truth of mathematical logic, and especially against the absolute and general truth of every-day arithmetic[15] and expressed a belief that proving the lack of contradictions among the consequences of a system is impossible "since the number of consequences of any given set of axioms is infinite".

Rodríguez-Consuegra claims that Russell's reference (in the letter to Henkin from 1963 quoted above) to the every-day arithmetic "was intended merely to reinforce our confidence in the truth of self-evident elementary axioms" (p. 235). He claims also that "the ultimate source of his [i.e. Russell's – R.M.] misinterpretation of Gödel might lie in his impression that Gödel's methods supposed a violation of the requirement that the syntax of a language, although it can be

---

15 Russell wrote: "It is maintained by those who hold this view that no systematic logical theory can be true of everything. Oddly enough, they never apply this opinion to elementary every-day arithmetic".

expressed (against Wittgenstein), this can be done only in a higher language (against Gödel)" (p. 236). Indeed Gödel showed in (1931a) that the syntax of a language can be expressed in the same language (via arithmetization) and indicated in this way how one can overcome one of the deepest beliefs of most of the members of the Vienna Circle (influenced by Wittgenstein). Russell had evidently difficulties in accepting Gödel's devices proposed in (1931a). He wrote in (1937):

> It is true, as Tarski (*Der Wahrheitsbegriff in den formalisierten Sprachen*, Lvov 1935) and Carnap (*Logical Syntax of Language*, Kegan Paul, 1937. – German, 1934) have proved, that in any given language there are things that cannot be said, but they can be said in a language of higher order. To say something about what cannot be said at all is not necessarily self-contradictory, but there is no reason known to me for supposing that there is any actual significant statement of this sort.

6. The above considerations show that the reactions to the discovery of the incompleteness phenomenon by Gödel were very different and represent a wide spectrum. On the one hand there were mathematicians who immediately grasped the importance of Gödel's results and even developed them further (as was in the case of von Neumann), on the other there were those who had great difficulties to understand and to accept the theorems. Those difficulties are now well understandable. In fact Gödel's results were revolutionary, they destroyed in a sense the convictions and expectations towards logic and the foundations of mathematics cherished by his contemporaries. Hence the resistance against them and the suspicions they must contain a contradiction or at most indicate new antinomies. All that shows and stresses the genius of Gödel and the greatness of his achievements.

Gödel's Incompleteness Theorems and Computer Science

In this essay some applications of Gödel's incompleteness theorems to discussions of problems of computer science will be presented. In particular the problem of relations between the mind and machine (arguments by J. J. C. Smart and J. R. Lucas) is discussed. Next Gödel's opinion on this issue is studied. Finally some interpretations of Gödel's incompleteness theorems from the point of view of the information theory are presented.

1. Gödel's Theorems and Artificial Intelligence

Gödel's incompleteness theorems[1] have been used as arguments in various discussions of mathematical and metamathematical as well as philosophical problems. Recently they have also appeared in discussions of problems connected with computer science, especially with artificial intelligence.

The origins of the latter are connected with A. Turing's paper "Computing Machinary and Intelligence" (1950). One of the main problems investigated in this domain is the question of whether machines can act in an intelligent way. Note that this problem is in fact very old and was considered already by René Descartes (1596–1650) and Julien Offray de La Mettrie (1709–1751) who discussed the question of whether a human being is a mechanical machine. *Descartes in Discours de la méthode pour bien conduire sa raison et chercher la vérité dans les sciences* (1637) answered this question negatively arguing that it is impossible to construct a (mechanical) machine which could perform acts that are characteristic for a human being. On the contrary La Mattrie claimed in *L'homme-machine* (1748) that thinking is a function of a body and that a human being can be treated as a machine.

The perspective of the eighteenth century thinkers was naturally restricted to mechanical machines. New aspects of the problem appeared with the introduction of symbolical machines (such as Turing machines) and modern computers. It can be now formulated in the form: does the human mind work as a formalized

---

1   Gödel's incompleteness theorems were proved in his famous paper (1931a). They can be found in various handbooks and monographs – see, e.g., Mendelson (1964), Smoryński (1977) or Murawski (1999b). Recall that Gödel's First Incompleteness Theorem states that any consistent recursively enumerable (recursively axiomatizable) first-order theory containing the arithmetic of natural numbers is essentially incomplete. Gödel's Second Incompleteness Theorem says that no such theory proves its own consistency (here some additional conditions on the formula that expresses the metamathematical notion of consistency must be satisfied).

system, can a computer program be treated as a model of the human mind. At the beginning of the sixties two papers appeared in which an attempt was made to answer this question using Gödel's theorems: Smart (1961) and Lucas (1961). They caused a long and interesting debate.[2]

Smart understands by a machine a Turing machine and is of the opinion that (cf. Smart 1961, p. 105):

> Recent tendencies in biology and psychology have made plausible, to some of us at any rate, the view that men are complicated mechanisms.

He uses in his argumentation Gödel's theorem and writes (pp. 105–106):[3]

> Consider a formal language L_0 adequate for elementary number theory. Gödel has shown that if elementary number theory is consistent there is some closed arithmetical sentence expressible in the symbolism of L_0 which can not be proved or disproved in L_0. By reasoning which makes use of the syntax language of L_0, however, we can show that this proposition which is undecidable within L_0 is in fact true. Corresponding to any formal language L_0 (with constructive rules of proof) we can specify some Turing machine (or computer) T_0 which if given enough time will produce a proof or disproof of any decidable closed arithmetical sentence of L_0. The machine will not of course ever produce for us a proof of a Gödelian undecidable proposition of L_0, for no such proof exists. A proof does exist in the language L_1, which is the language got by adjoining the syntax language of L_0 to L_0. To L_1 will correspond a machine T_1, which if given enough time will produce a proof of the original sentence undecidable in L_0. However in L_1 there will be another proposition undecidable within it but which we by using the syntax language of L_1 can show to be true. And so on. Consider the sequence of languages $L_0, L_1, L_2, \ldots$ and the sequence of machines $T_0, T_1, T_2, \ldots$ Whatever you like to mention there will always be some proposition which we can prove but it can not.

Smart's argument is now based on the claim that a machine can transform itself into another machine with greater logical power and in this way he comes to the conclusion (p. 106):

> If it [the human nervous system – R.M.] is a machine there must be some Turing machine which can do anything it can do.

Let us turn now to Lucas who at the very beginning of his paper (1961) claims that (p. 112):

---

2 The arguments of Lucas were criticized in particular by I. J. Good in (1967; 1969) and P. Benacerraf in (1967). Lucas defended his position in (1967; 1968). See also Hofstadter (1979) and Penrose (1989; 1994).
3 Observe that Smart uses here the notion of the syntax language introduced by Carnap; Gödel spoke about metamathematical considerations.

> Gödel's Theorem seems to me to prove that Mechanism is false, that is, that minds cannot be explained as machines.

His argumentation is as follows.[4] Lucas wants to show that mathematical abilities of a human being, e.g., his own abilities, are bigger than those of a machine, even if restricted to arithmetical propositions. If Mechanist says that a machine M is equivalent to Lucas, or rather to mathematical powers of Lucas, then Lucas produces the Gödel formula φ_M corresponding to the theory $T(M)$ consisting of arithmetical statements provable by M. It is inessential how we define the theory $T(M)$ – the only important thing is that $T(M)$ is semi-decidable (i.e., recursively enumerable). It is always the case if one assumes that M is equivalent to a Turing machine (or other standard notion). There are two possibilities now:

Case 1: the theory $T(M)$ is consistent.
Case 2: the theory $T(M)$ is inconsistent.

In the first case the sentence φ_M is unprovable by M, i.e., it is unprovable in the theory $T(M)$, but Lucas can prove it by showing that it is true (Lucas uses here Gödel's theorem together with the assumption that $T(M)$ is consistent). In the second case every sentence is provable by M. But such a provability is useless and Lucas has certainly another notion of provability and consequently is different from the machine M. So in any case Lucas has shown that the machine M cannot be equivalent to him.

In this way Lucas comes in (1961) to the following conclusion (p. 116):

> We are trying to produce a model of the mind which is mechanical – which is essentially "dead" – but the mind, being in fact "alive", can always go on better than any formal, ossified, dead, system can. Thanks to Gödel's theorem, the mind always has the last word.

Observe that the above procedure can be mechanized. Let us assume that all machines are listed in a sequence $M_1, M_2, M_3, \ldots$ It is easy to show that there is a recursive function g such that for every n

if $T(M_n)$ is consistent then $g(n)$ is the Gödel number of a (Gödel) sentence which is true and unprovable in $T(M_n)$. (∗)

An objection to the above argumentation can be made on the basis of (∗). Indeed, if a machine corresponding to the function g can simulate the procedure described above (and called sometimes "out-Gödeling of M") then machines are not really worse than we are. It can be replied that this machine can be out-Gödeled too. In fact, the aim is not to dominate all machines at once but rather each machine proposed by the Mechanist. Therefore Lucas describes the procedure as dialectical and he writes in (1968, p. 154):

---

4   A nice presentation and discussion of Lucas' argument can be found, e.g., in Krajewski (1993). See also Rucker (1982).

The argument is dialectical. It is an argument between two persons, not a proof sequence constructed by one.

Hence it is a game in which Lucas wins in every move.

Another objection that can be raised is that the procedure does not work if the program of a machine is not known. Hence the mathematical powers of Lucas might be equal to a machine but he would not be able to find an n in order to produce $g(n)$. This leads to the following possibility: perhaps we are a machine but we do not know which one. This can be answered as follows: the appropriate Gödel sentence exists in an objective way and can be used by us at any time. On the other hand if a Mechanist presents a number n then Lucas can produce $g(n)$. The sentence corresponding to it will be true provided the theory $T(M_n)$ is consistent.

But here the next problem appears: how do we know that $T(M_n)$ is consistent? The set $C = \{n : T(M_n) \text{ is consistent}\}$ is not recursive,[5] so the problem "is a given machine consistent" is undecidable. Consequently it is theoretically impossible to distinguish Case 1 and Case 2. To distinguish them requires nonrecursive skills, i.e., the procedure of out-Gödeling assumes a non-mechanical nature of Lucas. But this is the thesis that was supposed to be proved!

In this situation one can argue that either Case 1 or Case 2 holds and hence the whole argument still works. But now the procedure becomes nonconstructive and its dialectical character is doubtful.

Since Gödel sentences for inconsistent theories are false and contradict even the most elementary arithmetical truths, i.e., truths that can be formulated as formulas with bounded quantifiers only,[6] Lucas is not allowed to commit even a single mistake. Indeed, his argument in Case 2 depends entirely on his consistency and if he presented Gödel sentence as a sentence proving his superiority over machines, he would fall into inconsistency.

So the question if we are consistent turns out to be significant. Lucas in (1961) claims that the answer is negative and writes (p. 120):

> [...] are not men inconsistent too? Certainly women are, and politicians; and even male non-politicians contradict themselves sometimes, and a single inconsistency is enough to make a system inconsistent.

---

5   Assume that C is recursive. Then $C' = \{g(n) : n \in C\}$ is recursively enumerable. Hence $C' = T(M_k)$ for some k. Since all $g(n)$ for $n \in C$ are true, it follows that $k \in C$. By the definition $g(k)$ is unprovable in $T(M_k)$. But on the other hand $g(k) \in T(M_k)$, a contradiction.

6   Gödel's undecidable sentence is of the form $\forall x \varphi(x)$ where $\varphi(x)$ is a formula containing only atomic formulas, connectives and bounded quantifiers (denote the class of all such formulas φ by BQ). Recall that if ψ is any formula of the form $\exists x \varphi(x)$ where $\varphi \in$ BQ and ψ is true in the standard model of the arithmetic of natural numbers then ψ is provable in Peano arithmetic PA.

And he adds (p. 123):

> [...] if we find that no system containing simple arithmetic can be free of contradictions, we shall have to abandon not merely the whole of mathematics and the mathematical sciences, but the whole of thought.

The assumption that mathematics (and we ourselves as well!) are consistent is necessary to do mathematics. But it should be noted that it may happen that proofs of contradictions are too long to matter in practice and that it is possible to remain in safe contradiction-free areas all the time.[7] Gödel's Second Theorem seems to imply that our consistency cannot be proved in a mathematical, formal way even if we are consistent. In practice we believe that we are in principle consistent or rather that we can correct and remove any inconsistencies when they appear. Lucas puts it in the following way (1961, p. 122):

> [...] the mind does indeed try out dubious axioms and rules of inference; but if they are found to lead to contradiction, they are rejected altogether. We try out axioms and rules of inference provisionally-true: but we do not keep them, once they are found to lead to contradictions.

The belief that we are in principal consistent does not, however, assure that Lucas' argument is right. Indeed, G. Lee Bowie remarked in (1982) that whatever Lucas may claim about his own consistency, we know enough to prove that he is actually inconsistent. Lucas' procedure can be expressed as the recursiveness of the function g satisfying the condition $(*)$ formulated above. But the range of g, i.e., the set

$$A = \{\text{the sentence with Gödel number } g(n) : n \in \mathbb{N}\}$$

is inconsistent. This can be proved in the following way: Since g is recursive, the set A is recursively enumerable. Hence one can assume that it is generated by a machine M_k for some $k \in \mathbb{N}$. If A were consistent then by $(*)$ the sentence with Gödel number $g(k)$ would be unprovable in $T(M_k)$. Yet it is an element of A and consequently it belongs to $T(M_k)$. This is a contradiction.

It seems impossible to modify Lucas' argument to avoid the above criticism – cf. Krajewski (1983; 1993).

Though no convincing arguments in favour of or against the mechanistic thesis were proposed so far, it seems that Gödel's results support the moderate opinion of Nagel and Newman formulated in (1958, pp. 100–101) in the following way:[8]

---

7   This is the case with large computer programs. Note also that the infinitesimal calculus was developed for centuries on inconsistent foundations.
8   Though formulated almost forty years ago, when computers and artificial intelligence were at the beginnings of their development, it seems to be still valid.

> Given a definite problem, a machine [...] might be built for solving it; but no one such machine can be built for solving every problem. The human brain may, to be sure, have built-in limitations of its own, and there may be mathematical problems it is incapable of solving. But, even so, the brain appears to embody a structure of rules of operation which is far more powerful than the structure of currently conceived artificial machines. There is no immediate prospect of replacing the human mind by robots.

2. Gödel on the Mind-Body Problem

What was the opinion of Gödel concerning the problem of interrelations between his incompleteness theorems and the question whether our minds can be treated as machines? Note first that Gödel was looking for arguments that "laws of thought are not mechanical" – cf. Kreisel (1980, p. 216). In the 25th Josiah Willard Gibbs Lecture "Some basic theorems on the foundations of mathematics and their implications"[9] Gödel considered implications of his incompleteness theorems for the problem of relations between mind and machine. One of the implications is that the human mind is incapable of formulating or mechanizing all its mathematical intuitions. Therefore one can speak about certain incompletability of mathematics. Gödel wrote (p. 309):

> [...] *it* [Second Incompleteness Theorem – R.M.] makes *it impossible that someone should set up a certain well-defined system of axioms and rules and consistently make the following assertion about it: All of these axioms and rules I perceive (with mathematical certitude) to be correct, and moreover I believe that they contain all of mathematics.* If someone makes such a statement he contradicts himself. For if he perceives the axioms under consideration to be correct, he also perceives (with the same certainty) that they are consistent. Hence he has a mathematical insight not derivable from his axioms.

On the other hand Gödel noted that on the basis of what has been proved so far, it remains possible that (pp. 309–310):

> the human mind (in the realm of pure mathematics) *is* equivalent to a finite machine that, however, is unable to understand completely its own functioning.

This means that such a machine would be equivalent to mathematical intuition but on the other hand it would be impossible to prove that fact, or even to prove that the machine yields only correct theorems. This possibility follows from Gödel's Second Theorem under the assumption that our mind is consistent. Gödel treated the latter as granted.

---

9 Gödel delivered this lecture at the meeting of the American Mathematical Society at Brown University on 26th December 1951. It was first published in the volume III of his *Collected Works* – cf. Gödel (1951).

The second important implication is formulated by Gödel in the following way (p. 310):

> Either mathematics is incompletable in this sense, that its evident axioms can never be comprised in a finite rule, that is to say, the human mind (even within the realm of pure mathematics) infinitely surpasses the powers of any finite machine, or else there exist absolutely unsolvable diophantine problems [...] (where the case that both terms of the disjunction are true is not excluded, so that there are, strictly speaking, three alternatives).

Gödel claimed however that the history of mathematics forced us to reject the second alternative. His argumentation is summarized by Hao Wang in the following way (1974, pp. 324–325):

> If it were true it would mean that human reason is utterly irrational by asking questions it cannot answer, while asserting emphatically that only reason can answer them. Human reason would then be very imperfect and, in some sense, even inconsistent, in glaring contradiction to the fact that those parts of mathematics which have been systematically and completely developed (such as, e.g., the theory of 1st and 2nd degree Diophantine equations, the latter with two unknowns) show an amazing degree of beauty and perfection.

Certain 'rationalistic optimism' can be seen here. In this way Gödel shared Hilbert's belief expressed in (1926) in the words:

> In mathematics there is no *ignorabimus*,

and repeated at the end of his speech over the local radio station in Königsberg in September 1930, just after the conference during which Gödel announced his incompleteness result:

> Wir müssen wissen. Wir werden wissen.
> (We must know. We shall know.)

In Gödel's opinion the attempted proofs for the equivalence of the mind and machines were fallacious. This follows, e.g., from the following unpublished paragraph about Turing's alleged proof that every mental procedure for producing an infinite series of integers is equivalent to a mechanical procedure:[10]

> Turing, in *Proc. Lond. Math. Soc.* 42 (1936/1937), p. 250, gives an argument which is supposed to show that mental procedures cannot carry any further than mechanical

---

10 This paragraph was to be added as a footnote at the word 'mathematics' on p. 73, line 3 of the English translation of Gödel (1931a) in Davis (1965). It is quoted here according to Wang (1974, pp. 326–327).

procedures. However, this argument is inconclusive, because it depends on the supposition that a finite mind is capable of only a finite number of distinguished states. What Turing disregards completely is the fact that *mind, in its use, is not static, but constantly developing*. This is seen, e.g., from the infinite series of ever stronger axioms of infinity in set theory, each of which expresses a new idea or insight. A similar process takes place with regard to the primitive terms. E.g., the iterative concept of set became clear only in the past few decades. Several more primitive ideas now appear on the horizon, e.g., the self-reflexive concept of proper class. Therefore, although at each stage of the mind's development the number of its possible states is finite, there is no reason why this number should not converge to infinity in the course of its development. Now there may exist systematic methods of accelerating, specializing, and uniquely determining this development, e.g., by asking the right questions on the basis of a mechanical procedure. But it must be admitted that the precise definition of a procedure of this kind would require a substantial deepening of our understanding of the basic operations of the mind. Vaguely defined procedures of this kind, however, are known, e.g., the process of defining recursive well-orderings of integers representing larger and larger ordinals or the process of forming stronger and stronger axioms of infinity in set theory.

In discussions with Hao Wang (cf. Wang 1974, p. 326) Gödel added that Turing's argument became valid under two additional assumptions: (1) there is no mind separated from matter, and (2) the brain functions basically like a digital computer (or (2') the physical laws, in their observable consequences, have a finite limit of precision). Gödel treated (2) as very likely and (2') as practically certain. On the other hand he believed that (1) was (cf. Wang 1974, p. 326):

> a prejudice of our time, which will be disproved scientifically (perhaps by the fact that there aren't enough nerve cells to perform the observable operations of the mind).

3. Information Theory and Gödel's Incompleteness Theorems

Gödel's theorems were investigated and interpreted also from the point of view of the information theory. We cannot present here the appropriate technical considerations (they can be found, e.g., in Chaitin 1974; 1982). Information theory suggests that the phenomenon revealed by Gödel's incompleteness theorems is natural and widespread, not pathological and unusual. The results obtained in it lead to the following corollaries: if a certain effective (recursive) set of methods of reasoning has been made precise ahead then there is an upper bound to the complexity of theorems that can be proved in such a system. Consequently, if one wishes to obtain more complex theorems (in the sense of the information theory, i.e., theorems containing more information) then one will have to continually introduce new axioms and new methods. Neither the admissible methods and rules can be fixed and codified nor the concept of a correct mathematical proof can be defined once and for ever. Hence, progress in mathematics seems to be much like progress in the natural sciences than hitherto expected. Note that such a thesis was already stated by Gödel in (1944) where he wrote (pp. 127–128):

> The analogy between mathematics and a natural science is enlarged upon by Russell also in another respect [...] axioms need not be evident in themselves, but rather their justification lies (exactly as in physics) in the fact that they make it possible for these "sense perceptions" to be deduced [...] I think that [...] this view has been largely justified by subsequent developments, and it is to be expected that it will be still more so in the future. It has turned out that solution of certain arithmetical problems requires the use of assumptions essentially transcending arithmetic. [...] Furthermore it seems likely that for deciding certain questions of abstract set theory and even for certain related questions of the theory of real numbers new axioms based on some hitherto unknown idea will be necessary. Perhaps also the apparently unsurmountable difficulties which some other mathematical problems have been presenting for many years are due to the fact that the necessary axioms have not yet been found. Of course, under these circumstances mathematics may lose a good deal of its "absolute certainty"; but, under the influence of the modern criticism of the foundations, this has already happened to a large extent.

All such arguments are used by quasi-empiricism (cf., e.g., Putnam 1975) which claims that mathematical knowledge is not *a priori*, absolute and certain but it is rather quasi-empirical, fallible and probable, it is in fact much like natural sciences.

The Present State of Mechanized Deduction, and the Present Knowledge of Its Limitations

1. Introduction

In 1936 Alan Turing and Alonzo Church proved two theorems which seemed to have destroyed all hopes of establishing a method of mechanizing reasonings. Turing in (1936–1937) reduced the decidability problem for theories to the halting problem for abstract machines modelling the computability processes (and named after him) and proved that the latter is undecidable. Church (1936) solving Hilbert's original problem proved the undecidability of the full predicate logic and of various subclasses of it.

On the other hand results of Skolem and Herbrand (cf. Chapter 6 of Marciszewski and Murawski 1995) showed that if a theorem is true then this fact can be proved in a finite number of steps – but this is not the case if the theorem is not true (in this situation either one can prove in some cases the falsity of the given statement or the verification procedure does not halt). This semidecidability of the predicate logic was the source of hope and the basis of further searches for the mechanized deduction systems. Those studies were heavily stimulated by the appearance of computers in early fifties. There appeared the idea of applying them to the automatization of logic by using the mechanization procedures developed earlier. The appearance of computers stimulated also the search for new, more effective procedures.

In the sequel we shall describe the history of those researches. In Section 2 the early attempts of applying computers to prove theorems will be presented, in particular we shall tell about results of Davis, Newell–Shaw–Simon, Gilmore, Gelernter *et al.*, Hao Wang and Davis–Putnam. Section 3 will be devoted to resolution and unification algorithms of Prawitz and Robinson and to their modifications. They turned out to be crucial for the further development of the researches towards mechanization and automatization of reasonings. We will sketch them in Section 4.

All those studies led to the idea of automated theorem proving by which one means the use of a computer to prove non-numerical results, i.e., to determine their truth (validity). One can demand here either a simple statement "proved" (what is the case in decision procedures) or human readable proofs. We can distinguish also two modes of operation: (1) fully automated proof searches and (2) man-machine interaction proof searches.

Note that the studies of mechanization of reasonings and automated theorem proving were motivated by two different philosophies. The first one – which we shall call the logic approach – can be characterized "by the presence of a dominant logical system that is carefully delineated and essentially static over the development stage of the theorem proving system. Also, search control is clearly separable from the logic and can be regarded as sitting 'over' the logic. Control 'Heuristics' exist but are syntax-directed primarily" (cf. Loveland 1984, p. 3). The second philosophical viewpoint is called the human simulation approach. It is generally the antithesis of the first one. It can be characterized shortly by saying that "the thrust is obviously simulation of human problem solving techniques" (cf. Loveland 1984, p. 3). Of course the logic and human simulation approaches are not always clearly delineated. Nevertheless this distinction will help us to order the results and to consider the influence of each of the approaches on various particular systems.

In the last Section 5 we indicate some limitations of mechanized deduction and automated theorem proving. In particular we distinguish effective and feasible computability/decidability/deducibility and show on some examples the complexity of decision procedures for certain theories. Connections with the famous problem $P =?NP$ are also indicated (and some information on the origins of this problem is presented). On the other hand we discuss the speed-up phenomenon that appears when moving from the logic of the level n to the logic of the level $n+1$ and – following Boolos – give a simple examples of a formula which can be proved (in the usual sense) in the second-order logic but which has no proof that could be written down in the first-order logic.

2. First Mechanized Deduction Systems

The different philosophical backgrounds mentioned above can be spotted already in the first studies towards mechanization of reasonings and automated theorem proving. In 1954 Martin Davis wrote a program to prove theorems in additive arithmetic (this program was never published in a paper). It was developed on the computer "Johniac" in the Institute for Advanced Study and was a straight implementation of the classical Presburger's decision procedure for additive number theory (i.e. for the theory of non-negative integers in the language with zero, successor and addition only) (cf. Presburger 1929). The complexity of this decision procedure is very high and therefore the program proved only very simple facts (e.g. that the sum of two even numbers is also an even number – this was the first mathematical theorem in the history proved by a computer!).

The Presburger prover of M. Davis was an example of the logic approach. The second achievement in the field of automated theorem proving called the logic theorist should be included among examples of the human simulation approach. We mean here the program of A. Newell, J. S. Shaw and H. A. Simon presented in 1956 at the Dartmouth conference (cf. Newell *et al.* 1956). This program could prove some theorems in the propositional calculus of *Principia Mathematica* of A. N. Whitehead and B. Russell. Its goal was to mechanically simulate the deduction processes of humans as they prove theorems of the sentential calculus. Two methods were used: (1) substitution into established formulas to directly obtain a desired result and, failing that, (2) to find a subproblem whose proof represents progress towards proving the goal problem. This program was able to prove 38 theorems of *Principia*.

The Geometry Theorem-proving Machine of Gelernter and others from 1959 (cf. Gelernter *et al.* 1959, 1960) is also an example of the second approach. It applied the idea of M. Minsky that the diagram that traditionally accompanies plane geometry problems is a simple model for the theorem that could greatly prune the proof search. The program worked backwards, i.e., from the conclusion (goal) towards the premises creating new subgoals. The geometry model was used just to say which subgoals were true and enabled to drop the false ones. It should be noticed that this program was able to prove most high school exam problems within its domain and the running time was often comparable to high school student time.

Simultaneously with the Geometry Theorem-proving Machine two new efforts in the logic framework occurred. We mean here works of Gilmore and Hao Wang. They used methods derived from classical logic proof procedures and in this way rejected opinions that logical methods cannot provide a useful basis for automated theorem proving. Such opinions were rather popular at that time. They were founded on the fact that logic oriented methods were inefficient and on the fact that the methods of Newell, Shaw, Simon and Gelernter proved to be successful. The method of P. C. Gilmore was based on Beth's semantic tableau technique. It was probably the first working mechanized proof procedure for the predicate logic – it proved some theorems of modest difficulty (cf. Gilmore 1959, 1960).

In the summer 1958 Hao Wang developed the first logic oriented program of automated theorem proving of IBM and continued this work at Bell Labs in 1959–63 (cf. Wang 1960a, 1960b, 1961). Three programs were developed: (1) for propositional calculus, (2) for a decidable part of the predicate calculus and (3) for all of predicate calculus. Those programs were based on Gentzen–Herbrand methods, the last one proved about 350 theorems of *Principia Mathematica* (they were rather simple theorems of pure predicate calculus with identity).

During the Summer Institute for Symbolic Logic held at Cornell University, USA, in 1954 Abraham Robinson put forward, in his short lecture, the idea of considering the additional points, lines and circles – which must be used in a search for a solution of a geometrical problem – simply as elements of the so called Herbrand universe. This should enable us to abandon the geometrical constructs and to use directly Herbrand's methods.

This idea turned out to be very influential and significant. One of the first programs that realized it was implemented in 1960–62 by M. Davis and H. Putnam (cf. Davis–Putnam 1960). By Herbrand's theorem, the question of validity of a predicate calculus formula Z can be reduced to a series of validity questions about ever-expanding propositional formulas. More exactly one should consider so called Herbrand disjunctions $A_1 \vee \cdots \vee A_n$ (which can be effectively obtained from Z). It holds that Z is valid if and only if there exists an n such that the disjunction $A_1 \vee \cdots \vee A_n$ is valid. The formulas A_i are essentially substitution instances over an expanded term alphabet of Z with quantifiers removed. So one can test now $A_1 \vee \cdots \vee A_n$ for ever-increasing n for example by truth table and conclude that Z is valid if among formulas $A_1 \vee \cdots \vee A_n$ a tautology was found. But truth tables happen to be quite inefficient. The procedure of Davis and Putnam tried to overcome this difficulty. In fact they were considering unsatisfiability (instead of validity) of formulas and worked with conjunctive normal forms. Such a form is a conjunction of clauses, each clause being a disjunction of literals, i.e. atomic formulas (atoms) or their negations. The Davis–Putnam procedure can be described now as follows: "[it] made optimal use of simplification by cancellation due to one-literal clauses or because some literals might not have their complement (the same atomic formula but only one with a negation sign) in the formula. A simplified formula was split into two formulas so further simplification could recur anew" (cf. Loveland 1984).

3. Unification and Resolution

The procedure of Davis and Putnam described in the last section had some defects. The main one was the enumeration substitutions – prior to this point substitutions were determined by some enumeration scheme that covered every possibility. Prawitz (1960) realized that the only important substitutions create complementary literals. He found substitutions by deriving a set of identity conditions (equations) that will lead to contradictory propositional formula if the conditions are met. In this way one got a procedure of substituting Herbrand terms. It is called today unification.

M. Davis developed right away the idea and combining it with the procedure of Davis–Putnam implemented it in a computer program based on a so called Linked-Conjunct method (cf. Davis 1963). This program used conjunctive normal forms of formulas and the unification algorithm developed by D. McIlroy in November 1962 at the Bell-Telephone-Laboratories. It was the first program which overcame the weaknesses of Herbrand procedure and improved it just by using conjunctive normal form and the unification algorithm (cf. Davis 1963 and Chinlund et al. 1964).

Simultaneously at Argonne National Lab near Chicago a group of scientists (G. Robinson, D. Carson, J. A. Robinson, L. Wos) was working on computer programs proving theorems. They used methods based on Herbrand theorem recognizing their inefficiency. Studying papers of Davis–Putnam and Prawitz they came to the idea of trying to find a general machine-oriented logical principle which would unify their ideas in a single rule of inference. Such a rule was found in 1963–1964 by John Alan Robinson and published in (1965a). It is called the resolution principle and is today one of the most fundamental ideas in the field of automated theorem proving. Therefore we shall describe it now more exactly.

The resolution principle is applied to formulas in a special form called conjunctive normal form. Given a formula A of the language of predicate calculus we first transform it into prenex normal form, i.e. to the form $(Q_1x_1)(Q_2x_2)\ldots(Q_nx_n)B$ where Q_i $(i=1,\ldots,n)$ are universal or existential quantifiers and B is quantifier-free (B is called matrix). One can show that the prenex normal form of a formula A is logically equivalent to the formula A. As an example consider the following formula (stating that the function f is continuous):

$$\forall \varepsilon \{\varepsilon > 0 \longrightarrow \exists \delta [\delta > 0 \wedge \forall x \forall y (|x-y| < \delta \longrightarrow |f(x)-f(y)| < \varepsilon)]\}.$$

Its prenex normal form is:

$$\forall \varepsilon \exists \delta \forall x \forall y [\varepsilon > 0 \longrightarrow \delta > 0 \wedge (|x-y| < \delta \longrightarrow |f(x)-f(y)| < \varepsilon)].$$

Observe that prenex normal form of a formula is not uniquely determined but on the other hand all prenex normal forms of a given formula are logically equivalent.

The next step of the transformation of formulas we need is skolemization (cf. Section 6.5 of Marciszewski and Murawski 1995). Recall here only that it consists of dropping of all the quantifiers and of replacing every occurrence of each variable bound by an existential quantifier by a functional term containing the variables that in this formula are bound by universal quantifiers preceding the considered existential quantifier. A formula obtained in this way is said to be in a skolemized prenex normal form or in its functional form for satisfiability. The skolemized prenex normal form of the formula from our above example is the following:

$$\varepsilon > 0 \longrightarrow g(\varepsilon) > 0 \wedge (|x-y| < g(\varepsilon) \longrightarrow |f(x)-f(y)| < \varepsilon).$$

Note that the skolemized normal form of a formula is not equivalent to the given formula but it is satisfiable (inconsistent) if and only if the given formula is satisfiable (inconsistent).

The next step of getting the conjunctive normal form consists of the elimination of connectives $\longleftrightarrow$ and $\longrightarrow$ and moving all negation signs $\neg$ to the atoms. We proceed according to the following rules:

$A \longleftrightarrow B$ is logically equivalent to $(A \longrightarrow B) \wedge (B \longrightarrow A)$

$A \longrightarrow B$ is logically equivalent to $(\neg A \vee B)$

$\neg (A \wedge B)$ is logically equivalent to $(\neg A \vee \neg B)$

$\neg(A \vee B)$ is logically equivalent to $(\neg A \wedge \neg B)$
$\neg\neg A$ is logically equivalent to A.

A formula obtained in such a way is said to be in the negation normal form. For example the negation normal form of the formula considered above is:

$$\neg \varepsilon > 0 \vee \left[g(\varepsilon) > 0 \wedge (\neg |x-y| < g(\varepsilon) \vee |f(x) - f(y)| < \varepsilon) \right].$$

Note that atoms and negated atoms are usually called literals.

The last step of the considered transformation consists of the application of the following distributivity laws:

$$A \vee (B \wedge C) \longleftrightarrow (A \vee B) \wedge (A \vee C),$$
$$(A \wedge B) \vee C \longleftrightarrow (A \vee C) \wedge (B \vee C).$$

In this way the obtained formula is of the form of conjunctions of disjunctions of literals. Such a form is called the conjunctive normal form and the disjunctions of literals are called clauses. Clauses consisting of a single literal are called unit clauses. Clauses with only one positive literal are called Horn clauses. They correspond to formulas of the form $A_1 \wedge \ldots \wedge A_n \longrightarrow B$. To illustrate the last step note that the conjunctive normal form of our formula stating that a function f is continuous is the following:

$$(\neg \varepsilon > 0) \vee \left[g(\varepsilon) > 0 \wedge \neg |x-y| < g(\varepsilon) \right] \vee \left[g(\varepsilon) > 0 \wedge |f(x) - f(y)| < \varepsilon \right].$$

Note that if $(A_1^1 \vee \cdots \vee A_{n_1}^1) \wedge \ldots \wedge (A_1^l \vee \cdots \vee A_{n_l}^l)$ is a conjunctive normal form of a formula A then we write it sometimes also as: $\{A_1^1 \vee \cdots \vee A_{n_1}^1, \ldots, A_1^l \vee \cdots \vee A_{n_l}^l\}$ or as $\{\{A_1^1, \cdots, A_{n_1}^1\}, \ldots, \{A_1^l, \cdots, A_{n_l}^l\}\}$.

This form is called the clause form. In this way we have shown that for any formula A of the language of predicate calculus there exists a formula A' in conjunctive normal form such that the formula A is satisfiable (inconsistent) if and only if the formula A' is satisfiable (inconsistent).

Having described the needed form of formulas we can introduce now the resolution principle. First observe that a formula B is a logical consequence of formulas $A_1, \ldots, A_n$ if and only if the formula $A_1 \wedge \ldots \wedge A_n \wedge \neg B$ is inconsistent, i.e. unsatisfiable. Let $\square$ denote an empty clause (i.e. a contradiction) and let formulas $A_1, \ldots, A_n, \neg B$ be in conjunctive normal form. Hence to show that B follows logically from $A_1, \ldots, A_n$ it suffices to prove that $\square$ is contained in the set $\mathscr{S}$ of all clauses constituting $A_1, \ldots, A_n, \neg B$ or that $\square$ can be deduced from this set (the sense of the word 'deduce' will be explained below). This is the main idea of the method of resolution. Hence we can say that this method is a negative test calculus.

The resolution calculus introduced by J. A. Robinson (1965a) is a logical calculus in which one works only with formulas in clause form. It has no logical axioms and only one inference rule (the resolution rule). The simplest version of this rule has the following form: if C_1 and C_2 are two clauses such that C_1 contains a literal L_1 and C_2 contains a literal L_2 which is inconsistent with L_1 (i.e. L_1 and L_2 are complementary literals) then one obtains a new clause C consisting of all literals of C_1 except L_1 and all literals of C_2 except L_2. Symbolically it can be written as:

$$\begin{array}{ll} \text{clause } C_1: & K_1 \vee \cdots \vee K_n \vee L \\ \text{clause } C_2: & \underline{M_1 \vee \cdots \vee M_m \vee \neg L} \\ & K_1 \vee \cdots \vee K_n \vee M_1 \vee \cdots \vee M_m \end{array}$$

The clause $K_1 \vee \cdots \vee K_n \vee M_1 \vee \cdots \vee M_m$ is called the resolvent of C_1 and C_2 and clauses C_1 and C_2 are called parent clauses. We say that L_1 and L_2 are the literals resolved upon when the resolvent exists. Observe that the resolvent is a logical consequence of the parent clauses.

Let $\mathscr{S}$ be a set of clauses. A refutation (or a proof of unsatisfiability) of $\mathscr{S}$ is a finite sequence $C_1, \ldots, C_n$ of clauses such that (1) any C_i ($i = 1, \ldots, n$) either belongs to $\mathscr{S}$ or there exist C_j and C_k, $j, k < i$ such that C_i is a resolvent of C_j and C_k and (2) the last clause C_n is $\square$. This explains what we meant above by saying that $\square$ can be "deduced" from the set $\mathscr{S}$ of clauses.

We give now two examples.

A. We show that U is a logical consequence of formulas $P \longrightarrow S$, $S \longrightarrow U$ and P. Indeed it suffices to show that the formula $(P \longrightarrow S) \wedge (S \longrightarrow U) \wedge P \wedge \neg U$ is unsatisfiable (inconsistent). Writing it in conjunctive normal form we get

$$(\neg P \vee S) \wedge (\neg S \vee U) \wedge P \wedge \neg U.$$

Its clause form is $\{\neg P \vee S, \neg S \vee U, P, \neg U\}$. Denote this set of clauses by $\mathscr{S}$. The following sequence of formulas is a proof of unsatisfiability of $\mathscr{S}$:

(1) $\neg P \vee S$
(2) $\neg S \vee U$
(3) P
(4) $\neg U$
(5) $\neg P \vee U$ resolvent of (1) and (2)
(6) U resolvent of (3) and (5)
(7) $\square$ resolvent of (4) and (6).

Hence U is a logical consequence of formulas $P \longrightarrow S$, $S \longrightarrow U$ and P.

B. We show that the formula $\neg Q$ is a logical consequence of formulas $P \longrightarrow (\neg Q \vee (R \wedge S)), P, \neg S$. Hence consider the formula

$$\bigl(P \longrightarrow (\neg Q \vee (R \wedge S))\bigr) \wedge P \wedge \neg S \wedge Q.$$

Its clause form is

$$\mathscr{S} = \{\neg P \vee \neg Q \vee R, \neg P \vee \neg Q \vee S, P, \neg S, Q\}.$$

We have the following proof of unsatisfiability of $\mathscr{S}$:
(1) $\neg P \vee \neg Q \vee R$
(2) $\neg P \vee \neg Q \vee S$
(3) P
(4) $\neg S$
(5) Q
(6) $\neg P \vee \neg Q$ resolvent of (2) and (4)
(7) $\neg P$ resolvent of (5) and (6)
(8) $\square$ resolvent of (3) and (7).

So far we considered the simplest form of the resolution principle and its application in the propositional calculus. In the case of formulas containing variables the whole situation is more complicated.

First we describe a substitution device called unification (we shall do it following Chang–Lee 1973). By a substitution we mean a finite set of the form $\{t_1/v_1,\ldots,t_n/v_n\}$ where v_i are variables and t_i are terms different from v_i. An empty substitution will be denoted by ε. If $\theta = \{t_1/v_1,\ldots,t_n/v_n\}$ is a substitution and E is a formula then by $E\theta$ we denote the formula $E(v_1/t_1,\ldots,v_n/t_n)$. Observe that substitutions can be composed, i.e., if $\theta = \{t_1/x_1,\ldots,t_n/x_n\}$ and $\lambda = \{u_1/y_1,\ldots,u_m/y_m\}$ are two substitutions then by $\theta \circ \lambda$ we shall denote a composition of θ and λ and define it as a substitution obtained from the set $\{t_1\lambda/x_1,\ldots,t_n\lambda/x_n, u_1/y_1,\ldots,u_m/y_m\}$ by removing from it all the elements $t_j\lambda/x_j$ such that $t_j\lambda = x_j$ and all the elements u_i/y_i such that $y_i \in \{x_1,\ldots,x_n\}$. Note that the composition $\circ$ is associative and that ε is the left and right unit, i.e., $\varepsilon \circ \theta = \theta \circ \varepsilon = \varepsilon$.

A substitution θ is said to be a unifier of a set of formulas $\{E_1,\ldots,E_k\}$ if and only if $E_1\theta = E_2\theta = \cdots = E_k\theta$. If there exists a unifier of a set $\{E_1,\ldots,E_k\}$ then this set is said to be unifiable. A unifier σ of the set $\{E_1,\ldots,E_k\}$ is called a most general unifier if and only if for any unifier θ of this set there exists a substitution λ such that $\theta = \sigma \circ \lambda$. J. A. Robinson showed that for any set $\mathscr{S}$ of formulas there exists at most one most general unifier.

We shall describe now an algorithm of finding a most general unifier. It is called the unification algorithm.

Let $\mathscr{S}$ be a nonempty set of formulas. A disagreement set of $\mathscr{S}$ is defined as follows: one indicates the first (from the left) position such that there are two formulas from $\mathscr{S}$ which differ on this position. Then for every element E of $\mathscr{S}$ we write its subformula beginning with the symbol being written on this position. The set of these subformulas is just the disagreement set of $\mathscr{S}$. For example the disagreement set of the set $\{P(x, f(y,z)), P(x,a), P(x, g(h(k(x))))\}$ is $\{f(y,z), a, g(h(k(x)))\}$.

The unification algorithm can now be given by the following rules. Let a set $\mathscr{S}$ of formulas be given.

Step 1: $\mathscr{S}_0 = \mathscr{S}$, $\sigma_0 = \varepsilon$.

Step 2: If $\mathscr{S}_k$ is a unit clause then STOP and σ_k is the most general unifier of $\mathscr{S}$. Otherwise let D_k be the disagreement set of $\mathscr{S}_k$.

Step 3: If there are v_k and t_k in D_k such that v_k is a variable not occurring in t_k then move to Step 4. Otherwise STOP: $\mathscr{S}$ is not unifiable.

Step 4: Let $\sigma_{k+1} = \sigma_k \circ \{t_k/v_k\}$ and $\mathscr{S}_{k+1} = \mathscr{S}_k\{t_k/v_k\}$. (Observe that $\mathscr{S}_{k+1} = \mathscr{S} \sigma_{k+1}$.)

Note that the unification algorithm always halts when applied to a finite nonempty set of formulas. J. A. Robinson proved that if $\mathscr{S}$ is a finite nonempty unifiable set of formulas then the unification algorithm halts on Step 2 and the last σ_k is the most general unifier of $\mathscr{S}$.

We give some examples. First find a most general unifier of the set $\mathscr{S} = \{Q(f(a), g(x)), Q(y,y)\}$. We proceed as follows:

(1) $\sigma_0 = \varepsilon$, $\mathscr{S}_0 = \mathscr{S}$.
(2) Since $\mathscr{S}_0$ is not a unit clause we find a disagreement set D_0 of $\mathscr{S}_0$. We have $D_0 = \{f(a), y\}$. Hence $v_0 = y, t_0 = f(a)$.
(3) $\sigma_1 = \sigma_0 \circ \{t_0/v_0\} = \sigma_0 \circ \{f(a)/y\} = \{f(a)/y\}$,
$\mathscr{S}_1 = \mathscr{S}_0\{t_0/v_0\} = \{Q(f(a), g(x)), Q(f(a), f(a))\}$.
(4) $\mathscr{S}_1$ is not a unit clause and we have $D_1 = \{g(x), f(a)\}$. By Step 3 we conclude that the set $\mathscr{S}$ is not unifiable.

We give one more example.

Let $\mathscr{S} = \{P(a, x, f(g(x))), P(z, f(z), f(u))\}$. Find a most general unifier of $\mathscr{S}$. We proceed as follows:

(1) $\sigma_0 = \varepsilon$, $\mathscr{S}_0 = \mathscr{S}$.
(2) $D_0 = \{a, z\}$ and we have $v_0 = z, t_0 = a$.
(3) $\sigma_1 = \sigma_0 \circ \{t_0/v_0\} = \{a/z\}$,
$\mathscr{S}_1 = \mathscr{S}_0\{t_0/v_0\} = \{P(a, x, f(g(y))), P(a, f(a), f(u))\}$.
(4) $D_1 = \{x, f(a)\}$ and we put $v_1 = x, t_1 = f(a)$.
(5) $\sigma_2 = \sigma_1 \circ \{t_1/v_1\} = \{a/z, f(a)/x\}$,
$\mathscr{S}_2 = \mathscr{S}_1 \circ \{t_1/v_1\} = \{P(a, f(a), f(g(y))), P(a, f(a), f(u))\}$.
(6) $D_2 = \{g(y), u\}$ and we put $v_2 = u, t_2 = g(y)$.
(7) $\sigma_3 = \sigma_2 \circ \{t_2/v_2\} = \{a/z, f(a)/x, g(y)/u\}$,

$$\mathscr{S}_3 = \mathscr{S}_2\{t_2/v_2\} = \{P(a, f(a), f(g(y)))\}.$$

By Step 2 the set $\mathscr{S}$ is unifiable and σ_3 is the most general unifier of it.

We have to define two more notions to formulate at last the general form of the resolution rule. If two or more literals (with the same sign) of a clause C have a most general unifier σ then the clause $C\sigma$ is called a factor of C. Let C_1 and C_2 be two clauses which have no common variable and let L_1 and L_2 be two literals of C_1 and C_2, resp. If L_1 and $\neg L_2$ have a most general unifier σ then the clause $(C_1\sigma - L_1\sigma) \cup (C_2\sigma - L_2\sigma)$ is called a binary resolvent of C_1 and C_2. The clauses C_1 and C_2 are called parent clauses and we say that L_1 and L_2 are literals resolved upon. A resolvent of two parent clauses C_1 and C_2 is now defined as one of the following resolvents: (1) binary resolvent of C_1 and C_2, (2) binary resolvent of C_1 and a factor of C_2, (3) binary resolvent of C_2 and a factor of C_1, (4) binary resolvent of a factor C_1 and a factor of C_2.

Now we can define a resolution deduction. Let a set $\mathscr{S}$ of clauses and a clause A be given. A resolution deduction of A from $\mathscr{S}$ is a finite sequence $C_1, \ldots, C_n$ of clauses such that: (1) C_n is A, (2) for any $i, 1 \leq i \leq n$, C_i is either a member of $\mathscr{S}$ or there exist $j, k < i$ such that C_i is a resolvent of C_j and C_k (i.e., C_i is obtained from C_j and C_k by the resolution rule). A resolution deduction of the empty clause $\square$ from $\mathscr{S}$ is called a refutation (or a proof of unsatisfiability) of $\mathscr{S}$.

We shall give some examples (cf. Chang–Lee 1973).

1. Show that the formula $B =: \exists x(S(x) \wedge R(x))$ is a logical consequence of formulas $A_1 =: \forall x[P(x) \longrightarrow (Q(x) \wedge R(x))]$ and $A_2 =: \exists x(P(x) \wedge S(x))$. It suffices to show that the formula $A_1 \wedge A_2 \wedge \neg B$ is unsatisfiable. Transform the given formulas $A_1, A_2, \neg B$ into conjunctive normal form. We get the following formulas, resp.:

$$(\neg P(x) \vee Q(x)) \wedge (\neg P(x) \vee R(x)),$$
$$P(a) \wedge S(a),$$
$$\neg S(x) \vee \neg R(x).$$

We can construct now the following resolution deduction:

(1) $\neg P(x) \vee Q(x)$
(2) $\neg P(x) \vee R(x)$ $\Big\}$ from A_1
(3) $P(a)$
(4) $S(a)$ $\Big\}$ from A_2
(5) $\neg S(x) \vee \neg R(x)$ from $\neg B$
(6) $R(a)$ resolvent of (2) and (3)
(7) $\neg R(a)$ resolvent of (4) and (5)
(8) $\square$ resolvent of (6) and (7).

Hence we have shown that B is a logical consequence of A_1 and A_2.

2. We show that the formula $B =: \exists x(P(x) \land R(x))$ is a logical consequence of the following formulas:

$$A_1 =: \forall x[Q(x) \land \neg T(x) \longrightarrow \exists y(S(x,y) \land R(y))],$$
$$A_2 =: \exists x[P(x) \land Q(x) \land \forall y(S(x,y) \longrightarrow P(y))],$$
$$A_3 =: \forall x(P(x) \longrightarrow \neg T(x)).$$

Transforming the formulas A_1, A_2, A_3 and $\neg B$ into conjunctive normal form we get the following clauses:

(1) $\neg Q(x) \lor T(x) \lor S(x, f(x))$ } from A_1
(2) $\neg Q(x) \lor T(x) \lor R(f(x))$
(3) $P(a)$
(4) $Q(a)$ } from A_2
(5) $\neg S(a,y) \lor P(y)$
(6) $\neg P(x) \lor \neg T(x)$ from A_3
(7) $\neg P(x) \lor \neg R(x)$ from $\neg B$

The needed deduction of $\square$ can be the following:

(8) $\neg T(a)$ resolvent of (3) and (6)
(9) $T(a) \lor R(f(a))$ resolvent of (2) and (4)
(10) $R(f(a))$ resolvent of (8) and (9)
(11) $T(a) \lor S(a, f(a))$ resolvent of (1) and (4)
(12) $S(a, f(a))$ resolvent of (8) and (11)
(13) $P(f(a))$ resolvent of (5) and (12)
(14) $\neg R(f(a))$ resolvent of (7) and (13)
(15) $\square$ resolvent of (10) and (14).

The method of resolution has an important property called the refutation completeness. One can prove that a set $\mathscr{S}$ of clauses is unsatisfiable if and only if there exists a resolution deduction of the empty clause $\square$ from $\mathscr{S}$. The proof of this fact uses semantical trees and Herbrand's theorem. Hence the method gives a semidecidability of the predicate logic: if a given formula is a logical consequence of a given set $\mathscr{S}$ of formulas then by a systematic application of the resolution rule we get in a finite number of steps the empty clause $\square$, but if it is not a logical consequence then sometimes one can decide it after a finite number of steps but in general the procedure does not halt. On the other hand the considered method does not give a procedure of finding a formal proof of a formula B on the basis of formulas $A_1, \ldots, A_n$ in the case when B is a logical consequence of $A_1, \ldots, A_n$.

Recall that we used in the resolution most general unifiers, i.e., most general substitutions that allow the equality of literals. This guarantees the elimination of branching of search due to different possible substitutions that equate those atoms but lead to different clauses. Therefore the method of resolution is simple, elegant and powerful. But the world is not so perfectly beautiful – this method has also some defects. If one generates from a given set of clauses new clauses by the method of resolution then they accumulate at a rapid rate. Indeed given a set $\mathscr{S}$ of clauses one can obtain new clauses systematically using the level-saturation method which is described by the following equations: $\mathscr{S}^0 = \mathscr{S}, \mathscr{S}^{n+1} = \{$resolvents of C_1 and C_2: $C_1 \in \mathscr{S}^0 \cup \cdots \cup \mathscr{S}^n, C_2 \in \mathscr{S}^n\}, n = 1, 2, \ldots$ In this way we get all the resolvents of all pairs of elements of $\mathscr{S}$, add them to $\mathscr{S}$, further we calculate all the resolvents of elements of this new set etc. till we come to the empty clause $\square$. Among clauses generated in this procedure there are many irrelevant ones and the total number of clauses grows rapidly. Hence the idea of improving the method of resolution by finding restrictions and strategies to control the growth of the number of clauses. We shall tell here only about some of the proposed improvements (Loveland 1978 summarizes in Appendix twenty five such improvements but more exist). By a restriction of resolution we mean a variant for which some clauses generated by the basic resolution procedure are not generated. A strategy of resolution only rearranges the order of generation to get likely useful clauses earlier.

One of the earliest refinements of resolution is unit-preference introduced in Wos–Carson–Robinson (1964). This strategy guarantees that the resolvent is shorter than the longer parent clauses. In the same paper L. Wos, D. F. Robinson and G. A. Carson introduced the set-of-support restriction. It can be described as follows: one chooses a subset $\mathscr{T}$ (called a support) of a given set $\mathscr{S}$ of clauses and then two clauses from $\mathscr{S} - \mathscr{T}$ are never resolved together. This means that every resolvent has in its deduction history some clause of $\mathscr{T}$. In practice $\mathscr{T}$ is usually chosen to be a (sub)set of the clauses special to the considered problem.

J. A. Robinson in (1965b) introduced a restriction called hyperresolution. It restricts resolutions to where one parent clause contains only positive literals and any resolvent containing a negative literal is immediately used in all permitted resolutions and then discarded.

D. W. Loveland (1970) and D. Luckham (1970) proposed another restriction called linear resolution (Luckham used the name "ancestry-filter form"). It constrains the deduction so that a new clause is always derived from the preceding clause of the deduction by resolving against an earlier clause of the deduction. In this way one is always seeking to transform the last clause obtained into a clause closer to the goal clause. This method was further developed by R. Anderson and W. W. Bledsoe (1970). R. Yates, B. Raphael and T. Hart (1970), R. Reiter (1971), D. W. Loveland (1972) and R. Kowalski and D. Kuehner (1971).

D. W. Loveland (1968) and (1969) introduced a procedure which is not a resolution procedure but is close to a very restricted form of linear resolution –

it is called model elimination. R. Kowalski and D. Kuehner (1971) provided the translation of it into a very restricted linear resolution format called SL-resolution. It was used by A. Colmerauer and P. Roussel to an early version of a programming language Prolog.

There appeared a number of resolution refinements which reduce multiple derivations of the same clause by ordering literals in clauses. An example of this type of procedures is the method called locking or lock-resolution introduced by R. S. Boyer (1971). Its main idea is to use indices to order literals in clauses from a given set of clauses. The occurrences of literals are indexed by integers. Then resolution need be done using only the lowest indexed integer of each clause.

We should mention here also the semantic resolution of J. R. Slagle (1967) which generalizes hyperresolution of J. A. Robinson (1965b), resolution with renaming of B. Meltzer (1966) and the set-of-support restriction of L. Wos, G. A. Robinson and D. F. Carson (1965).

All types of refinements of resolution given above are refutation complete. There exist also two forms of resolution restrictions which are incomplete. We mean here unit clause resolution and input clause resolution. The former was introduced by L. Wos, D. F. Carson and G. A. Robinson (1964). It permits resolution only when one parent is a unit clause, i.e., it consists of one literal. The input clause resolution was introduced by C. L. Chang (1970). It is a restricted form of linear resolution where one parent is always an input (given) clause. Chang proved that the unit clause and input clause resolutions are of equal power – they are complete over the class of Horn formulas.

Most mathematical theories contain among its symbols the equality symbol and among its axioms – the equality axioms. The immediate application of the usual resolution procedure generates in this case a lot of undesired clauses. Hence L. Wos and G. A. Robinson (1970) introduced a procedure called paramodulation. This is the equality replacement rule with unification. It replaces all equality axioms except certain reflexivity axioms for functions. When paramodulation is restricted to replacement of the (usually) shorter term by the longer term with no instantiation allowed in the formula incorporating the replacement then one uses the term demodulation (cf. Wos–Robinson–Carson–Shalla 1967). It should be mentioned that there are also two other systems treating equality: system introduced by E. E. Sibert (1969) and E-resolution introduced by J. B. Morris (1969).

To finish this section we should add that almost simultaneosly with Robinson's invention of resolution, J. Ju. Maslov in the USSR introduced a proof procedure very close to resolution in spirit. His method is called the inverse method and it is a test for validity rather than unsatisfiability (cf. Maslov 1964, 1971 and Kuehner 1971).

4. Development of Mechanized Deduction After 1965

The last section was devoted to the method of resolution and to its refinements. This approach to the mechanization and automatization of reasonings was dominating in the sixties. Nevertheless there were developed also other methods. In this section we shall discuss them briefly and sketch further development of the researches in the field after 1965.

We should begin with the Semi Automated Mathematics (SAM) project that spanned 1963 to 1967. It belongs to the human simulation approach (cf. Section 1). In the framework of this project a succession of systems designed to interact with a mathematician was developed. They used many sorted ω-order logic with equality and λ-notation. The system SAM I was a proof checker but the theorem proving power continued to increase through SAM V which had substantial automatic capability. Only a part of the work in this project got recorded in the literature – cf. J. R. Guard–F. C. Oglesby–J. H. Bennett–L. G. Settle (1969).

Another project belonging to the human simulation approach developed in the mid-sixties was ADEPT, the Ph. D. thesis of L. M. Norton (cf. Norton 1966). It was a heuristic prover for group theory.

The dominant position of the resolution methods brought sharp criticism from some researchers. Their main argument was that there cannot be a unique procedure which would suffice to realize (to simulate) the real intelligence. They stressed the necessity of using many components. One of those critics was M. Minsky from MIT. In 1970 C. Hewitt, a Ph. D. student at MIT wrote a dissertation on a new programmic language called PLANNER (cf. Hewitt 1971). Its goal was to structure a theorem prover system in such a way that locally distributed knowledge could be represented at various positions of the proving program. In fact it was not a theorem prover *per se*, but a language in which a "user" was to write his own theorem prover, specifically tailored to the problem domain at hand. This language was never fully implemented, only a subset of it was realized (microPLANNER).

About the same time another effort in human simulation approach was undertaken by A. Nevins (cf. Nevins 1974, 1975a, 1975b). He built in fact at least two provers that were able to prove theorems which most resolution provers could not touch. For example Nevins could prove fully automatically that:

$$x^3 = e \longrightarrow f(f(a,b),b) = e.$$

where $f(x,y) = xyx^{-1}y^{-1}$. This result is much harder than the implication: "$x^2 = e \longrightarrow$ the group is Abelian" which constituted the limit of capability of ADEPT.

We should mention also works of the group of scientists gathered around W. W. Bledsoe at the University of Texas which proved to be very important. They were working not for single uniform rule of inference for the whole mathematics but were seeking specific methods for particular domains of mathematics such as analysis, set theory or nonstandard analysis (cf. Bledsoe 1983, 1984).

So far we spoke about studies in seventies of the automated theorem proving which could be classified as human simulation approach. It does not mean that this was the only direction. There were also some new ideas within the logic approach. We should say here first of D. E. Knuth and P. B. Bendix (1970). They used the idea of rewrite rules – a device familiar to logicians. It is a replacement rule and allow to replace the left hand side by the right hand side at any occurrence of the left hand side. In equational theories one converts equations just to rewrite rules. Those rules enable us to reduce terms and to equate them. Knuth and Bendix proposed an algorithm which for a class of equational theories gave a complete set of rewrite rules – i.e., a set of rewrite rules sufficient to check the truth of every equation of the theory by demanding that equal terms reduce to the same normal form. An example of a theory to which the algorithm applies is the theory of groups (while the theory of Abelian groups is not). There is an open problem connected with this algorithm: what equational theories have complete sets of rewrite rules?

Recall that the unification algorithm (described in the previous section) permits computation of a most general substitution for variables to make atomic formulas identical. Working with special theories, usually equational theories, one can simplify the procedure given by unification and resolution by introducing special unification. Its main idea is that the usual unification is augmented by equations or rewrite rules obtained from axioms of the considered theory. This idea was first formalized by G. D. Plotkin (1972) and further developed by M. E. Stickel (1981, 1985) and M. Livesey and J. Siekmann (1976). The idea of theory resolution of Stickel can be summarized as follows: since the resolution rule enlarges the whole number of steps, it is desirable to find macrorules in which certain sequences of steps could be performed in one step.

Another example of results which should be classified as logic oriented is the system of R. Overbeek developed later by S. Winker, E. Lusk, B. Smith and L. Wos and named AURA (Automated Reasoning Assistant). It was based on the old (i.e., coming from the sixties) ideas of unit preference, set-of-support for resolution, paramodulation and demodulation to which hyperresolution as well as more flexibility in demodulation and preprocessors for preparation of input from a variety of formats were added.

Around 1973 there appeared a very interesting effort different from the resolution approach and the strongly human oriented prover of Bledsoe. We mean here the Computational Logic Theorem Prover of R. S. Boyer and J. S. Moore (1975, 1979, 1981). This system uses the language of quantifier-free first order logic with equality and includes a general induction principle among the inference rules. It can be used to work within traditional mathematics (e.g. number theory) as well as to prove properties of programs and algorithms (so called proofs of correctness).

We should tell also about graph representation and about prover for systems of higher order. The former is based on the idea of enriching the structure of basic data with additional information, e.g. by representing the potential resolution steps in the graph structure. Literals or clauses and possible complementary literals form vertices of graphs which are connected by edges. This approach was introduced by R. Kowalski (1975) (cf. also S. Sickel 1976, R. E. Shostak 1976 and P. Andrews 1976).

The first proving system for higher order logics was developed by a group of scientists working under the direction of J. R. Guard in the early sixties. The studies were continued by W. E. Gould (1966), G. D. Huet (1975) and D. C. Jensen and T. Pietrzykowski (1976). The most important and influential group of people working in this direction is today at the Carnegie-Melon University (its chief is P. B. Andrews). They developed in the late seventies a theorem prover for type theory (TPS) (cf. Andrews 1981 and Miller, Cohen, Andrews 1982). It can prove for example Cantor's theorem as well as numerous first order theorems.

We shall finish this survey of activities in the field of mechanization and automatization of reasonings in sixties and seventies by mentioning an ambitious theorem prover now being developed at the University of Karlsruhe and named Markgraf Karl Refutation Procedure (cf. K. Bläsius, N. Eininger, J. Siekmann, G. Smolka, A. Herold, C. Walther 1981).

5. Some Limitations

Having described so far the positive achievements in the field of mechanized deduction and automated theorem proving let us turn now to the discussion of some (essential) limitations in this process.

It is a trivial observation that an automatic theorem prover would have wide application if it operated effectively enough. Hence it is useful to distinguish effective and feasible computability (decidability). Both are intuitive (nonformal) concepts. Recall that a problem is said to be effectively decidable (a function is effectively computable) if and only if there exists a definite mechanical procedure which can solve in a finite number of prescribed steps every instance of the problem (calculate the value of the function for any given arguments). It is believed that the concept of recursive computability (or equivalently of Turing machine computability) is an adequate mathematical formalization (counterpart) of this concept – this is stated by Church's Thesis. On the other hand by "feasible" one means "computable in practice" or "computable in the real world". As a mathematical model of this intuitive notion one can consider the concept of polynomial time computability, i.e., computability by a deterministic Turing machine in the time bounded by a polynomial of the size of the input, hence by a deterministic Turing machine that needs a number of steps bounded by a polynomial of the size of the input. Denote by P the class of predicates (problems) recognizable (solvable) in a polynomial time. A closely related class is the class *NP* of problems recognizable in non-deterministic polynomial time. There is an open problem whether problems in *NP* are feasibly decidable or are polynomial time decidable – it is shortly denoted as $P =?NP$ and is a central problem in the contemporary computer science.

It seems that this problem was stated (in a certain sense) for the first time in a letter of Kurt Gödel to John von Neumann from 20th March 1956 (cf. Gödel 2003b, p. 373–375). Gödel was thinking about computational complexity of Turing machine computations and asked von Neumann about the computational complexity of a problem (which is in fact a *NP* complete problem; note that we are using the modern terminology, in Gödel's letter the problem is not referred to as *NP* complete) about proof systems and wondered if it could be solved in linear or possibly quadratic time. He asked how hard is it (computationally) to decide if a statement has a proof of length n in a formal system (it is of course a question about the fundamental nature of mathematics). It is worth quoting here some fragments of Gödel's letter:

> Obviously, it is easy to construct a Turing machine that allows us to decide, for each formula F of the restricted functional calculus and every natural number n, whether F has a proof of length n [length = number of symbols]. Let $\psi(F,n)$ be the number of steps required for the machine to do that, and let $\varphi(n) = \max_F \psi(F,n)$. The question is, how rapidly does $\varphi(n)$ grow for an optimal machine? It is possible to show that $\varphi(n) \geqslant Kn$. If there really were a machine with $\varphi(n) \sim Kn$ (or even just $\sim Kn^2$) then that would have consequences of the greatest significance. Namely, this would clearly mean that the thinking of a mathematician in the case of yes-or-no questions could be completely[1] replaced by machine, in spite of the unsolvability of the Entscheidungsproblem. n would merely have to be chosen so large that, when the machine does not provide a result, it also does not make any sense to think about the problem. Now it seems to me to be quite within the realm of possibility that $\varphi(n)$ grows so slowly. For 1.) $\varphi(n) \geqslant Kn$ seems to be the only estimate obtainable by generalizing the proof of the unsolvability of the Entscheidungsproblem; 2.) $\varphi(n) \sim Kn$ (or $\sim Kn^2$) just means that the number of steps when compared to pure trial and error can be reduced from N to $\log N$ (or $\log N^2$). Such significant reductions are definitely involved in the case of other finitist problems, e.g., when computing the quadratic remainder symbol by repeated application of the law of reciprocity. It would be interesting to know what the case would be, e.g., in determining whether a number is prime, and how significantly *in general* for finitist combinatorial problems the number of steps can be reduced when compared to pure trial and error.

To formulate the problems raised by Gödel more clearly let us think of first-order predicate logic formalized in one of the usual Hilbert-style systems with a finite set of axiom schemata and *modus ponens* and generalization as the only rules of inference. Let the symbol $\vdash_n \varphi$ denote that φ has a first-order proof of $\leqslant n$ symbols. Gödel was asking about the difficulty of answering questions of

---
[1] Except for the formulation of axioms.

the form "$\vdash_n \varphi$?". Let $A = \{\langle \varphi, 0^n \rangle : \vdash_n \varphi\}$. Gödel's questions is now: is the set A recognizable on a (multitape) Turing machine in time $O(n)$ or in time $O(n^2)$. It can be shown that there is an effective algorithm for deciding membership in A and that the set A is in fact NP-complete (it is NP-complete even for propositional logic – cf. Buss 1995).

It is interesting that Gödel was thinking in his letter about problems related to $P = ?NP$ well before these classes were widely stated (cf. Hartmanis 1989 for the discussion of Gödel's letter). Note also that Gödel treated linear or quadratic time computability as corresponding to feasible computation and did not realize the importance of polynomial time as a mathematical model of feasible computability.

The problem $P = ?NP$ belongs to the most famous open problems in computer science. On the other hand it is known today that even in the case of simple mathematical theories the decision procedures are of high complexity. The following theorems can serve as examples indicating the measure of the complexity.

Let F be a function defined in the following way:

$$F(n,1) = 2^n, \qquad F(n,m+1) = 2^{F(n,m)}.$$

Denote by $l(\varphi)$ the length of a formula φ, i.e., the number of (logical and nonlogical) symbols occurring in φ.

It has been shown that:
– (Meyer 1975) Let $f(n)$ be the function

$$F(n, [dn]).$$

There exists a constant $d > 0$ such that for any Turing machine P deciding the weak second-order monadic theory of one successor WS1S there exist infinitely many sentences φ with the property that the machine P needs more than $f(l(\varphi))$ steps to decide whether $\varphi \in$ WS1S or not.

– (Meyer 1975) The complexity of the theory of linear order is at least $F(n, [dn])$ for a certain positive constant d, i.e., to decide a formula of the length n one needs at least $F(n, [dn])$ steps.

– (Fisher and Rabin 1974) A decision procedure for Presburger arithmetic is of the complexity at least $2^{2^{cn}}$ for a certain constant $c > 0$, i.e., to decide whether a formula φ of the length n is a theorem of Presburger arithmetic one needs at least $2^{2^{cn}}$ steps.

– (Fisher and Rabin 1974) The complexity of the theory $Th(\langle \mathbb{N}, \cdot \rangle)$ is at least $F(cn, 3)$, i.e., $2^{2^{2^{cn}}}$ for a certain constant $c > 0$.

So one sees that even for such simple theories as $Th(\langle \mathbb{N}, + \rangle)$ or $Th(\langle \mathbb{N}, \cdot \rangle)$ decision procedures are of exponential complexity. Add also that the decision procedure for the theory of real numbers is doubly exponential in the number of quantifier blocks (cf. Heintz et al. 1989) and that even deciding first-order sentences for the ordered group $\langle \mathbb{R}, +, < \rangle$ is exponentially hard (cf. Fisher et al. 1974).

On the other hand if one moves from the level of first-order logic to higher systems (or generally from the level n to the level $n+1$) the complexity of proofs (their lengths) can be reduced. This observation was made by Gödel in (1936). Having already shown that a logic S_{n+1} of a higher order could prove formulas that a logic S_n of a lower order could not prove, in this abstract he considered the question of formulas that can be proved in both the weaker and the stronger logics. He stated that if the length of a proof is defined to be the number of lines in it, there are formulas that can be proved in both S_n and S_{n+1} but that have a proof in S_{n+1} much shorter than their shortest proof in S_n. This speed-up can be by arbitrary function computable in S_n, i.e., for any function F computable in S_n there exist infinitely many formulas φ such that if k is the length of a shortest proof of φ in S_n and l is the length of a shortest proof of φ in S_{n+1} then $k > F(l)$. Hence "passing to the logic of the next higher order has the effect, not only of making provable certain propositions that were not provable before, but also of making it possible to shorten, by an extraordinary amount, infinitely many of the proofs already available" (Gödel 1936).

Gödel did not give a proof of his result. An analogous result (taking the length of a proof to be its Gödel number rather than the number of lines in it) was given by Mostowski in (1952). Similar results were also proved by Ehrenfeucht and Mycielski (1971) and by Parikh (1971). Statman in (1978) has shown that there is no function F provably recursive in second-order arithmetic such that whenever a first-order formula φ is derivable in a certain standard system of second-order logic with length $\leqslant l$ then φ is derivable in a certain standard system of first-order logic with length $\leqslant F(l)$.

Note that the problem $P =?NP$ discussed above can be thought of as a speed-up question.

A nice illustration of the phenomenon indicated by Gödel was given by Boolos in (1987). He considered there the following set of axioms (in the language with function symbols F and S and a unary predicate D):

(1) $\forall n F(n,1) = S(1)$,

(2) $\forall x F(1, S(x)) = SS(F(1,x))$,

(3) $\forall n \forall x F(S(n), S(x)) = F(n, F(S(n), x))$,

(4) $D(1)$,

(5) $\forall x [D(x) \longrightarrow D(S(x))]$.

In the intended interpretation of the above formulas the variables range over the natural numbers, 1 denotes the number one and S is the successor function. There is no particular interpretation intended for D. By this interpretation F denotes an Ackermann-style function $f: \mathbb{N} \times \mathbb{N} \to \mathbb{N}$ defined by the following equations: $f(1,x) = 2x$, $f(n,1) = 2$ and $f(n+1, x+1) = f(n, f(n+1, x))$. This is a rapidly growing function. In fact one can easily show that

$$f(1,x) = 2x,$$
$$f(2,x) = 2^x,$$
$$f(3,x) = 2^{2^{\cdot^{\cdot^{2}}}} \quad \text{(the value of a stack of } x \text{ 2's),}$$
$$f(4,1) = 2,$$
$$f(4,2) = 4,$$
$$f(4,3) = 2^{2^{\cdot^{\cdot^{2}}}} \quad (= 64K),$$
$$f(4,4) = 2^{2^{\cdot^{\cdot^{2}}}} \quad (64 \text{ K of 2's in all)}.$$

In second-order logic one can deduce from the given set of axioms that

$$D(F(SSSS(1), SSSS(1))).$$

The usage of the second-order logic is essential here (one uses, e.g., the comprehension principle). The proof can be found in (Boolos 1987, Appendix). This formula can be also proved in a first-order system but any derivation of it must contain at least $f(4,4)$ symbols (!) – details of the proof of this (metatheoretical) statement can be found in Appendix to (Boolos, 1987).[2] Hence it is impossible to write down such a proof, no actual or conceivable creature or device could do it. There are simply far too many symbols in any such derivation. It shows that first-order logic is in a certain sense practically incomplete. Boolos says even in (1987, p. 135) that "no standard first-order logical system can be taken to be a satisfactory idealization of the psychological mechanisms or processes, whatever they might be, whereby we recognize (first-order!) logical consequences. "Cognitive scientists" ought to be suspicious of the view that logic as it appears in logic texts adequately represents the whole of the science of valid inference." We can add that this thesis should be taken into account not only by cognitive scientists but also by specialists in mechanized deduction and automated theorem proving (as well as in the artificial intelligence).

---

2   There is another example of a similar result: H. Friedman has shown (cf. Nerode and Harrington 1984) that a certain "finitization" of a combinatorial theorem due to J. Kruskal concerning embeddings of trees can be proved in Zermelo–Fraenkel set theory with the axiom of choice (ZFC) in a few pages but not in the system of second-order arithmetic called ATR (for Arithmetic Transfinite Recursion) in under $f(3, 1000)$ pages.

6. Final Remarks

In the previous sections we have presented the recent period of the history of efforts to find an automated theorem prover. They were stimulated by the appearance of computers which made possible the practical realization of earlier ideas. The emphasis was put on the resolution procedure and the unification algorithm and their modifications because they proved to be the most influential ideas.

As was proved by Turing, Gödel and Church there exists no universal automated theorem prover for the whole mathematics – and even more, there are no such provers for most mathematical theories. Proving theorems in mathematics and logic is too complex a task for total automation because it requires insight, deep thought and much knowledge and experience. Nevertheless the semidecidability of mathematical theories was a sufficient motivation for looking for weaker theorem provers. We have described those efforts in the previous sections.

What does one expect from an automated theorem prover? First of all one obtains a certain unification of reasonings and their automatization. Having that one can shift the burden of proof finding from a mathematician and a logician to the computer. In this way we are also assured that faulty proofs would never occur. Are such automated theorem provers clever than people? Of course they can proceed quicker than a human being. But can they discover new mathematical results? The answer is YES. Some open questions have been answered in this way within finitely axiomatizable theories. For example S. Winker, L. Wos and E. Lusk (1981) answered positively the following open question: does there exist a finite semigroup which simultaneously admits of a nontrivial antiautomorphism without admitting a nontrivial involution? The progress in more complex theories such as analysis or set theory is slower, but there are also provers being able to prove some nontrivial theorems such as for example Cantor's theorem stating that a set has more subsets than elements (cf. P. Andrews, D. A. Miller, E. L. Cohen, F. Pfenning 1984) and various theorems in introductory analysis. The latter includes limit theorems of calculus such as
 – the sum, product and composition of two continuous functions is continuous,
 – differentiable functions are continuous,
 – a uniformly continuous function is continuous,
as well as theorems of intermediate analysis (on the real numbers) such as
 – Bolzano–Weierstrass theorem,
 – if the function f is continuous on the compact set S then f is uniformly continuous on S,
 – if f is continuous on the compact set S then $f[S]$ is compact,
 – intermediate value theorem
(cf. Bledsoe 1984).

All those achievements can be treated as partial realizations and fulfilments of Leibniz's dreams of *characteristica universalis* and *calculus ratiocinator*. They are still far from what Leibniz did expect but they prove that a certain progress in the mechanization and automatization of reasonings and generally human thought has been made. On the other hand one should be aware of some limitations indicated above.

On Proofs of the Consistency of Arithmetic

1. The main aim and purpose of Hilbert's program was to defend the integrity of classical mathematics (referring to the actual infinity) by showing that it is safe and free of any inconsistencies. This problem was formulated by him for the first time in his lecture at the Second International Congress of Mathematicians held in Paris in August 1900 (cf. Hilbert 1901). Among twenty three problems Hilbert mentioned under number 2 the problem of proving the consistency of axioms of arithmetic (under the name "arithmetic" Hilbert meant number theory and analysis).

Hilbert returned to the problem of justification of mathematics in lectures and papers, especially in the twentieth,[1] where he tried to describe and to explain the problem more precisely (in particular the methods allowed to be used) and simultaneously presented the partial solutions obtained by his students.

Hilbert distinguished between the unproblematic, finitistic part of mathematics and the infinitistic part that needed justification. Finitistic mathematics deals with so called real sentences, which are completely meaningful because they refer only to given concrete objects. Infinitistic mathematics on the other hand deals with so called ideal sentences that contain reference to infinite totalities. It should be justified by finitistic methods – only they can give it security (*Sicherheit*). Hilbert proposed to base mathematics on finitistic mathematics via proof theory (*Beweistheorie*). It should be shown that proofs which use ideal elements in order to prove results in the real part of mathematics always yield correct results, more exactly, that (1) finitistic mathematics is conservative over finitistic mathematics with respect to real sentences and (2) the infinitistic mathematics is consistent. This should be done by using finitistic methods only.

2. It seems that the first result in this direction was obtained by Wilhelm Ackermann in 1924. In his paper "Begründung des „tertium non datur" mittels der Hilbertschen Theorie der Widerspruchsfreiheit" (cf. Ackermann 1924/1925) Ackermann gave a finitistic proof of the consistency of arithmetic of natural numbers without the axiom (scheme) of induction. In fact it was a much weaker system than the usual systems of arithmetic but the paper provided the first attempt to solve the problem of consistency. Add that Ackermann used in (1924) a formalism with Hilbert's ε-functions.

3. Next attempt to solve the second Hilbert's problem was the paper by Janos (later Johann, John) von Neumann "Zur Hilbertschen Beweistheorie" published

---

[1] More information on this can be found for example in Mancosu (1998).

in 1927. He used another formalism than that in Ackermann (1924) and, similarly as Ackermann, proved in fact the consistency of a fragment of arithmetic of natural numbers obtained by putting some restrictions on the induction. It is worth mentioning here that in the introductory section of von Neumann's paper a nice and precise formulation of aims and methods of Hilbert's proof theory was given. It indicated how was at that time the state of affairs and how Hilbert's program was understood. Therefore we shall quote the appropriate passages.

Von Neumann writes that the essential tasks of proof theory are (cf. von Neumann 1927, pp. 256–257):

> I. First of all one wants to give a proof of the consistency of the classical mathematics. Under 'classical mathematics' one means the mathematics in the sense in which it was understood before the begin of the criticism of set theory. All set-theoretic methods essentially belong to it but not the proper abstract set theory. [...]
> II. To this end the whole language and proving machinery of the classical mathematics should be formalized in an absolutely strong way. The formalism cannot be too narrow.
> III. Then one must prove the consistency of this system, i.e., one should show that certain formulas of the formalism just described can never be "proved".
> IV. One should always strongly distinguish here between various types of "proving": between formal ("mathematical") proving in a given formal system and contents ("metamathematical") proving [of statements] about the system. Whereas the former one is an arbitrarily defined logical game (which should to a large extent be analogues to the classical mathematics), the latter is a chain of directly evident contents insights. Hence this "contents proving" must proceed according to the intuitionistic logic of Brouwer and Weyl. Proof theory should so to speak construct classical mathematics on the intuitionistic base and in this way lead the strict intuitionism ad absurdum.[2]

---

2   "I. In erster Linie wird der Nachweis der Widerspruchsfreiheit der klassischen Mathematik angestrebt. Unter „klassischer Mathematik" wird dabei die Mathematik in demjenigen Sinne verstanden, wie sie bis zum Auftreten der Kritiker der Mengenlehre anerkannt war. Alle mengentheoretischen Methoden gehören im wesentlichen zu ihr, nicht aber die eigentliche abstrakte Mengenlehre. [...]
II. Zu diesem Zwecke muß der ganze Aussagen- und Beweisapparat der klassischen Mathematik absolut streng formalisiert werden. Der Formalismus darf keinesfalls zu eng sein.
III. Sodann muß die Widerspruchsfreiheit dieses Systems nachgewiesen werden, d.h. es muß gezeigt werden, daß gewisse Aussagen „Formeln" innerhalb des beschriebenen Formalismus niemals „bewiesen" werden können.
IV. Hierbei muß stets scharf zwischen verschiedenen Arten des „Beweisens" unterschieden werden: Dem formalistischen („mathematischen") Beweisen innerhalb des formalen Systems, und dem inhaltlichen („metamathematischen") Beweisen über das System. Während das erstere ein willkürlich definiertes logisches Spiel ist (das freilich mit der klassischen Mathematik weitgehend analog sein muß), ist das letztere eine Verkettung unmittelbar evidenter inhaltlicher Einsichten. Dieses „inhaltliche Beweisen" muß also ganz im Sinne der Brouwer-Weylschen intuitionistischen Logik verlaufen: Die Beweistheorie soll sozusagen auf intuitionistischer Basis die klassische Mathematik aufbauen und den strikten Intuitionismus so ad absurdum führen."

Note that von Neumann identifies here finitistic methods with intuitionistic ones. This was then current among members of the Hilbert's school. The distinction between those two notions was to be made explicit a few years later – cf. (Hilbert and Bernays 1934, pp. 34 and 43) and (Bernays 1934, 1935b, 1941).

4. In 1930 Kurt Gödel obtained a result which undermined Hilbert's program. Gödel proved that any consistent theory extending the arithmetic of natural numbers and based on a recursive set of axioms is incomplete (this result is called today Gödel's First Incompleteness Theorem). This result was announced for the first time by Gödel during a conference in Königsberg in September 1930.

It seems that the only participant of the conference in Königsberg who immediately grasped the meaning of Gödel's theorem and understood it was J. von Neumann. After Gödel's talk he had a long discussion with him and asked him about details of the proof. Soon after coming back from the conference to Berlin he wrote a letter to Gödel (on 20th November 1930) in which he announced that he had received a remarkable corollary from Gödel's First Theorem, namely a theorem on the unprovability of the consistency of arithmetic in arithmetic itself. In the meantime Gödel developed his Second Incompleteness Theorem and included it in his paper "Über formal unentscheidbare Sätze der 'Principia Mathematica' und verwandter Systeme. I" (cf. Gödel 1931a). In this situation von Neumann decided to leave the priority of the discovery to Gödel.

In fact in Gödel (1931a) one finds only a statement of the theorem on the unprovability of consistency (called today Gödel's Second Incompleteness Theorem) and a remark that it can be proved by formalizing the proof of the first theorem. Gödel promised also there to publish the full proof in the second part of the paper which would be ready soon. But this second part was never written and Gödel published in fact no proof of his second theorem. Moreover, his remark on the proof was not correct. The first proof of the theorem on the unprovability of consistency appeared in the second volume of Hilbert and Bernay's monograph *Grundlagen der Mathematik* (1939). It has turned out that the way in which the metamathematical sentence "the theory T is consistent" is formalized in the formal language of T is significant here. Hilbert and Bernays formulated certain so called derivability conditions for formulas representing in T the metamathematical notion of provability in T (in fact those conditions require certain internal properties of provability to be formally derivable in T). If those conditions are fulfilled then the second incompleteness theorem holds.

Hilbert-Bernay's conditions were not elegant. A useful and elegant form of them was given by M. H. Löb in 1954 (cf. Löb 1955). It was also shown that there exist formal translations of the sentence "T is consistent" which are provable in *T* and for which the second incompleteness theorem fails. Examples of such formulas were given by J. B. Rosser and A. Mostowski.³

Those results weakened in a sense (the metamathematical and philosophical meaning of) Gödel's Second Incompleteness Theorem. In fact this theorem does not say simply that Peano arithmetic, if consistent, cannot prove its own consistency (and similarly for any consistent extension of it). It turns out that the way in which the metamathematical property of consistency is expressed in the language of the considered theory plays here the crucial rôle. The crude numerical adequacy in the sense of strong representability is not enough here – one needs in fact that the formal representation "reflects" the very structure of the notion of provability (cf. Feferman 1960). Nevertheless Gödel's theorem indicated certain limitations of formalized systems and showed that certain corrections in Hilbert's program are necessary.

In spite of those new circumstances Hilbert defended the very idea of his program. In the Preface to the first volume of *Grundlagen der Mathematik* he wrote:

> [...] the occasionally held opinion that from the results of Gödel follows the non-executability of my Proof Theory, is shown to be erroneous. This result shows indeed only that for more advanced consistency proofs one must use the finite standpoint in a deeper way than is necessary for the consideration of elementary formalisms.⁴

5. Through von Neumann about Gödel's incompleteness theorems learned (in November 1930) Jacques Herbrand. He found them to be of great interest. They also stimulated him to reflect on the nature of intuitionistic proofs and of schemes for the recursive definition of functions. In a letter to Gödel of 7th April 1931 Herbrand suggested the idea of extending the schemes for the recursive definition of functions. His remarks inspired Gödel to formulate the notion of general recursive function (in the lectures he gave at Princeton in 1934 – cf. Gödel 1934).

From the point of view of the present paper however more important is Herbrand's paper "Sur la non-contradiction de l'arithmétique" published in 1931 already after the Gödel's "Über formal unentscheidbare Sätze ...". Herbrand probably started to write his paper before Gödel's paper reached him (the manuscript sent for publication to the *Journal für reine und angewandte Mathematik* was dated "Göttingen, 14 July 1931"; it was sent just before Herbrand left for a vacation

---

3 For technical as well as philosophical and historical information on Gödel's theorems see, e.g., Murawski (1999b).

4 "[...] die zeitweilig aufgekommene Meinung, aus gewissen neueren Ergebnissen von Gödel folge die Undurchführbarkeit meiner Beweistheorie, als irrtümlich erwiesen ist. Jenes Ergebnis zeigt in der Tat auch nur, daß man für die weitergehenden Widerspruchsfreiheitsbeweise den finiten Standpunkt in einer schärferen Weise ausnutzen muß, als dieses bei der Betrachtung der elementaren Formalismen erforderlich ist."

trip in the Alps, and was received on 27 July 1931 – on that day Herbrand was killed in a fall). Nevertheless he had opportunity to examine Gödel's results (in particular his second theorem) and in the last section of his paper he was dealing with them.

Herbrand's paper presents a proof of the consistency of a fragment of arithmetic of natural numbers. It was certainly intended to be a contribution to the realization of Hilbert's program. The fragment considered by Herbrand is arithmetic with induction for formulas containing no bounded variables and induction for formulas containing bounded variables but containing no function symbols except eventually the successor function. The proof uses Herbrand's fundamental theorem[5] (section 1 consists of a very clear presentation of this theorem).

It is worth noting here that Herbrand, similarly as von Neumann (see above), uses the name "intuitionistic" to describe methods which are allowed in the metamathematics, hence finitistic methods. This identification was then current in Hilbert's school.

The key trick of Herbrand's proof of the consistency of the indicated fragment of arithmetic is the elimination of the induction axiom scheme through the introduction of functions. The definition conditions for those functions are such that, for every set of arguments, a well-determined number can be proved in a finitary way to be the value of the function. It should be noted that those functions are (general) recursive functions. This is in fact the first appearance of the notion of a general recursive function as opposed to primitive recursive (cf. Gödel's definition of general recursive functions from 1934 "suggested by Herbrand" – see Gödel 1934, p. 26).

As indicated above, in the last section of his paper (1931b) Herbrand considered the problem of connections between his result and Gödel's theorem on the unprovability of consistency. He explains very clearly why the latter does not hold for the fragment of arithmetic he considers. The reason is that the metamathematical description of the system cannot be projected into the system itself (because the system is too weak).

6. First proof of the consistency of the arithmetic of natural numbers was given by Gerhard Gentzen in the paper "Die Widerspruchsfreiheit der reinen Zahlentheorie" (1936) (cf. also his paper "Neue Fassung des Widerspruchsfreiheitsbeweises für die reine Zahlentheorie" from 1938). According to Gödel's Second Incompleteness Theorem a proof of the consistency of the full arithmetic of natural

---

5 This theorem contains a reduction (in a certain sense) of predicate logic to propositional logic, more exactly it shows that a formula is derivable in the axiomatic system of quantification logic if and only if its negation has a truth-functionally inconsistent expansion. Herbrand intended to prove this theorem by finitistic means. The theorem was contained in Chapter 4 of his doctoral dissertation presented to the Sorbonne in 1930 and published in the same year – cf. Herbrand (1930).

numbers should use means stronger than those available in the arithmetic itself (modulo the restrictions concerning the way of expressing in the formal language the property of consistency). Indeed the analysis of Gentzen's proof shows that it is just in the concept of a reduction process applied by Gentzen in (1936) that the transgression of the methods formalizable in the formal system under consideration comes about. By assigning ordinals to the derivations one sees that the transfinite induction up to ε_0 suffices for the proof.[6]

It is worth noting here that the first version of Gentzen's consistency proof was submitted in 1935 but was withdrawn after criticism directed against the means used in the proof which were considered to be too strong. Gentzen took care of the criticism and modified his original proof before it was published (the modified proof was published in the paper (1936)). Fortunately the text of the original proof was preserved in galley proof. It became publicly known because of the paper by Bernays (1970) and was recently published in the name of Gentzen (cf. Gentzen 1974). Bernays remarks in (1970) that Gentzen's original proof was certainly easier to follow than the first published proof and at least as easy to follow as the second Gentzen consistency proof from (1938).

7. Gentzen's proof was apparently accepted by Hilbert and Bernays in the second volume of *Grundlagen der Mathematik* (1939). Indeed in the Preface Bernays wrote there (p. VII):

> In any case one can say on the basis of Gentzen's proof that the short-lived failure of proof theory was caused solely by the whimsicality of the methodological demand put on it.[7]

In the same Preface it was also announced that W. Ackermann is working on extending his earlier consistency proof (published in 1927) along the lines indicated by Gentzen, i.e., by applying the transfinite induction. Indeed, in 1940 appeared Ackermann's paper "Zur Widerspruchsfreiheit der Zahlentheorie" in which the consistency of the full arithmetic of natural numbers was proved by using methods from his paper (1927) and the transfinite induction.

Since then other proofs along Gentzen's lines have been published. One should mention here among others papers by Lorenzen (1951), Schütte (1951, 1960) and Hlodovskii (1959).

---

6   The countable ordinal ε_0 is defined as the smallest ordinal ε such that $\omega^\varepsilon = \varepsilon$ or as the limit of the sequence $\omega, \omega^\omega, \omega^{\omega^\omega}, ...$

7   "Jedenfalls kann schon auf Grund des Gentzenschen Beweises die Auffassung vertreten werden, daß das zeitweilige Fiasko der Beweistheorie lediglich durch eine Überspannung der methodischen Anforderung verschuldet war, die man an die Theorie gestellt hat."

Decidability vs. Undecidability. Logico-Philosophico-Historical Remarks

1. Origin of the Decidability Problem

It was David Hilbert with whom one should connect the beginnings of researches on the decidability – he drew attention of mathematicians to this problem and made it into a central problem of mathematical logic. He called it *das Entscheidungsproblem* (what literally means: "the decision problem").[1] It appeared in a sense already in his famous lecture at the Congress of Mathematicians in Paris in August 1900. Hilbert proposed there a list of 23 most important problems of mathematics which should be solved in the future. Problem X was (cf. Hilbert 1901):

> Given a diophantine equation with any number of unknown quantities and with rational integral numerical coefficients: *To devise a process according to which it can be determined by a finite number of operations whether the equation is solvable in rational integers.*[2]

In the first quarter of the 20th century Hilbert formulated and developed a research program, called today Hilbert program. Its aim was the justification of the classical mathematics. In this context appeared the decidability problem (closely connected with the completeness problem), that is the problem of finding an effective method which would enable us to decide in a finite number of prescribed steps whether a given formula is a theorem of a considered (formalized) theory. First – not quite clear and explicite – formulation of this problem can be found in Hilbert's paper "Axiomatisches Denken" (1918).[3] The decidability problem was formulated in a direct and explicite way by H. Behmann in his *Habilitationsschrift* in 1922 (cf. Behmann 1922) where he wrote:

> A well defined general procedure should be given which in the case of any given statement formulated with the help of purely logical means, would enable us to decidein

---

[1] It seems that the word *das Entscheidungsproblem* appeared for the first time in the talk "Das Entscheidungsproblem der mathematischen Logik" of H. Behmann given at the meeting of Deutsche Mathematiker-Vereiningung in Göttingen in May 1921 – cf. (1921), p. 21.

[2] "Eine diophantische Gleichung mit irgendwelchen Unbekannten und mit ganzen rationalen Zahlkoeffizienten sei vorgelegt: *man soll ein Verfahren angeben, nach welchem sich mittels einer endlichen Anzahl von Operationen entscheiden läßt, ob die Gleichung in ganzen rationalen Zahlen lösbar ist.*"

[3] One should admit that similar problems can be also found by Schröder in (1895) and by Löwenheim in (1915).

a finite number of steps whether it is true or false or at least this aim would be realized within the – precisely fixed – framework in which its realization is really possible.[4]

Hilbert and Ackermann formulated the decidability problem in the book *Grundzüge der theoretischen Logik* in Chapter "Das Entscheidungsproblem im Funktionalkalkül und seine Bedeutung" in the following way (cf. 1928, p. 73):

> The Entscheidungsproblem is solved when we know a procedure that allows for any given logical expression to decide by finitely many operations its validity or satisfiability.[5]

They distinguished there various aspects of *Entscheidungsproblem*:
– the *satisfiability problem* (or the *consistency problem*): given a formula, decide if it is consistent,
– the *validity problem*: given a formula, decide if it is valid,
– the *provability problem*: given a formula, decide if it is provable (in a given system).

In the first-order logic all those aspects are equivalent – this follows from the completeness theorem of Gödel . Moreover, by Deduction Theorem in the case of finitely axiomatizable theories it suffices to investigate only the system of the first-order logic and not theories in general (in the case of non-finitely axiomatizable theories it does not suffice). Add that the decision problem for a system of first-order logic is called the *classical decision problem*.

Hilbert and Ackermann were of the opinion that the *Entscheidungsproblem* is the main problem of mathematical logic. In (1928) they wrote:

> The *Entscheidungsproblem* must be considered the main problem of mathematical logic. [...] The solution of the *Entscheidungsproblem* is [an issue] of the fundamental significance for the theory of all domains whose propositions could be developed on the basis of a finite number of axioms.[6]

This conviction was based just on Deduction Theorem – indeed, a decision procedure for first-order logic would generate (via this theorem) a decision procedure for any (finitely axiomatizable) first-order theory.

---

4 "Es soll eine ganz bestimmte allgemeine Vorschrift angegeben werden, die über die Richtigkeit oder Falschheit einer beliebig vorgelegten mit rein logischen Mitteln darstellbaren Behauptung nach einer endlichen Anzahl von Schritten zu entscheiden gestattet, oder zum mindesten dieses Ziel innerhalb derjenigen – genau festzulegenden – Grenzen verwirklicht werden, innerhalb deren seine Verwirklichung tatsächlich möglich ist."

5 "Das Entscheidungsproblem ist gelöst, wenn man ein Verfahren kennt, das bei einem vorgelegten logischen Ausdruck durch endlich viele Operationen die Entscheidung über die Allegemeingültigkeit bzw. Erfüllbarkeit erlaubt."

6 "Das Entscheidungsproblem muss als das Hauptproblem der mathematischen Logik bezeichnet werden. [...] Die Lösung des Entscheidungsproblems ist für die Theorie aller Gebiete, deren Sätze überhaupt einer logischen Entwickelbarkeit aus endlich vielen Axiomen fähig sind, von grundsätzlicher Wichtigkeit."

Other logicians shared the conviction of the importance of the *Entscheidungsproblem*. P. Bernays and M. Schönfinkel wrote in (1928):

> The central problem of mathematical logic, which is also most closely related to the question of axiomatics, is the *Entscheidungsproblem*.[7]

J. Herbrand begins the paper (1929) with the words:

> We could consider the fundamental problem of mathematics to be the following. Problem A: What is the necessary and sufficient condition for a theorem to be true in a given theory having only finite number of hypotheses?

Herbrand finished the paper (1930) with the words:

> The solution of this problem [i.e., the decision problem – R.M.] would yield a general method in mathematics and would enable mathematical logic to play with respect to classical mathematics the role that analytic geometry plays with respect to ordinary geometry.[8]

In (1931a) Herbrand added:

> In a sense it [the classical decision problem – R.M.] is the most general problem of mathematics.

F. P. Ramsey wrote in (1930, p. 264) that this paper was:

> concerned with a special case of one of the leading problems in mathematical logic, the problem of finding a regular procedure to determine the truth or falsity of any given logical formula.

The roots of the decision problem can be traced while back by those philosophers who were interested in a general method of problem solving. One should mention here first of all the medieval thinker Raimundus Lullus and his *ars magna* as well as Descartes' idea of *mathesis universalis* and Gottfried Wilhelm Leibniz with his *characteristica universalis* and *calculus ratiocinator*. The realization of the latter idea should provide a method of mechanical solving of any scientific problem (expressed in the symbolic language). Partial realization of this idea (in fact restricted to mathematics) was found by the mathematical logic at the end of the 19th and the beginning of the 20th century.

---

7   "Das zentrale Problem der mathematischen Logik, welches auch mit den Fragen der Axiomatik im engsten Zusammenhang steht, ist das Entscheidungsproblem."

8   "La solution de ce problème fournirait une méthode générale en Mathématique, et permettrait de faire jouer à la logique mathématique, vis-à-vis de Mathématique classique, le même rôle que la géométrie analytique vis-à-vis de la géométrie ordinaire."

One should note that Leibniz distinguished between two different versions of *ars magna*:
- *ars inveniendi* which finds all true scientific statements,
- *ars iudicandi* which allows one to decide whether any given scientific statement is true or not.

In fact in the framework of the first-order logic an *ars inveniendi* exists: the collection of all valid first-order formulas is recursively enumerable, hence there is an algorithm that lists all valid formulas. On the other hand, the classical decision problem can be viewed as the *ars iudicandi* problem in the first-order framework. It can be formulated in the following way: Does there exist an algorithm that decides the validity of any given first-order formula?

It is worth noticing that some logicians felt sceptical about the possibility of finding such an algorithm. Among them was J. von Neumann who wrote in (1927, pp. 11–12):

> It appears thus that there is no way of finding the general criterion for deciding whether or not a well-formed formula a is provable. (We cannot, however, at the moment demonstrate this. Indeed, we have no clue as to how such a proof of undecidability would go.) [...] The undecidability is even *the condition sine qua non* for the contemporary practice of mathematics, using as it does heuristic methods, to make any sense. The very day on which the undecidability would cease to exist, so would mathematics as we now understand it; it would be replaced by an absolutely mechanical prescription, by means of which anyone could decide the provability or unprovability of any given sentence.
>
> Thus we have to take the position; it is generally undecidable, whether a given well-formed formula is provable or not. The only thing we can do is [...] to construct an arbitrary number of provable formulas. [...] In this way, we can establish for many well-formed formulas that they are provable. But in this way we never succeed to establish that a well-formed formula is not provable.[9]

---

9 "Es scheint also, daß es keinen Weg gibt, um das allgemeine Entscheidungskriterium dafür, ob eine gegebene Normalformel a beweisbar ist, aufzufinden. (Nachweisen können wir freilich gegenwärtig nichts. Es ist auch gar kein Anhaltspunkt dafür vorhanden, wie ein solcher Unentscheidbarkeitsbeweis zu führen wäre.) [...] Und die Unentscheidbarkeit ist sogar die *Conditio sine qua non* dafür, daß es überhaupt einen Sinn habe, mit den heutigen heuristischen Methoden Mathematik zu treiben. An dem Tage, an dem die Unentscheidbarkeit aufhörte, würde auch die Mathematik im heutigen Sinne aufhören zu existieren; an ihre Stelle würde eine absolut mechanische Vorschrift treten, mit deren Hilfe jedermann von jeder gegebenen Aussage entscheiden könnte, ob diese beweisen werden kann oder nicht. Wir müssen uns also auf den Standpunkt stellen: Es ist allgemein unentscheidbar, ob eine gegebene Normalformel beweisbar ist oder nicht. Das einzige, was wir tun können, ist [...], beliebig viele beweisbare Normalformeln aufzustellen. [...] Auf diese Art können wir von vielen Normalformeln feststellen, daß sie beweisbar sind. Aber auf diesem Weg kann uns niemals die Feststellung gelingen, daß eine Normalformel nicht beweisbar ist."

J. Herbrand in an appendix to his (1931a) wrote:

> Note finally that, although at present it seems unlikely that the decision problem can be solved, it has not yet been proved that it is impossible to do so.

2. First (Negative) Results

Note that before the 1930s some positive answers to the decision problem for particular theories have been obtained (we shall say more about those results later). However, the classical decision problem (i.e., the decision problem for the first-order logic) was unsolved. Notice also that to prove that there is no effective procedure to decide the formulas of the first-order logic one needs a precise definition of the notion of an algorithm and of an effective method (in the case of positive solutions one does not need a precise general definition). In fact, in the 1930s such definitions have been given by Church, Gödel, Turing, Herbrand, Kleene. The Church–Turing thesis formulated in 1936 stated that those precise definitions are adequate with respect to the intuitive notion of an effective procedure.[10]

The method of an arithmetization of syntax introduced by Gödel enabled also to formulate precisely the very decision problem. This was done by Alfred Tarski in (1953). Tarski proposed the following definitions:

– a (first-order) theory T is said to be *decidable* if and only if the set of (Gödel numbers of) theorems of T is recursive,

– a (first-order) theory T is said to be *undecidable* if and only if the set of (Gödel numbers of) theorems of T is not recursive,

– a (first-order) theory T is said to be *essentially undecidable* if and only if T is undecidable and every consistent extension T' of T (in the same language as T) is undecidable.

The first result concerning the *Entscheidungsproblem* in a strict formulation was the theorem due to A. Church from 1936 (cf. Church 1936) providing the negative solution of the decision problem for the first-order predicate calculus. In fact Church has proved that the set of all valid formulas of the first-order logic is not effectively decidable. A similar result was obtained a bit later by A. Turing. The method used was similar in both cases: it was shown that a certain undecidable combinatorial problem can be represented in the first-order logic, hence the latter is undecidable. In fact Church has shown that the set of provable formulas (theorems) of the first-order logic is not λ-definable. The undecidability of the first-order logic is then a corollary via the completeness theorem (due to Gödel, 1929) and the Church–Turing thesis.

---

10 On Church–Turing thesis, its history and epistemological status see Murawski (2004).

The result of Church has been later "sharpened", i.e., it has been shown that the *Entscheidungsproblem* has a negative solution for a fragment of the first-order logic, viz. for the first-order predicate calculus in a language containing at least one binary predicate – this was done by Kalmár (1936). This contrasted with the earlier result by L. Löwenheim (1915), Th. Skolem (1919) and H. Behmann (1922) on the decidability of the classical monadic first-order predicate calculus. Note that the intuitionistic monadic first-order predicate calculus is not decidable!

The undecidability result of Church implied also undecidability of the second-order logic (add that – as noticed above – Hilbert and Ackermann were talking about logic as such not distinguishing the order of it). Today one knows that the first-order logic being undecidable is semi-decidable, i.e., the set of (Gödel numbers of) its theorems is recursively enumerable whereas the second-order logic is not even semi-decidable (this follows from Gödel's incompleteness theorem).

3. Studies on (Un)decidability

Church's result was a beginning of very intensive studies on the problem of (un)decidability. During a long time those problems were treated as central in the mathematical logic and the foundations of mathematics. Together with investigations on the complexity of decision procedures a new group of problems appeared, viz. the problem of how complicated the possible decision procedures can be.

To systematize the presentation of the results obtained let us distinguish studies on the (un)decidability of: (a) concrete mathematical theories, (b) fragments of the first-order logic (both are connected via Deduction Theorem), and (c) problems in the computation theory. Those investigations contributed to the development of both mathematical logic and the recursion theory. The literature is very large here. In what follows examples of most important results in the indicated fields will be provided and some remarks on the methods used in the proofs will be given.

3.1. (Un)decidability of Mathematical Theories

Investigations on the decidability of mathematical theories wsere carried out long before the theorem of Church – in fact to prove the decidability of a theory one does not need a precise definition of decidability itself (only for a negative result such a definition is necessary!).

The main methods used (nowadays) in proving the decidability of (mathematical) theories are the following:[11]
  (i) elimination of quantifiers,
  (ii) modeltheoretic method,
  (iii) method of interpretation.

The first two methods are based on a theorem due to A. Janiczak (1950) and stating that if a theory T is consistent, complete and (recursively) axiomatizable then T is decidable. Methods (i) and (ii) are used just to show that a considered theory is complete and to obtain in this way its decidability.

The method of elimination of quantifiers consists of indicating a set of a certain class Φ of formulas in the language of the considered theory (called basic formulas) such that (a) every formula of Φ is decidable, (b) any formula of T is T-equivalent to a Boolean combination of some formulas form Φ and (c) the decidability of basic formulas implies the decidability of the Boolean combinations of them.

The method was initiated by L. Löwenheim (1915) and used in fully-developed form by Th. Skolem (1919) and C. H. Langford (1927). It was also intensively studied at the seminar led by A. Tarski at Warsaw University in 1927–29. Tarski and his students used this method to characterize definability and to prove the decidability of particular mathematical theories (cf. Murawski 1996)). It was also used to describe and classify all complete extensions of a given theory. The elimination of quantifiers became there *the* method and a paradigm of how logicians should study axiomatic theories. The very name of the method comes from Tarski.

With the help of the method of elimination of quantifiers the decidability of various theories has been established, in particular the following theories have been shown to be decidable:
  – the arithmetic of addition (Presburger arithmetic) (Presburger 1930),
  – elementary theory of identity (Löwenheim 1915),
  – theory of finitely many sets (Löwenheim 1915),
  – theory of discrete order DO (Langford 1927),
  – theory of linear order in the set of rationals (Tarski 1936),
  – theory of algebraically closed fields ACF (1949),
  – theory of Boolean algebras (Tarski 1949a),
  – theory of real numbers (Tarski 1951; Cohen 1969),
  – theory of Abelian groups (Szmielew 1949a, 1949b),
  – theory of well order (Donner, Mostowski, Tarski 1978).[12]

---

11 Detailed information on the methods used in proving the decidability or the undecidability of theories together with examples and references to the literature can be found, e.g., in Murawski (1999b).
12 On the (dramatic) history of this result see Murawski (1996).

Note that Presburger arithmetic T_+ is a first-order theory in the language $L(T_+)$ with $0, S, +$ as the non-logical constants and based on the following non-logical axioms:

$$0 \neq S(x),$$
$$S(x) = S(y) \to x = y,$$
$$x + 0 = x,$$
$$x + S(y) = S(x+y),$$

induction scheme for formulas φ of the language $L(T_+)$.

Notice that results on the decidability of the first-order theory of successor (Herbrand 1928) and of the theory of multiplication (with successor but without addition!) (Skolem 1930) has been also obtained. Note that in contrast to those results the arithmetic of successor, addition and multiplication is essentially undecidable!

The second indicated method, i.e., the modeltheoretic method is usually used to show (by methods of model theory) that a given theory is complete or to study systematically all complete extensions of it. Sometimes a combination of this method and the method of elimination of quantifiers is used. Using those methods one has proved the decidability of the following theories:
 – theory of linear dense order without the first and last element DNO,
 – theory of algebraically closed fields of a given characteristic,
 – theory of p-adic fields (Ax and Kochen 1965a, 1965b, 1966),
 – theory of all finite fields (Ax 1968),
 – theory of real closed fields RCF (Tarski 1949b),
 – theory of linearly ordered sets (Ehrenfeucht 1959 – result announced only; Läuchli and Leonard 1966).

The last indicated method of proving decidability, i.e., the method of interpretation can be briefly described as follows. Let a decidable theory T_0 formalized in a language L_0 be given. We are asking if another given theory T formalized in a language L is decidable. To answer this question one defines a computable (recursive) function f mapping formulas of the language L on formulas of the language L_0 such that if φ is a sentence of L then $f(\varphi)$ is a sentence of L_0 and $T \vdash \varphi$ if and only if $T_0 \vdash f(\varphi)$. This gives us a decision procedure for the theory T.

By this method the decidability of, e.g., the second-order monadic theory of one successor S1S (Büchi 1962) and of the weak second-order theory of one successor WS1S (Büchi 1960; Elgot 1961) have been proved.

In the case of decidable theories one can ask the question: how complex is a decision procedure? To indicate some answers and to show that decisions procedures are usually very complicated (mostly of exponential complexity) and hence not applicable practically, let us mention the following results:

– A decision procedure for Presburger arithmetic is of the complexity at least $2^{2^{cn}}$ for a certain constant $c > 0$, i.e. to decide whether a formula φ of the length n is a theorem of Presburger arithmetic one needs at least $2^{2^{cn}}$ steps [Fisher and Rabin 1974].

– The complexity of the theory of multiplication is at least $2^{2^{2^{cn}}}$ for a certain constant $c > 0$ [Fisher and Rabin 1974].

The problem whether there are decidable theories with practically applicable decision procedures is still open. It is connected with the famous problem of whether $P = NP$, which is nowadays the central problem of the recursion theory and of the complexity theory.

Let us turn now to proofs of the undecidability. Main methods of proving the undecidability of a theory are the following:

(i) the method based directly on the ideas of Gödel's proof of the incompleteness theorem,

(ii) the method of interpretation.

The method (i) is based on the theorem stating that if all recursive relations are strongly representable in a theory T then T is undecidable (moreover, the set of Gödel numbers of theorems of T and the set of Gödel numbers of negations of theorems of T are not recursively separable). This method can be applied only in the case of theories which have built-in an appropriate fragment of the arithmetic of natural numbers.

The method (ii) has been mostly developed by A. Tarski. Generally speaking it consists of showing that a known undecidable theory T_1 can be interpreted (embedded) into a theory T_2 under question. If it is so then the theory T_2 is undecidable. Here are some examples of undecidable theories:

– Peano arithmetic (i.e., the arithmetic of natural numbers in the language with $0, S, +, \cdot$ as non-logical constants),

– theory of rings (Tarski 1951),

– theory of ordered fields (R. M. Robinson 1951),

– theory of lattices (Tarski 1949c),

– predicate calculus with at least one binary predicate (Kalmár 1936),

– theory of partial order (Tarski),

– theory of two equivalence relations whose intersection is the identity relation, theory of two equivalence relations, theory of one equivalence relation and one bijection (Janiczak 1953),

– theory of groups (Tarski 1953),

– theory of rationals with $+$ and $\cdot$ (J. Robinson 1949).

The tenth problem of Hilbert mentioned at the very beginning of this story on (un)decidability was solved in the 1970s by Y. Matiyasevich who using some earlier results of M. Davis, H. Putnam and J. Robinson showed that a relation R is recursively enumerable if and only if R is diophantine. Since there are recursively enumerable relations which are not recursive, it follows that not every diophantine relation is recursive, hence the tenth problem of Hilbert has a negative solution, i.e., there is no effective method of deciding whether a given diophantine equation has solutions or not.

3.2. (Un)decidability of Fragments of the First-Order Logic

Since the first-order predicate calculus is undecidable (as shown by Church), one can ask whether given fragments of it are decidable or not. This problem, called the classical decision problem, has been studied intensively. Nowadays this field of problems can be treated as closed (cf. the monograph Börger *et al.* 1997).

The investigated fragments of the first-order logic are usually described with the help of prefixes. Let us explain this on an example: so $[\forall\exists\forall, (\omega, 1), (0)]$ denotes the class of all formulas in the prenex form with the quantifier prefix $\forall\exists\forall$ in the language with infinitely many unary predicates, one binary predicate and no function symbols. The symbol $[\forall\exists\forall, (\omega, 1), (0)]_=$ denotes similar class of formulas but now in the language there is the identity $=$ symbol.

As an example of results obtained in studies of (un)decidability of fragments of the first-order logic let us say that the following fragments are undecidable:[13]
- $[\exists\forall\exists\forall, (0, 3), (0)]$ (Büchi 1962),
- $[\forall\exists\forall, (0, \infty)(0)]$ (Kahr, Moore, Wang 1962),
- $[\forall^3\exists, (0, \infty), (0)]$ (Gödel 1933c),
- $[\exists^\infty\forall^2\exists^2\forall^\infty, (0, 1), (0)]$ (Kalmár 1932).

3.3. (Un)decidability in the Computation Theory

Problems of decidability have been studied also with respect to the computation theory. The best known result is here the undecidability of the halting problem. The question is: can it be effectively decided whether a given Turing machine stops at a given input x? The problem can be reformulated in the following way: let (φ_x) be the effective enumeration of all recursive functions. We are asking whether the set $\{\langle x, y\rangle : \varphi_x(y)\downarrow\}$ is recursive, i.e., whether it can be effectively decided if the function φ_x is defined for an argument y? The answer to this problem is negative. Hence the halting problem is undecidable.

Here are some examples of other undecidable problems from the computation theory:
- is $\varphi_x = 0$?

---

13 Let us add that in this field worked also and received some interesting results Polish logician József Pepis. He was active at the Jan Kazimierz University in Lvov. In August 1941 Pepis was killed by Gestapo.

- is $\varphi_x = \varphi_y$?
- does $y \in \text{dom}(\varphi_x)$?
- does $y \in \text{rng}(\varphi_x)$?
- is $\varphi_x(x) = 0$?
- does $\varphi_x(x)\downarrow$?
- does $\varphi_x(y) = 0$?

And here are some other examples in the language of Turing machines. The following problems are undecidable:
- does a given Turing machine M stop on all inputs?
- does for a given Turing machine M and a given input x there exist an y such that $M(x) = y$?
- does the computation of the Turing machine M on the input x use all the states of M?
- does for given Turing machine M and for x and y hold $M(x) = y$?

4. Conclusions

As shown above most mathematical theories are undecidable. This means that for such theories sets of Gödel numbers of their theorems are not recursive, hence not definable (strongly representable) in Peano arithmetic. This indicates some limits in defining notions in formal systems. An interesting comment to this was given by W. V. O. Quine who said that those systems wanted to swallow a greater piece of ontology than they were able to digest. On the other hand, the cardinality argument shows that this phenomenon is quite normal: in fact, there are uncountably many subsets of the set of natural numbers while the set of formulas of the language of Peano arithmetic, hence the set of definable subsets of the set of natural numbers, is countable. What is surprising here is that among sets of natural numbers that are not definable (not representable) in arithmetic are sets of Gödel numbers of theorems of most mathematical theories.

The fact that most mathematical theories are undecidable should not be astonishing. Indeed, problems formulated in decidable theories are not any longer scientific problems – they can be solved (at least theoretically) in a mechanical way. On the other hand, since the decision procedures are usually of an exponential complexity, they are not practically applicable. So one comes to the conclusion that the mind of a mathematician cannot be replaced by a machine (even in the case of a decidable theory). But why does the human mind overcome a machine? Does the reason for that lie in the fact that a human being is able to perform infinite operations? And maybe it does not work algorithmically but in a creative way, it can move to higher levels (to use higher types) and in this way find solutions unaccessible at lower levels?

The final (and in a sense optimistic) conclusion can be that there will always be open problems mathematicians can work on, there will always be a need for a creative thinking in mathematics. Tarski put it in the following amusing way (cf. 1995, p. 166):

> I have no doubt that many mathematicians experienced a profound feeling of relief when they heard of this result. Perhaps sometimes in their sleepless nights they thought with horror of the moment when some wicked metamathematician would find a positive solution of the problem, and design a machine which would enable us to solve any mathematical problems in a purely mechanical way, so that any further creative mathematical thought would become a worthless hobby. The danger is now over, that such a robot will ever be created; mathematicians have regained their *raison d'être* and can sleep quietly.

Undefinability of Truth. The Problem of the Priority: Tarski vs. Gödel

The essay is devoted to the discussion of some philosophical and historical problems connected with the theorem on the undefinability of the notion of truth. In particular the problem of the priority of proving this theorem will be considered. It is claimed that Tarski obtained this theorem independently though he made clear his indebtedness to Gödel's methods. On the other hand Gödel was aware of the formal undefinability of truth in 1931 but he did not publish this result. Reasons for that are also considered.

The theorem on the undefinability of truth was published by Alfred Tarski in his famous paper *Pojęcie prawdy w językach nauk dedukcyjnych* (1933a) (German translation – 1936, English translation – 1956a). It was numbered as Theorem I (β) and stated that (cf. Tarski 1956a, p. 247):

> Assuming that the class of all provable sentences of the metatheory is consistent, it is impossible to construct an adequate definition of truth in the sense of convention **T** on the basis of the metatheory.

It was followed by a description of the idea of the proof and then by a sketch of the proof. The theorem was proved by the diagonalization, hence it is closely connected with methods and results developed by Kurt Gödel in the paper "Über formal unentscheidbare Sätze der *Principia Mathematica* und verwandter Systeme. I" (1931a). In this way the problem of the priority and anticipation arises here. This is the main problem studied in the present paper.

1. Tarski and the Undefinability of Truth

Tarski made clear his indebtedness to Gödel's methods. In footnote 88 on pp. 96–97 of (1933a) (footnote 1 on p. 247 of the English translation 1956a) where the idea of the proof of the theorem is sketched he wrote:

> We owe the method used here to Gödel, who employed it for other purposes in his recently published work, Gödel, K. (22)[1] [...]. This exceedingly important and interesting article is not directly connected with the theme of our work – it deals with strictly methodological problems: the consistency and completeness of deductive systems; nevertheless we shall be able to use the methods and in part also the results of Gödel's investigations for our purpose.

---

1 This is Gödel's paper (1931a) – my remark, R.M.

I take this opportunity of mentioning that Th. I and the sketch of its proof was only added to the present work after it had already gone to press. At the time the work was presented at the Warsaw Society of Sciences (21 March 1931), Gödel's article – as far as I know – had not yet appeared. In this place therefore I had originally expressed, instead of positive results, only certain suppositions in the same direction, which were based partly on my own investigations and partly on the short report, Gödel, K. (21),[2] which had been published some months previously.

In *Historische Bemerkungen* added at the end of (1936a) (Historical Notes at the end of 1956a) Tarski wrote explicitly:

In the one place in which my work is connected with the ideas of Gödel – in the negative solution of the problem of the definition of truth for the case where the metalanguage is not richer than the language investigated – I have naturally expressly emphasized this fact (cf. p. 247, footnote); it may be mentioned that the result so reached, which very much completed my work, was the only one subsequently added to the otherwise already finished investigation.

On the other hand Tarski strongly emphasized the fact that his results were obtained independently. In particular in Historical Notes he wrote:

I may say quite generally that all my methods and results, with the exception of those at places where I have expressly emphasized this – cf. footnotes, pp. 154 and 247 – were obtained by me quite independently. [...]
I should like to emphasize the independence of my investigations regarding the following points of detail: [...] (5) the discussion on pp. 184 f. on the interpretation of the metasystem in arithmetic, which already contain the so called 'method of arithmetizing the metalanguage' which was developed far more completely and quite independently by Gödel.

Tarski did not claim any priority for Gödel's own results. In footnote 2 to the paper „Einige Betrachtungen über die Begriffe der ω-Widerspruchsfreiheit und der ω-Vollständigkeit" (1933b) Tarski wrote:

Already, in the year 1927 [...] I also communicated the example of a consistent and yet not ω-consistent system which I give in the present article in a slightly altered form. Naturally it is not hereby claimed that I already knew then the results later obtained by Gödel or had even foreseen them. On the contrary, I had personally felt that the publication of the work of Gödel cited above[3] was a most exciting scientific event.

Though Tarski saw the similarity of the system P used by Gödel in the paper „Über formal unentscheidbare Sätze ..." (1931a) and his own system used in

---

2   This is Gödel's paper (1930b) – my remark, R.M.
3   I.e., Gödel (1931a) – my remark, R.M.

(1933a) he adds at the end of footnote 88 on pp. 96–97 of (1933a) (footnote 1 on pp. 247–248 of the English translation 1956a) the remark that

> [...] the abstract character of the methods used by Gödel renders the validity of his results independent to a high degree of the specific peculiarities of the science investigated.

It is known that Tarski's theorem on undefinability of truth implies the existence of undecidable sentences, hence Gödel's first incompleteness theorem (see, e.g., Murawski 1999b). So one can ask whether Tarski was close to this theorem. S. R. Givant writes in connection with this question that (see Givant 1991, p. 25):

> Tarski once wondered out loud to me whether he might not have obtained his theorem [i.e., the theorem on the existence of undecidable sentences – R. M.] before Gödel's work appeared, had he had, in the period 1925–1931, a position that allowed him to devote himself more to research. At any rate he lamented the fate that so limited the time and energy available to him for creative mathematics during the years when he was at the hight of his mathematical powers.

2. Gödel and the Undefinability of Truth

Having considered the problem of the priority of proving the undefinability of truth from "Tarski's side" let us consider it now from "Gödel's side". The first fact is that Gödel was aware of the formal undefinability of the notion of truth in 1931. This follows from a letter of Paul Bernays to Gödel of 3rd May 1931 (cf. Dawson 1985a). Earlier, on 20th April, Bernays sent Gödel a letter in which he wrote about his inability to see why a truth predicate could not be formally defined in number theory and he proposed even a candidate for such a definition. In the letter of 3rd May he recognized his error.

Further evidence for the discussed thesis is the letter of Gödel to Ernst Zermelo of 12th October 1931 (cf. Gödel 2003, pp. 426 and 428; see also Grattan--Guinness 1979 and Dawson 1985b). Gödel wrote in it:

> Now it becomes quite clearly manifest that in the definition for K^* a new concept, namely the concept 'correct formula', or, respectively, the class of correct formulas, occurs. This concept, however, may not, without further ado, be traced back to a combinatorial property of formulas (but rather rests upon the meaning of the symbols), and therefore may not be traced back in arithmetized metamathematics to simple arithmetical concepts; or, in other words: The class of correct formulas is *not* expressible by means of a class sign of the given system[4] (hence neither is the class K^* defined from it). The situation

---

[4] More precisely stated, it is of course a question of the class of those *numbers* that are assigned to correct formulas.

is quite otherwise for the concept 'provable formula' (respectively, the class of provable formulas, which occurs in the definition of K). The property of a formula, that it is provable, is a purely combinatorial (formal) one, in that it does *not* depend on the meaning of the symbols.[5]

Note that Gödel used here the term 'richtige Formel' (correct formula) instead of 'wahre Formel' (true formula). The reasons of that will be discussed below.

Later in the considered letter to Zermelo, Gödel showed that the undefinability of truth implies the existence of undecidable sentences. He wrote (cf. Gödel 2003, pp. 426 and 428; 427 and 429:

> In connection with what has been said, one can moreover also carry out my proof as follows: The class W of correct formulas *is never* coextensive with a class sign of that same system (for the assumption that that is the case leads to a contradiction). The class B of provable formulas *is* coextensive with a class sign of that same system (as one can show in detail); consequently B and W can not be coextensive with each other. But because $B \subseteq W$, $B \subset W$ holds, i.e., there is a correct formula A that is not provable. Because A is correct, not-A is also not provable, i.e., A is undecidable. This proof has, however, the disadvantage that it furnishes no construction of the undecidable statement and it not intuitionistically unobjectionable.[6]

Important for our discussion are also Gödel's own reports of how he arrived at his incompleteness theorems. Hao Wang, on the basis of his discussions with Gödel, reports this in the following way (see 1981):

---

5 „Jetzt zeigt sich ganz deutlich, dass in der Definition für K^* ein neuer Begriff, nämlich der Begriff 'richtige Formel' bzw. die Klasse der richtigen Formeln vorkommt. Dieser Begriff lässt sich aber *nicht* ohne weiteres auf eine kombinatorische Eigenschaft der Formeln zurückführen (sondern stützt sich auf die Bedeutung der Zeichen) und lässt sich daher in der arithmetisierten Metamathematik nicht auf einfache arithmetische Begriffe zurückführen; oder anders ausgedrückt: Die Klasse der richtigen Formeln ist *nicht* durch ein Klassenzeichen des gegebenen Systems ausdrückbar [Fussnote: genauer gesprochen, handelt es sich natürlich um die Klasse derjenigen *Zahlen*, welche richtigen Formeln zugeordnet sind] (daher auch nicht die daraus definierte Klasse K^*). Ganz anders steht es mit dem Begriff 'beweisbare Formel' (bzw. der Klasse der beweisbaren Formel, welche in der Definition von K vorkommt). Die Eigenschaft einer Formel, beweisbar zu sein, ist eine rein kombinatorische (formale), bei der es auf die Bedeutung der Zeichen *nicht* ankommt."
6 „Im Anschluss an das Gesagte kann man übrigens meinen Beweis auch so führen: Die Klasse W der richtigen Formeln *ist niemals* mit einem Klassenzeichen desselben Systems umfangsgleich (denn die Annahme, dass dies der Fall sei, führt auf einen Widerspruch). Die Klasse B der beweisbaren Formeln *ist* mit einem Klassenzeichen desselben Systems umfangsgleich (wie man ausführlich zeigen kann); folglich können B und W nicht miteinander umfangsgleich sein. Weil aber $B \subseteq W$, so gilt $B \subset W$ d.h. es gibt eine richtige Formel A, die nicht beweisbar ist. Weil A richtig ist, so ist auch non-A nicht beweisbar, d.h. A ist unentscheidbar. Dieser Beweis hat aber den Nachteil, dass er keine Konstruktion des unentscheidbaren Satzes liefert und intuitionistisch nicht einwandfrei ist."

[Gödel] represented real numbers by formulas [...] of number theory and found he had to use the concept of truth for sentences in number theory in order to verify the comprehension axiom for analysis. He quickly ran into the paradoxes (in particular, the Liar and Richard's) connected with truth and definability. He realized that truth in number theory cannot be defined in number theory and therefore his plan [...] did not work.

Gödel himself wrote on his discovery in a draft reply to letter dated 27th May 1970 from Yossef Balas, then a student at the University of Northern Iowa (cf. Wang 1987, pp. 84–85). Gödel indicated there that it was precisely his recognition of the contrast between the formal definability of provability and the formal undefinability of truth that led him to his discovery of incompleteness. One finds also there the following statement:

> [...] long before, I had found the *correct* solution of the semantic paradoxes in the fact that truth in a language cannot be defined in itself.

Note also that Gödel was convinced of the objectivity of the concept of mathematical truth. In a latter to Hao Wang (cf. Wang 1974, p. 9) he wrote:

> I may add that my objectivist conception of mathematics and metamathematics in general, and of transfinite reasoning in particular, was fundamental also to my other work in logic. How indeed could one think of *expressing* metamathematics *in* the mathematical systems themselves, if the latter are considered to consist of meaningless symbols which acquire some substitute of meaning only *through* metamathematics. [...] it should be noted that the heuristic principle of my construction of undecidable number theoretical propositions in the formal systems of mathematics is the highly transfinite concept of 'objective mathematical truth' as *opposed* to that of 'demonstrability' (cf. M. Davis, *The Undecidable*, New York 1965, p. 64 where I explain the heuristic argument by which I arrive at the incompleteness results), with which it was generally confused before my own and Tarski's work.

In this situation one should ask why Gödel did not mention the undefinability of truth in his writings. In fact, Gödel even avoided the terms 'true' and 'truth' as well as the very concept of being true (compare the remark above on his usage of the term 'richtige Formel' instead of 'wahre Formel'). In the paper „Über formal unentscheidbare Sätze ..." (1931) the concept of a true formula occurs only at the end of Section 1 where Gödel explains the main idea of the proof of the first incompleteness theorem (but again the term 'inhaltlich richtige Formel' and not the term 'wahre Formel' appears here). Indeed, talking about the construction of a formula which should express its own unprovability invokes the interpretation of the formal system. At the very end of the introductory section one finds the following remarks (see Gödel 1931a, pp. 175–176 and Gödel 1986, p. 151):

> The method of proof just explained can clearly be applied to any formal system that, first, when interpreted as representing a system of notions and propositions, has at its

disposal sufficient means of expression to define the notions occurring in the argument above (in particular, the notion 'provable formula') and in which, second, every provable formula is true in the interpretation considered. The purpose of carrying out the above proof with full precision in what follows is, among other things, to replace the second of the assumptions just mentioned by a purely formal and much weaker one.[7]

On the other hand the term 'truth' occurred in Gödel's lectures on the incompleteness theorems at the Institute for Advanced Study in Princeton in the spring of 1934.[8] He discussed there, among other things, the relation between the existence of undecidable propositions and the possibility of defining the concept 'true (false) sentence' of a given language in the language itself. Considering the relation of his arguments to the paradoxes, in particular to the paradox of "The Liar", Gödel indicates that the paradox disappears when one notes that the notion 'false statement in a language B' cannot be expressed in B. Even more, the paradox can be considered as a proof that 'false statement in B' cannot be expressed in B. In the footnote 25 (added to the version published in Davis 1965) Gödel wrote:

> For a closer examination of this fact see A. Tarski's papers published in: *Trav. Soc. Sci. Lettr. de Varsovie*, Cl. III, No. 34, 1933 (Polish) (translated in: *Logic, Semantics, Metamathematics. Papers from 1923 to 1938 by A. Tarski*, see in particular p. 247 ff.) and in Philosophy and Phenom. Res. 4 (1944), p. 341–376. In these two papers the concept of truth relating to sentences of a language is discussed systematically. See also: R. Carnap, *Mon. Hefte f. Math. u. Phys.* 4 (1934), p. 263.

The reasons for the incompleteness results were also explicitly mentioned in Gödel's reply to a letter of A. W. Burks. This reply is quoted in von Neumann's *Theory of Self-Reproducing Automata* (1966), pp. 55–56. Gödel wrote:

> I think the theorem of mine which von Neumann refers to is not that on the existence of undecidable propositions or that on the length of proofs but rather the fact that a complete epistemological description of a language A cannot be given in the same language A, because the concept of truth of sentences of A cannot be defined in A. It is this theorem which is the true reason for the existence of the undecidable propositions in the formal systems containing arithmetic. I did not, however, formulate it explicitly in my paper of 1931 but only in my Princeton lectures of 1934. The same theorem was

---

7 „Die eben auseinandergesetzte Beweismethode läßt sich offenbar auf jedes formale System anwenden, das erstens inhaltlich gedeutet über genügend Ausdrucksmittel verfügt, um die in der obigen Überlegung vorkommenden Begriffe (insbesondere den Begriff 'beweisbare Formel') zu definieren, und in dem zweitens jede beweisbare Formel auch inhaltlich richtig ist. Die nun folgende exakte Durchführung des obigen Beweises wird unter anderem die Aufgabe haben, die zweite der eben angeführten Voraussetzungen durch eine rein formale und weit schwächere zu ersetzen."
8 Notes of Gödel's lectures taken by S. C. Kleene and J. B. Rosser were published in Davis's book in 1965 (cf. Gödel 1934).

proved by Tarski in his paper on the concept of truth published in 1933 in *Act. Soc. Sci. Lit. Vars.*, translated on pp. 152–278 of *Logic, Semantics and Metamathematics*.

What were the reasons of avoiding the concept of truth by Gödel? An answer can be found in a crossed-out passage of a draft of Gödel's reply to a letter of the student Yossef Balas (mentioned already above). Gödel wrote there:

> However in consequence of the philosophical prejudices of our times 1. nobody was looking for a relative consistency proof because [it] was considered axiomatic that a consistency proof must be finitary in order to make sense, 2. a concept of objective mathematical truth as opposed to demonstrability was viewed with greatest suspicion and widely rejected as meaningless.

Hence it leads us to the conclusion formulated by S. Feferman in (1984) in the following way:

> [...] Gödel feared that work assuming such a concept [i.e., the concept of mathematical truth – R. M.] would be rejected by the foundational establishment, dominated as it was by Hilbert's ideas. Thus he sought to extract results from it which would make perfectly good sense even to those who eschewed all non-finitary methods in mathematics.

On the other hand Gödel was aware of limitations of the formalists' approach. In a letter to Hao Wang (cf. Wang 1974, p. 9) he wrote:

> [...] formalists considered formal demonstrability to be an *analysis* of the concept of mathematical truth and, therefore were of course not in a position to *distinguish* the two.

Note that A. Tarski was free of such limitations. In fact in the Lvov-Warsaw School no restrictive initial preconditions were assumed before the proper investigation could start. The main demands were clarity, anti-speculativeness and scepticism towards many fundamental problems of traditional philosophy. The principal method that should be used was logical analysis. The Lvov-Warsaw School was not so radical in its criticism of metaphysics as the Vienna Circe (see, for example, Woleński 1989 and 1995a).

Tarski pointed out on many occasions that mathematical and logical research should not be restricted by any general philosophical views. In particular he wrote in (1930):

> In conclusion it should be noted that no particular philosophical standpoint regarding the foundations of mathematics is presupposed in the present work.

And in (1954) he wrote:

> As an essential contribution of the Polish school to the development of metamathematics one can regard the fact that from the very beginning it admitted into metamathematical research all fruitful methods, whether finitary or not.

Hence Tarski, though indicating his sympathies with nominalism, freely used in his logical and mathematical studies the abstract and general notions that a nominalist seeks to avoid.

3. Definition of Truth

Considering the problem of the priority of Tarski and Gödel in proving the undefinability of the notion of truth one should mention still another thing. In fact Tarski showed not only the undefinability but – and this is his main merit here – he gave the precise inductive definition of satisfiability and truth. In connection with this one should ask whether Gödel saw the necessity to give an analysis of the concept of truth (note that in his doctoral dissertation „Über die Vollständigkeit des Logikkalküls" (1929) and in his paper „Die Vollständigkeit der Axiome des logischen Funktionenkalküls" (1930a) on the completeness of the first-order predicate calculus the notion of the validity was understood in an informal way what was in fact a long tradition – cf. Löwenheim and Skolem). The answer is affirmative. Indeed, in a letter to R. Carnap of 11th September 1932 he wrote (quotation after Köhler 1991):

> On the basis of this idea I will give in the second part of my work a definition of [the concept] 'true' and I am of the opinion that one cannot do it in another way and that the higher calculus of induction cannot be grasped semantically [i.e., at that time – syntactically].[9]

Köhler explains in (1991) that 'II. Teil meiner Arbeit' means here the joint project of Gödel together with A. Heyting to write a survey of the current investigations in mathematical logic for Springer-Verlag (Berlin). Heyting wrote his part while the part by Gödel was never written (the reasons were his problems with the health). One can assume that Gödel planned to develop there a theory of truth based on the set theory.

4. Conclusions

The above considerations lead to the following conclusions. Tarski made clear his indebtedness to Gödel's methods in proving the theorem on the undefinability of truth, but on the other hand he strongly emphasized the fact that his results had been obtained independently. Gödel was aware of the formal undefinability of

---

9   „Ich werde auf Grund dieses Gedankens im II. Teil meiner Arbeit eine Definition für 'wahre' geben und ich bin der Meinung, daß sich die Sache anders nicht machen läßt und daß man den höheren Induktionenkalkül nicht semantisch [d.h. damals syntaktisch!] auffassen kann."

the notion of truth in 1931. In fact it was precisely his recognition of the contrast between the formal definability of provability and the formal undefinability of truth that led him to his discovery of incompleteness. Gödel did not mention the undefinability of truth in his writings, he even avoided the terms 'truth' and 'true', because he feared that work assuming such a concept would be rejected by foundational establishment dominated by Hilbert's ideas. Tarski was free of such limitations. In fact, in the Lvov-Warsaw School no restrictive initial preconditions were assumed before the proper investigations could start.

Troubles With (the Concept of) Truth in Mathematics

Several concepts of truth, several approaches to this concept have been proposed in the logic: coherence theory, correspondence theory, pragmatist theory, redundancy theory and semantic theory. The last one due to Tarski is probably the most influential and most widely accepted theory of truth – though not free of critiques. Tarski hoped that his definition will "catch hold of the actual meaning of an old notion" (Tarski 1944). Since according to him the "old" notion of truth is ambiguous and even doubtfully coherent, he restricted his concern to what he called the "classical Aristotelian conception of truth" as expressed in Aristotle's dictum:

> To say of what is that it is not, or of what is not that it is, is false, while to say of what is that it is, or of what is not that it is not, is true.

Tarski's theory falls into two parts: he provided, first, adequacy conditions, i.e. conditions which any acceptable definition of truth ought to fulfil; and then a definition of truth for a specified formal language.

The question of the philosophical significance of Tarski's theory of truth is a hard one. It has been criticized both for saying too little and for saying too much. For example Black wrote in (1948, p. 260):

> the neutrality of Tarski's definition with respect to the competing philosophical theories of truth is sufficient to demonstrate its lack of philosophical relevance.

On the other hand Mackie (1973, p. 40) said that

> The Tarskian theory [...] belongs to factual rather than conceptual analysis [...]. Tarski's theory has plenty of meat to it, whereas a correct conceptual analysis of truth has very little.

Tarski himself was modest about the epistemological pretensions of his theory. Though he was convinced that his concept of satisfaction and truth is a contribution to the philosophical problem of truth (cf. his famous paper 1933a), on the other hand he emphasized that his conception is philosophically neutral. In (1944) he wrote:

> we may accept the semantic conception of truth without giving up any epistemological attitude we may have had, we may remain naive realists or idealists, empiricists or metaphysicians. [...] The semantic conception is completely neutral toward all these issues.

Despite of these controversies it is the fact that just Tarski's theory of truth has been accepted in the foundations of mathematics. Hence we shall not discuss the philosophical problems connected with it but we shall indicate some other problems – of a metamathematical and foundational character.

* * *

Tarski provided in (1933a) a definition (in a non-formalized metasystem) of satisfaction and truth and, on the other hand, proved a theorem on the undefinability of the concept of truth for a formalized language L in L itself. It was stated as Theorem I (β) and said that:[1]

> Assuming that the class of all provable sentences of the metatheory is consistent, it is impossible to construct an adequate definition of truth in the sense of convention **T** on the basis of the metatheory.

It was followed by a description of the idea of the proof and then by a sketch of the proof. Note that the theorem was proved by diagonalization.

To fix our attention and to be more precise let us restrict ourselves to Peano arithmetic. This is a first-order theory formalized in the language L(PA) with the following nonlogical symbols: $0, S, +, \cdot$ and based on the following nonlogical axioms:

(A1) $S(x) = S(y) \rightarrow x = y$,
(A2) $\neg(0 = S(x))$,
(A3) $x + 0 = x$,
(A4) $x + S(y) = S(x + y)$,
(A5) $x \cdot 0 = 0$,
(A6) $x \cdot S(y) = x \cdot y + x$,
(A7) $\varphi(0) \land \forall x[\varphi(x) \rightarrow \varphi(S(x))] \longrightarrow \forall x \varphi(x)$,

where φ is any formula of the language L(PA).

Fix an arithmetization of the language L(PA) and denote by $\ulcorner \varphi \urcorner$ the Gödel number of a formula φ by the given arithmetization.[2] Let $\bar{n}$ be the term $\underbrace{S \ldots S(0)}_{n}$ denoting the natural number n.

The strong version of Tarski's theorem (i.e., the version without parameters) can be now formulated in the following way.

---

[1] Cf. Tarski (1965, p. 247).
[2] Detailed information on the arithmetization and on the arithmetical counterparts of various metamathematical notions can be found, e.g., in Mendelson (1964), Shoenfield (1967) or Murawski (1999b).

Theorem 1 (Tarski 1933a). *If Peano arithmetic* PA *is consistent then there exists no formula* **St**(x) *of the language* L(PA) *being the definition of truth for formulas of* L(PA), *i.e., such a formula* **St**(x) *that for any sentence* ψ *of the language* L(PA)

$$\text{PA} \vdash \psi \equiv \textbf{St}(\ulcorner\psi\urcorner).$$

Let $\mathfrak{N}_0$ be the standard interpretation of the language of Peano arithmetic, i.e., $\mathfrak{N}_0 = \langle \mathbb{N}, 0, S, +, \cdot \rangle$ where $\mathbb{N}$ is the set of natural numbers, 0 is the number zero, S is the successor function and $+$ and $\cdot$ are addition and multiplication of natural numbers, resp. The structure $\mathfrak{N}_0$ is called the standard model of PA. Tarski's theorem states that there exists no formula **St** of the language L(PA) such that for any sentence ψ of L(PA), PA $\vdash \psi \equiv \textbf{St}(\ulcorner\psi\urcorner)$, hence in particular there exists no formula **St** such that for any sentence ψ of L(PA), $\mathfrak{N}_0 \models \psi$ if and only if $\mathfrak{N}_0 \models \textbf{St}(\ulcorner\psi\urcorner)$, i.e., there is no definition (in the language of L(PA)) of the set of (Gödel numbers of) those sentences of L(PA) which are true in the domain of natural numbers (= in the standard model $\mathfrak{N}_0$). Consequently the notion of truth for arithmetic of natural numbers, i.e., the set

$$\{\ulcorner\varphi\urcorner : \varphi \text{ is a sentence of L(PA) \& } \mathfrak{N}_0 \models \varphi\}$$

is not an arithmetical set. This contrasts with the fact that the notion of provability for arithmetic, i.e., the set

$$\{\ulcorner\varphi\urcorner : \varphi \text{ is a sentence of L(PA) \& PA} \vdash \varphi\}$$

is an arithmetical set, in fact it is recursively enumerable. This indicates the gap between provability and truth. On the other hand one can show that the notion of truth for arithmetic is hyperarithmetical, i.e., it belongs to the class Δ_1^1.[3]

Tarski's theorem can be easily generalized to theories extending Peano arithmetic PA. In fact the following theorem holds.

Theorem 2. *Let* T *be any consistent first-order theory extending Peano arithmetic* PA *and let* $\mathfrak{M}$ *be any model of* T. *Then the set* $Th(\mathfrak{M}) = \{\ulcorner\psi\urcorner : \mathfrak{M} \models \psi\}$, *i.e., the set of Gödel numbers of all sentences true in* $\mathfrak{M}$, *is not definable in* $\mathfrak{M}$.

Note also that in the above theorems only the notion of truth, i.e., of satisfaction of sentences, was considered. One can generalize them of course to the case of satisfaction of formulas with free variables. Let T be an extension of PA (the language L(T) can also be an extension of the language L(PA)).

Definition 3. *A binary predicate* **S** *of the language of* L(T) *is said to be a satisfaction predicate for the theory* PA *in the sense* (A) *if and only if for for every*

---

[3] Cf. Mostowski (1949–1950). For information on the hyperarithmetical hierarchy see Rogers (1967) or Shoenfield (1967).

formula φ of L(PA) *all free variables of which occur among variables* $x_1,\ldots,x_n$ *and any natural numbers* $k_1,\ldots,k_n$:

$$T \vdash \varphi(\overline{k_1},\ldots,\overline{k_n}) \equiv S(\overline{\ulcorner\varphi\urcorner},\overline{\langle k_1,\ldots,k_n\rangle}).$$

Definition 4. *A binary predicate* **S** *of the language of* L(T) *is said to be a satisfaction predicate for the theory* PA *in the sense* (B) *if and only if for every formula* φ *of* L(PA) *all free variables of which occur among variables* $x_1,\ldots,x_n$:

$$T \vdash \forall x\{\mathbf{Seq}(x) \wedge \mathbf{lh}(x) = \overline{n} \rightarrow [\varphi((x)_1,\ldots,(x)_n) \equiv \mathbf{S}(\overline{\ulcorner\varphi\urcorner},x)]\}.$$

Definition 5. *A binary predicate* **S** *of the language of* L(T) *is said to be a satisfaction predicate for the theory* PA *in the sense* (C) *if and only if the following formulas are provable in* T:

$$\mathbf{S}(u,v) \rightarrow \mathbf{Form}(u) \wedge \mathbf{Seq}(v) \wedge \mathbf{lh}(v) = \mathbf{F}(u),$$

$$\mathbf{Term}(t_1) \wedge \mathbf{Term}(t_2) \wedge u = \overline{\langle SN(=),t_1,t_2\rangle} \rightarrow$$
$$\rightarrow [\mathbf{S}(u,v) \equiv \mathbf{val}(t_1,v|\mathbf{F}(t_1)) = \mathbf{val}(t_2,v|\mathbf{F}(t_2))].$$

$$u = \overline{\langle SN(\neg),u_1\rangle} \wedge \mathbf{Form}(u_1) \rightarrow [\mathbf{S}(u,v) \equiv \neg\mathbf{S}(u_1,v)],$$

$$u = \overline{\langle SN(\vee),u_1,u_2\rangle} \wedge \mathbf{Form}(u_1) \wedge \mathbf{Form}(u_2) \rightarrow$$
$$\rightarrow [\mathbf{S}(u,v) \equiv \mathbf{S}(u_1,v|\mathbf{F}(u_1)) \vee \mathbf{S}(u_2,v|\mathbf{F}(u_2))],$$

$$u = \overline{\langle SN(\exists),\ulcorner x_k\urcorner,u_1\rangle} \wedge \mathbf{Form}(u_1) \wedge \neg\mathbf{Fr}(u_1,2k) \rightarrow$$
$$\rightarrow [\mathbf{S}(u,v) \equiv \mathbf{S}(u_1,v)],$$

$$u = \overline{\langle SN(\exists),\ulcorner x_k\urcorner,u_1\rangle} \wedge \mathbf{Form}(u_1) \wedge \mathbf{Fr}(u_1,2k) \rightarrow$$
$$\rightarrow \left[\mathbf{S}(u,v) \equiv \exists x \mathbf{S}\left(u_1,v * \binom{k}{x}\right)\right]$$

where $v * \binom{k}{x}$ *denotes a sequence number* w *such that*

$$lh(w) = max(lh(v),k),$$
$$\forall i < lh(v)[i \neq k \rightarrow (w)_i = (v)_i],$$
$$(w)_k = x,$$
$$\forall i\, [lh(v) < i < k \rightarrow (w)_i = 0].$$

Add that we adopt here the following convention: if R is a recursive relation then by **R** we denote a formula of the language L(PA) strongly representing R in PA.

Note that Tarski considered in (1933a) the notion of a satisfaction predicate in sense (A). Observe also that if **S** is a satisfaction predicate in the sense (C) then it is a satisfaction predicate in the sense (B) (this follows by induction) and if **S** is a satisfaction predicate in the sense (B) then it is a satisfaction predicate in the sense (A) (this is obvious from the definitions). So Tarski's theorem on undefinability of truth implies that there is no satisfaction predicate for PA in the sense (A) definable in PA. Hence there are no satisfaction predicates in the sense (B) or (C) for PA definable in PA.

A connection between the notion of satisfaction and the notion of consistency is indicated by the following theorem.[4]

Theorem 6. *Let* T *be an extension of Peano arithmetic* PA *such that induction (with respect to all formulas of the language* L(T)*) holds in* T. *If* **S** *is a satisfaction predicate for* PA *in the theory* T *in the sense* (C) *then* T *proves the consistency of* PA, *i.e.,* T ⊢ Con$_{PA}$, *where* Con$_{PA}$ *denotes the formula* $\neg \mathbf{Pr}(\ulcorner 0 = \bar{1} \urcorner)$.

Note that the last theorem does not hold for **S** being a satisfaction predicate in the sense (A) or (B).

As mentioned above Tarski used in his undefinability theorem Gödel's method of diagonalization. From a historico-philosophical point of view it should be noted that Tarski made clear his indebtedness to Gödel's methods but on the other hand he strongly emphasized the fact that his results had been obtained independently. Gödel was aware of the formal undefinability of the notion of truth in 1931. In fact it was precisely his recognition of the contrast between the formal definability of provability and the formal undefinability of truth that led him to his discovery of incompleteness. Gödel did not mention the undefinability of truth in his writings, he even avoided the terms "truth" and "true", because he feared that work assuming such a concept would be rejected by foundational establishment dominated by Hilbert's ideas. Tarski was free of such limitations. In fact, in the Lvov-Warsaw School no restrictive initial preconditions were assumed before the proper investigations could start. Note also that Gödel had no precise definition of the concept of truth.[5]

Having shown that the notion of truth for Peano arithmetic PA cannot be defined in PA itself one should ask *where* it can be defined. We have here two possibilities: (1) one can consider an appropriate extension of PA (possibly weak) in which the notion can be defined and (2) one can extend the language L(PA) by adding a new binary predicate **S** (called satisfaction class) and characterizing it axiomatically by adding to Peano arithmetic PA (as new axioms) sentences given above in the definition of a satisfaction predicate in the sense (C). Note that since those axioms form a finite set of axioms one can write them as a single

---

[4] The proof of this theorem can be found, e.g., in Murawski's (1999b).
[5] More information on this problem can be found in Woleński (1991) and Murawski (1998).

formula of the language L(PA) ∪ **S** (denote it as "**S** is a satisfaction class"). Let us consider both those possibilities.

A natural extension of PA which can be considered in our context is the so-called second-order arithmetic A_2^-. This is a first-order (!) system formalized in a language with two sorts of variables: number variables $x, y, z, \ldots$ and set variables $X, Y, Z, \ldots$ Its nonlogical constants are those of Peano arithmetic, i.e., 0, $S, +, \cdot$ as well as symbols for all primitive recursive functions and the membership relation $\in$. Nonlogical axioms of A_2^- are the following:

(1) axioms of PA without the axiom scheme of induction,
(2) (extensionality) $\forall x (x \in X \equiv x \in Y) \to X = Y$,
(3) (induction axiom)

$$0 \in X \land \forall x (x \in X \to Sx \in X) \to \forall x (x \in X),$$

(4) recursive definitional equations for primitive recursive functions,
(5) (axiom scheme of comprehension)

$$\exists X \forall x [x \in X \equiv \varphi(x, \ldots)],$$

where φ is any formula of the language of A_2^- (possibly with free number- or set-variables) in which X does not occur free.

If Γ is a class of formulas of the language $L(A_2^-)$ then we denote by $A_2^- | \Gamma$ the subsystem of A_2^- obtained by restricting the comprehension axiom to formulas belonging to the class Γ. Later we shall consider in particular the system $A_2^- | \Sigma_1^1$ where Σ_1^1 is the class of formulas of the form $\exists X \varphi(X, \ldots)$ where φ is an arithmetical formula, i.e., a formula containing possibly any quantifiers bounding number-variables and no quantifier over set-variables. One can proved the following theorems:[6]

Theorem 7. *Second order arithmetic A_2^- proves the existence of the satisfaction predicate in the sense (C) for Peano arithmetic. Moreover, this can be proved in the fragment $A_2^- | \Sigma_1^1$ of A_2^-.*

Using this theorem and Theorem 6 we obtain

Theorem 8. $A_2^- | \Sigma_1^1 \vdash \text{Con}_{\text{PA}}$.

In this way we showed that the notion of truth (in fact the notion of satisfaction in the sense (C)) for L(PA) can be defined in the theory $A_2^- | \Sigma_1^1$.

It turns out that, in contrast with Tarski's theorem, the notion of satisfaction and truth for certain fragments of the language L(PA) can be defined in Peano

---
[6] Proofs of those theorems can be found in Murawski (1999a, 1999b).

arithmetic itself. To formulate precisely appropriate results a hierarchy of formulas of the language L(PA) similar to the arithmetical hierarchy of relations is needed. Let $\Sigma_0^0 = \Pi_0^0 = \Delta_0^0$ be the smallest class of formulas of the language L(PA) containing atomic formulas and closed under connectives and bounded quantifiers. We define Σ_{n+1}^0 to be the set of all formulas equivalent (in PA) to formulas of the form $\exists x \psi$ for $\psi \in \Pi_n^0$ and Π_{n+1}^0 to be the set of all formulas equivalent (in PA) to formulas of the form $\forall x \psi$ for $\psi \in \Sigma_n^0$. We put also Δ_n^0 to be the set of all formulas equivalent (in PA) to a Σ_n^0 formula and to a Π_n^0 formula.

One can show that there exist formulas $Sat_{\Delta_n^0}$, $Sat_{\Sigma_n^0}$ and $Sat_{\Pi_n^0}$ of L(PA) which are definitions of satisfaction for, resp., Δ_n^0, Σ_n^0 and Π_n^0 formulas ($n \in \mathbb{N}$). Moreover, the formula $Sat_{\Delta_n^0}$ can be written as both Σ_1^0 and Π_1^0 formula. Hence one can say that there exists a Δ_1^0 definition of satisfaction for Δ_0^0 formulas of L(PA). Consequently the formulas $Sat_{\Sigma_n^0}$ and $Sat_{\Pi_n^0}$ are, resp., Σ_n^0 and Π_n^0 definitions of satisfaction for Σ_n^0 and Π_n^0 formulas of L(PA) ($n \in \mathbb{N}$). One can also show that the appropriate properties of those formulas (corresponding to the metamathematical properties of the appropriate notions of satisfaction) can be proved in Peano arithmetic PA.[7] Let further $Tr_{\Sigma_n^0}$ and $Tr_{\Pi_n^0}$ denote truth predicates for Σ_n^0 and Π_n^0 sentences.[8] In the sequel we shall identify formulas defining satisfaction and truth and their extensions in the standard model $\mathfrak{N}_0$.

It turns out that the (partial) truth (in the standard model $\mathfrak{N}_0$), i.e., truth for Σ_n^0 formulas can be approximated by iterations of the so-called ω-rule. Hence one can say that the (infinitary) ω-rule enables us to express, to reach the partial truth. To be more precise let us introduce the following hierarchies. Let T be any first-order theory in the language L(PA) of Peano arithmetic. The first hierarchy is defined as follows:

$T^0 = T$,

$T^{\alpha + \frac{1}{2}} = T^\alpha \cup \{\varphi : \varphi \text{ is of the form } \forall x \psi(x) \text{ and } \psi(\bar{n}) \in T^\alpha \text{ for every } n \in \mathbb{N}\}$,

$T^{\alpha+1} =$ the smallest set of formulas containing $T^{\alpha + \frac{1}{2}}$ and closed under the rules of inference of PA,

$T^\lambda = \bigcup_{\alpha < \lambda} T^\alpha$ for λ limit.

The second hierarchy is defined so (cf. Niebergall 1996):

$T^{(0)} = T$,

$T^{(\alpha + \frac{1}{2})} = T^{(\alpha)} \cup \{\varphi : \varphi \text{ is of the form } \forall x \psi(x) \text{ and } \psi(x) \in \Sigma_{2\alpha+1}^0$
and $\psi(\bar{n}) \in T^{(\alpha)}$ for every $n \in \mathbb{N}\}$,

---

7   In fact it can be proved even in the fragment of Peano arithmetic with induction for Σ_1^0 formulas only. Cf. Kaye (1991), Murawski (1999b) and Hájek–Pudlák (1993).
8   Construction of $Sat_{\Sigma_n^0}$ and $Sat_{\Pi_n^0}$ can be found in Kaye (1991) and Murawski (1999b).

$T^{(\alpha+1)}$ = the smallest set of formulas containing $T^{(\alpha+\frac{1}{2})}$ and closed under the rules of inference of PA,

$T^{(\lambda)} = \bigcup_{\alpha<\lambda} T^{(\alpha)}$ for λ limit.

Hence the ω-rule is now applied at stage n to Σ^0_{2n+1} formulas only.
The last hierarchy is the following one (cf. Niebergall 1996):

$(\Sigma^k T)^0 = T,$

$(\Sigma^k T)^{\alpha+\frac{1}{2}} = (\Sigma^k T)^\alpha \cup \{\varphi : \varphi$ is of the form $\forall x \psi(x)$ and $\psi(x) \in \Sigma^0_k$
and $\psi(\bar{n}) \in (\Sigma^k T)^\alpha$ for every $n \in \mathbb{N}\},$

$(\Sigma^k T)^{\alpha+1}$ = the smallest set of formulas containing $(\Sigma^k T)^{\alpha+\frac{1}{2}}$
and closed under the rules of inference of PA,

$(\Sigma^k T)^\lambda = \bigcup_{\alpha<\lambda} (\Sigma^k T)^\alpha$ for λ limit.

We still need one notion – reflection principle. So let T be a theory whose set of (Gödel numbers of) theorems is strongly representable in PA. Denote by $RFN(T)$ the uniform reflection principle for T, i.e., the scheme

$$\forall x Pr_T(\ulcorner \varphi(x) \urcorner) \to \forall x \varphi(x)$$

for $\varphi(x)$ formula of L(T) with at most one free variable. If one restricts the class of formulas to a class Γ (for example Σ^0_k or Π^0_k) then one obtains $RFN_\Gamma(T)$.
The local reflection principle for T denoted by $Rfn(T)$ is the following scheme

$$Pr_T(\ulcorner \varphi \urcorner) \to \varphi$$

for φ closed.
We have now the following facts:[9]

Theorem 9. (1) $PA^n \supseteq PA + Tr_{\Sigma^0_{2n+1}}$, *i.e., for every* $n \in \mathbb{N}$ *the theory* PA^n *is complete with respect to* Σ^0_{2n+1} *sentences.*
(2) (Niebergall 1996) *For any* $n \in \mathbb{N}$, $PA^{(n)} = PA + Tr_{\Sigma^0_{2n+1}}$.
(3) (Niebergall 1996) *For any* $n \in \mathbb{N}$, $(\Sigma^k PA)^{n+1} = PA + Tr_{\Sigma^0_{k+2}}$ *if* $k \leq 2n$.
(4) (Niebergall 1996) *For any* $n \in \mathbb{N}$, $PA^{n+1} = PA + Tr_{\Sigma^0_{2n+3}} + RFN(PA^n)$.
(5) (Niebergall 1996) *For any* $n \in \mathbb{N}$, $(PA + Tr_{\Sigma^0_k})^n = PA^n + Tr_{\Sigma^0_{k+2n}}$.

---

9   They are only examples of theorems that should indicate the character of results.

(6) (Feferman 1962) *For a suitable class of ordinals:* (a) *iterating* $T \to T + Con_T$ *or* $T \to T + Rfn$ (T) *one has* $\bigcup PA_\alpha = PA + Tr_{\Pi_1^0}$; (b) *iterating* $T \to T + RFN(T)$ *one obtains* $\bigcup PA_\alpha = Th(\mathfrak{N}_0) =$ *all true sentences of arithmetic.*

Turn now to the second possibility indicated above, i.e., to the axiomatic characterization of satisfaction and truth. Recall that one extends now the language L(PA) by adding a new binary predicate **S** (called a satisfaction class; denote the new language by L_S) and characterizing it axiomatically by adding to Peano arithmetic PA (as new axioms) sentences given above in the definition of a satisfaction predicate in the sense (C). Note that since those axioms form a finite set of axioms one can write them as a single formula of the language L(PA) ∪ **S** (denote it as "**S** is a satisfaction class"). One can add certain additional axioms stating that **S** has special properties. Two such properties are significant: being full and being inductive. A satisfaction class **S** is said to be full if and only if it decides every formula on any valuation. And **S** is said to be inductive if and only if the induction principle holds for all formulas of the extended language L_S. If Γ is a class of formulas of L_S and one requires that the induction principle holds for all formulas of Γ only then **S** is called Γ-inductive. Denote by $\Gamma - PA(S)$ the theory PA + "**S** is a full Γ-inductive satisfaction class" and by PA(**S**) the theory PA + "**S** is a full inductive satisfaction class".

There arises a question whether theories of the type $\Gamma - PA(S)$ or the theory PA(**S**) are consistent, i.e., whether they have models. Note that if $\langle \mathfrak{M}, S \rangle$ is a model of such an extension of PA then $\mathfrak{M}$ is a model of PA and S is a satisfaction predicate for L(PA) over the model $\mathfrak{M}$ (S is called a satisfaction class over the model $\mathfrak{M}$). It turns out that not over every model $\mathfrak{M}$ of PA one can define a predicate S such that the structure $\langle \mathfrak{M}, S \rangle$ is a model of PA + "**S** is a satisfaction class", i.e., not for every model of PA the notion of satisfaction (truth) (satisfying the natural Tarski's conditions) exists. The crucial property of a model $\mathfrak{M}$ needed here is recursive saturation defined as follows:

Definition 10. *A model* $\mathfrak{M} \models PA$ *is said to be recursively saturated if and only if for every recursive type* Θ *over the model* $\mathfrak{M}$, *if* Θ *is consistent over* $\mathfrak{M}$ *then* Θ *is realized in* $\mathfrak{M}$.

In fact the following theorem holds:

Theorem 11. *For any countable model* $\mathfrak{M}$ *of PA the following conditions are equivalent:*
  (a) $\mathfrak{M}$ *is recursively saturated,*
  (b) $\mathfrak{M}$ *has a satisfaction class,*
  (c) $\mathfrak{M}$ *has a full satisfaction class,*
  (d) $\mathfrak{M}$ *has an inductive satisfaction class.*

From this theorem it follows also that the theories:

PA + "**S** is a satisfaction class",
PA + "**S** is a full satisfaction class",
PA + "**S** is an inductive satisfaction class"

are all conservative extensions of PA, i.e., one can prove in those theories exactly the same theorems about natural numbers (i.e., formulas of the language L(PA)) as in Peano arithmetic PA. Hence the addition of a new notion, i.e., of a notion of a satisfaction (truth), with properties indicated above does not increase the proof-theoretical power of a theory with respect to sentences of the language L(PA). On the other hand the assumption that a satisfaction class is full and Δ_0^0-inductive gives a nonconservative extension of PA! In fact one can prove in this theory, i.e., in Δ_0^0-PA(**S**) the consistency of PA.

This leads us to the problem when does there exist a model of a theory of the type $\Gamma - $PA(**S**) for Γ such that $\Delta_0^0 \subseteq \Gamma$, i.e., when for a model $\mathfrak{M}$ of PA does there exist a full Γ-inductive satisfaction class over $\mathfrak{M}$? The answer is: the model $\mathfrak{M}$ must be recursively saturated and must satisfy certain extension of Peano arithmetic. Those extensions can be characterized in the language of consistency of appropriate ω-logics or of appropriate transfinite induction.

Consider the following sequence of formulas of the language L(PA) (one uses here arithmetization):

$\Gamma_0(\varphi) = $ "PA $\vdash \varphi$",
$\Gamma_{n+\frac{1}{2}}(\varphi) = $ "φ *is of the form* $\eta \vee \forall z\ \psi(z)$ *and* $\forall z\ \Gamma_n(\eta \vee \psi(S^z 0))$",
$\Gamma_{n+1}(\varphi) = $ "*there exists a proof of the formula* φ
    based on PA $\cup \{\psi : \Gamma_{n+\frac{1}{2}}(\psi)\}$".

Observe that in this system of ω-logic only the application of the ω-rule increases the degree of complexity of a proof.

Theorem 12 (Kotlarski 1986). *Let $\mathfrak{M}$ be a countable recursively saturated model of PA. Then there exists a full Δ_0^0-inductive satisfaction class over $\mathfrak{M}$ if and only if for any $n \in \mathbb{N}$: $\mathfrak{M} \models \neg \Gamma_n(0 = 1)$.*

It can also be proved (cf. Kotlarski 1986) that the theory $\Delta_0^0 - $PA(**S**) is equal to the theory

PA $+ $ **S** is a full satisfaction class $+ \forall \varphi[($PA $\vdash \varphi) \rightarrow $ **S**$(\varphi)]$.

The last sentence can be read as: "**S** makes all theorems of PA true". It is equivalent to the Δ_0^0-inductiveness of the satisfaction class **S**.

The system of ω-logic described above can be iterated in the transfinite. So let us fix a "natural" system of notations for ordinals $< \varepsilon_0$ (one gets it by Cantor's

Normal Form Theorem). By transfinite induction on $\alpha < \varepsilon_0$ we define theories T^α and formulas Γ_n^α in the following way:

$$T^0 = \text{PA},$$
$$\Gamma_0^0(\varphi) = \text{``PA} \vdash \varphi\text{''},$$
$$\Gamma_0^\alpha(\varphi) = \text{``}T^\alpha \vdash \varphi\text{''},$$
$$\Gamma_{n+\frac{1}{2}}^\alpha(\varphi) = \text{``}\varphi \text{ is of the form } \eta \vee \forall z\, \psi(z) \text{ and } \forall z\, \Gamma_n^\alpha(\eta \vee \psi(z))\text{''},$$
$$\Gamma_{n+1}^\alpha(\varphi) = \text{``}T^\alpha \cup \Gamma_{n+\frac{1}{2}}^\alpha \vdash \varphi\text{''},$$
$$T^{\alpha+1} = T^\alpha \cup \{\neg \Gamma_n^\alpha(0=1) : n \in \mathbb{N}\},$$
$$T^\lambda = \bigcup_{\alpha<\lambda} T^\alpha,\ \lambda \text{ limit.}$$

Using Recursion Theorem one can formalize those definitions in PA. Define now for an ordinal α a sequence $\omega_m(\alpha)$ in the following way: $\omega_0(\alpha) = \alpha$, $\omega_{m+1}(\alpha) = \omega^{\omega_m(\alpha)}$. The following theorem holds.

Theorem 13 (Kotlarski and Ratajczyk 1990a). (1) *Let m be a natural number and let $\mathfrak{M} \models$ PA be countable and recursively saturated. Then there exists a full Σ_m^0-inductive satisfaction class over $\mathfrak{M}$ if and only if for every $k \in \mathbb{N}$: $\mathfrak{M} \models \neg\Gamma_k^{\omega_m(k)}(0=1)$.*

(2) *Let $\mathfrak{M}$ be a countable and recursively saturated model of PA. Then there exists a full inductive satisfaction class over $\mathfrak{M}$ if and only if for every $n \in \mathbb{N}$:*

$$\mathfrak{M} \models \neg\Gamma_n^{\omega_n}(0=1)$$

where $\omega_n = \omega_n(\omega)$.

Let now $TI(\rho)$, where ρ is an ordinal, denote the scheme of transfinite induction up to ρ. Then the following theorem holds.

Theorem 14 (Kotlarski and Ratajczyk 1990b). *Let $\mathfrak{M}$ be a countable and recursively saturated model of PA and let m be a natural number. Then*

(1) *there exists a full Σ_m^0-inductive satisfaction class over the model $\mathfrak{M}$ if and only if for every $k \in \mathbb{N}$, $\mathfrak{M}$ satisfies the transfinite induction up to $\varepsilon_{\omega_m(k)}$, i.e., $\mathfrak{M} \models TI(\varepsilon_{\omega_m(k)})$,*

(2) *there exists a full inductive satisfaction class over the model $\mathfrak{M}$ if and only if for every $k \in \mathbb{N}$, $\mathfrak{M}$ satisfies the transfinite induction up to ε_{ω_k}, i.e., $\mathfrak{M} \models TI(\varepsilon_{\omega_k})$.*

The above theorems show that not always a full Γ-inductive satisfaction class does exist. In fact a given model of PA must satisfy additional conditions. Those conditions indicate connections between satisfaction (truth) on the one hand and transfinite induction and consistency of certain ω-logics on the other.

They do this also in another way. Let T be an extension of Peano arithmetic PA. Define a theory PA^T in the following way:

$$\text{PA}^T = \{\varphi \in L(\text{PA}) : T \vdash \varphi\}.$$

Hence theorems of PA^T are those sentences of the language L(PA) of Peano arithmetic (hence sentences about natural numbers) which can be proved in the stronger theory T.

Let now T be a theory of the type of PA(S) or its fragment. How do theories PA^T look like? The answer is provided by the following theorem.

Theorem 15. (i) (Kotlarski 1986) $\text{PA}^{\Delta_0^0 - \text{PA}(S)} = \text{PA} \cup \{\neg \Gamma_n(0=1) : n \in \mathbb{N}\}$.
(ii) (Kotlarski and Ratajczyk 1990a) *Let m be a natural number. Then*

$$\text{PA}^{\Sigma_m^0 - \text{PA}(S)} = \text{PA} \cup \{\neg \Gamma_k^{\omega_m(k)}(0=1) : k \in \mathbb{N}\}.$$
$$\text{PA}^{\text{PA}(S)} = \text{PA} \cup \{\neg \Gamma_n^{\omega_n}(0=1) : n \in \mathbb{N}\}$$

where $\omega_n = \omega_n(\omega)$.
(iii) (Kotlarski and Ratajczyk 1990b) *Let m be a natural number. Then*

$$\text{PA}^{\Sigma_m^0 - \text{PA}(S)} = \text{PA} \cup \{TI(\varepsilon_{\omega_m(k)}) : k \in \mathbb{N}\}.$$
$$\text{PA}^{\text{PA}(S)} = \text{PA} \cup \{TI(\varepsilon_{\omega_k}) : k \in \mathbb{N}\}.$$

This theorem shows that what can be proved about natural numbers using Peano axioms and the notion of satisfaction (truth) that is assumed to be full and Σ_m^0-inductive is exactly the same as what can be proved in PA plus transfinite induction for ordinals $\varepsilon_{\omega_m(k)}$ (for all $k \in \mathbb{N}$) or in PA plus appropriate consistency statements. Similarly for PA plus full inductive satisfaction (truth) on the one hand and PA plus transfinite induction for ordinals ε_{ω_k} (for all $k \in \mathbb{N}$) or PA plus appropriate consistency statements on the other. It shows also that by adding to PA the notion of a satisfaction (truth) and assuming that it is full and makes all theorems of PA true one obtains a theory with exactly the same theorems about natural numbers as by taking PA augmented with a concept of a full and Δ_0^0-inductive satisfaction (truth) or PA plus appropriate consistency statements. So (the usage of) satisfaction (truth) can be in a certain sense approximated by transfinite induction or by adding certain consistency statements concerning appropriate systems of ω-logic. Recall also that if T is PA + "S is a full satisfaction class" or PA + "S is an inductive satisfaction class" then PA^T = PA. Hence only the assumption that satisfaction (truth) is inductive and full gives new information about natural numbers.

Next problem is the problem of uniqueness: so assume that over a given model $\mathfrak{M}$ there exists a satisfaction class. Is it determined uniquely, i.e., does $\mathfrak{M}$ admit exactly one satisfaction class or do there exist a variety of them over $\mathfrak{M}$? In other words: if a theory in the language L_S extending PA has a model then can it posses also other models with the same fixed part corresponding to the language L(PA)? The answer is given by the following theorems.

Theorem 16 (Krajewski 1976). *For any countable model $\mathfrak{M}$ of PA which admits a full satisfaction class S there exists a countable model $\mathfrak{M}_1$ such that* (1) $\mathfrak{M}_1 \equiv \mathfrak{M}$ *and* (2) *there exist $2^{\aleph_0}$ full satisfaction classes over the model $\mathfrak{M}_1$ which are mutually inconsistent on sentences and* $(\mathfrak{M}, S) \equiv (\mathfrak{M}_1, S_\alpha)$ *for* $\alpha < 2^{\aleph_0}$.

Theorem 17 (Kossak 1985). *If there exists a full inductive satisfaction class over a countable model $\mathfrak{M}$ then*

(1) *there exist $2^{\aleph_0}$ full inductive satisfaction classes over $\mathfrak{M}$ which are pairwise elementarily inequivalent, i.e., such that*

$$(\mathfrak{M}, S_{\alpha_1}) \not\equiv (\mathfrak{M}, S_{\alpha_2})$$

for $\alpha_1 < \alpha_2 < 2^{\aleph_0}$,

(2) *there exist $2^{\aleph_0}$ full inductive satisfaction classes over $\mathfrak{M}$ which are elementarily equivalent but pairwise nonisomorphic.*

Explain that two satisfaction classes over a given model $\mathfrak{M}$ are said to be mutually inconsistent on sentences if and only if there exists a (nonstandard) sentence φ (i.e., a formula without free variables) of the language Form($\mathfrak{M}$) such that $S_1(\varphi, \emptyset)$ and $S_2(\neg\varphi, \emptyset)$ or *vice versa*. Hence one of satisfaction classes over $\mathfrak{M}$ (i.e., one of the notions of satisfaction for the language Form($\mathfrak{M}$)) says that the sentence φ is true and the other says that φ is false! In general, satisfaction classes S_1 and S_2 over $\mathfrak{M}$ are said to be mutually inconsistent if and only if there exists a formula φ in the sense of the model $\mathfrak{M}$ and an M-valuation a for the formula φ such that $S_1(\varphi, a)$ and $S_2(\neg\varphi, a)$ or *vice versa*. Hence one of satisfaction classes says that the formula φ is satisfied on the valuation a and the other one says that φ is not satisfied on a!

Add that two models are said to be elementarily equivalent if and only if they satisfy exactly the same sentences (i.e., formulas without free variables).

The above theorems show that the axiomatic characterization of satisfaction and truth is non-unique. The reason is that Tarski's conditions put on satisfaction classes are too weak and do not uniquely determine the satisfaction and truth. What more, they admit various interpretations, even mutually inconsistent on sentences! Hence the classical principle of bivalency is not any longer valued for nonstandard languages. Moreover, one can find mutually inconsistent satisfaction classes being elementarily equivalent, i.e., having the same elementary properties in the language L(PA) with predicate **S**.

* * *

Let us turn to conclusions. As Gaifman (2004, p. 15) wrote:

> Intended interpretations are closely related to realistic conceptions of mathematical theories. By subscribing to the standard model of natural numbers, we are committing ourselves to the objective truth or falsity of number-theoretic statements, where these are usually taken as statements of first-order arithmetic. The standard model is supposed to provide truth-values for these statements.

Deductive systems can only yield recursively enumerable sets of theorems and therefore they can only partially capture truth in the standard model. Even more, the truth in the standard model is not arithmetically definable.

On the other hand there are nonstandard (hence unintended) models (not only for Peano arithmetic but even for the theory of the standard model $\mathfrak{N}_0$). This shows an essential shortcoming of a formalized approach: the failure to fully determine the intended model.

An attempt to define arithmetical truth (truth for arithmetic) in a higher order theory, for example in the second-order arithmetic or its appropriate fragment where its existence can be proved, does not give a satisfactory solution. Indeed second-order arithmetic as a deductive system is incomplete and, additionally, there appears the problem of nonstandard models and interpretations.

So we are forced to attempt to characterize the concept of truth (for PA or for other theories) in an axiomatic way. But here again we encounter the phenomenon of nonstandardness. In fact, considering a nonstandard[10] model $\langle \mathfrak{M}, S \rangle$ for the theory $\Gamma - \mathrm{PA}(\mathbf{S})$ or its fragment we have that $\mathfrak{M}$ is a nonstandard model of PA and S is the appropriate satisfaction class over $\mathfrak{M}$, hence the satisfaction class for formulas of the language $\mathrm{Form}(\mathfrak{M})$ consisting of all those elements of the universe M (standard and nonstandard numbers) that (from the point of view of $\mathfrak{M}$) are (i.e., behave like) formulas (identified here with their Gödel numbers). Among them there are also nonstandard formulas, i.e., objects that formally behave like formulas but have no proper metamathematical meaning (they are formulas from the point of view of the world of $\mathfrak{M}$, but not from the point of view of the real metamathematical world). Of course $\mathrm{L}(\mathrm{PA}) \subseteq \mathrm{Form}(\mathcal{M})$ and

$$S_{tr} = \{(\ulcorner \varphi \urcorner, a) : \varphi \text{ standard formula of } \mathrm{L}(\mathrm{PA}),$$
$$\text{a M-}valuation\ for\ \varphi,\ \mathfrak{M} \models \varphi[a]\} \subseteq S.$$

But this "real" satisfaction S_{tr} (and consequently also "real" truth) cannot be arithmetically defined in ("cut" from) the satisfaction class S. Indeed, the notion of being standard is not arithmetically definable.

---

10 It is impossible to exclude nonstandard models and to restrict ourselves to the standard one only since the latter cannot be characterized arithmetically (in an axiomatic way).

Theories of the type $\Gamma - \mathrm{PA}(\mathbf{S})$ have a rich variety of models. But on the other hand not every model $\mathfrak{M}$ of PA can be extended to a model $\langle \mathfrak{M}, S \rangle$ of $\Gamma - \mathrm{PA}(\mathbf{S})$ – indeed, the structure $\mathfrak{M}$ must satisfy appropriate conditions that can be characterized in the language of consistency of certain systems of ω-logic or of the transfinite induction. This shows also that the usage of satisfaction (truth) in proving theorems about natural numbers (i.e., proving properties of natural numbers in theories of the type $\mathrm{PA}^{\Gamma-\mathrm{PA}(S)}$) can be in a certain sense approximated by transfinite induction or by adding certain consistency statements concerning appropriate systems of ω-logic.

Moreover, even for a fixed model $\mathfrak{M}$ of Peano arithmetic for which there exists a satisfaction class, the concept of satisfaction and truth cannot be uniquely determined and, even worse, not always can be defined in such a way that the required (and expected because useful) nice metamathematical properties would be satisfied. There is no uniqueness and no bivalency (for nonstandard models). But nonstandard models and nonstandard languages (generated by such models and by axiomatic approach to the concept of truth) turn out to be useful and to have an impressive spectrum of applications. In particular they can be used to establish properties of deductive systems, provide insight into fragments of Peano arithmetic as well as into (second-order) expansions of it. They can also serve as a heuristic guide for behavior of the infinity (one can code by nonstandard objects appropriate infinite sets, in particular infinite sets of standard formulas).

Note also that considering satisfaction classes and truth for the language of Peano arithmetic and attempting to characterize them axiomatically we use the whole time at the metatheoretical level Tarski's definition with respect to structures of the type $\langle \mathfrak{M}, S \rangle$ and the latter is understood as being defined in a non-formalized metasystem.

A general moral of our considerations is that semantics needs infinitistic means and methods. Hence finitistic tools and means proposed by Hilbert in his program are essentially insufficient.

Part III
Philosophy of Mathematics in Poland

The Philosophy of Hoene-Wroński

15th August 1803, the catholic Feast of the Assumption played a significant rôle in the scientific biography of Hoene-Wroński (1776–1853). On that day, during a ball on the occasion of birthday of the First Consul, Wroński experienced an illumination. In remembrance of this he assumed a new name – Maria.

Two problems stood in the center of Wroński's interests:
 – the creation of a new philosophy – achrematic philosophy[1] which overcoming the world of things will reach the Absolute and the creation principles in order to deduce from them a logically consistent theory of the whole reality,
 – a profound reconstruction of the system of science, a beginning of which should be a reform of mathematics (the latter should consists of deducing all domains of mathematics from a unique general principle, namely from "the absolute law of algorithm").

We are interested here mainly in the first problem.

1. Predecessors and Contemporaries

Philosophical system of Hoene-Wroński developed at the turn of the 18th and 19th centuries is usually counted as a messianistic philosophy. This philosophy was established under the influence of German thinkers such as Kant , Fichte, Schelling and Hegel. But it has various specific features making it different than German thought. It was rather spiritualistic than idealistic, it maintained the existence of a personalistic God, was convinced of the eternity of soul, of the absolute superiority of spiritual powers over the corporal ones.

Hoene-Wroński belonged to the earlier generation of Polish messianists, he was in a sense a forerunner. Others were active between 1830 and 1863. One should mention here Bronisław Trentowski (1808–1869), Józef Gołuchowski (1797–1858), August Count Cieszkowski (1814–1894), Karol Libelt (1807–1875), Józef Kremer (1806–1875). They were unanimous with respect to fundamental theses but differed strongly in details. The unique center of this trend was Paris but they worked in fact in isolation: Trentowski in Germany, Gołuchowski in a village in the Congress Kingdom of Poland[2], Cieszkowski and Libelt in Great Poland, Kremer in Cracow. There were professional philosophers among them but an important rôle was played also by poets: Mickiewicz, Słowacki and Krasiński as well as persons actively engaged in religious work such as Towiański. What was

---
1 From the Greek *chrema* – thing.
2 Congress Kingdom of Poland was a puppet state under Russian imperial rule from 1814 to 1915, constitutionally in personal union.

common for them were their messianistic views. They differed in argumentation as well as in results and conclusions. The three great national poets wanted to create Polish philosophy, scholars – an absolute philosophy. The latter knew quite well European philosophy and referred to it, the former – spontaneously developed their own systems. Some of them were rationalists, the others – mystics.

Their common feature was that they attributed to philosophy not only a cognitive function but also a significant rôle in the realization of the reform of life and in the liberation of the mankind. Their philosophy was filled with the belief in the metaphysical meaning of nation and with the conviction that a man can fulfill his/her vocation only in the communion of spirits formed by a nation, that just nations determine the development of the mankind. Polish nation has a special task – the task of being a Messiah among other nations. According to them a nation forms an intermediate link between an individual and God.

2. Philosophical System of Hoene-Wroński

The starting point of the philosophical system of Hoene-Wroński was the system of Immanuel Kant. In fact he wanted to study philosophy under the direction of Kant – unfortunately when he decided to do this, Kant retired already. In the Marseille period among first works of Wroński one finds the work *Philosophie critique decouverteé par Kant, fondée sur le dernier principe du savoir* (1803) – a philosophical dissertation on the philosophy of Kant sent to Kościuszko[3] and Dąbrowski[4]. Unfortunately the whole edition of the book has been destroyed in a fire of a printing-house. It was in fact the first extensive work written in France and devoted to Kant – French philosophers got to know the achievements of the master from Königsberg with a delay and rather unwillingly being convinced of their own superiority.

According to Wroński, Kant made a mistake treating the knowledge (the reason) and the objective being as something heterogeneous instead of as homogeneous. This led consequently to dualism of the knowledge and being, of phenomenological and noumenal worlds. To overcome this limitation one should derive both from a higher principle: from the Absolute. One can see here also the influence of Schelling whose system tends towards absolute idealism and which – according to Hoene-Wroński – was an anticipation of the absolute knowledge.

---

3   Tadeusz Kościuszko (1746–1817), Polish national hero, general and a leader of the 1794 uprising against the Russian Empire. He fought in the American Revolutionary War on the side of Washington. In recognition of his service he became in 1783 a naturalized citizen of United States.
4   Henryk Dąbrowski (1755–1818), Polish general and national hero, organizer of Polish Legions during the Napoleonic Wars.

The starting point of Hoene-Wroński was the assumption about the rationality of the world. If the world were not rational then theoretical investigations as well as technology would be impossible. But the rational order of the world can be explained only by assuming that it exists and develops according to a certain law. Hoene-Wroński claimed that he has discovered this law. He called it "the law of creation" (*loi de création*). He spoke about it in a mysterious way. It formed the kern of his philosophy. It made possible not only to get to know the structure of the reality but also to describe and to rule the processes. It should provide the method of accurate cognition and activity.

Hoene-Wroński distinguished chrematic and achrematic philosophy. The source of the former is the world of created things. It looks for the conditions of their existence till the absolutely independent condition, i.e., the Absolute which overcomes the world of created objects. The whole philosophy existing so far belongs to this trend. The starting point of the second type of philosophy, i.e., of the achrematic philosophy, comes beyond the sphere of things. It is independent of experience and it forms the world as a set of conditions present in the mind of the creator of the universe. It transcends the world of created things and provides an analysis of fundamental principles. It reaches the most important principles of any being and eliminates the gap between the knowledge and the being proposing their synthesis in a direct inspection.[5]

The mind of a human being and the mind of the Absolute are of the same nature. But since the mind of a man has some limitations caused by our physical being, hence it possesses smaller power and smaller domain of activity. The mind acts – both in a human being and in the Absolute according to the law of creation. The mind does not discover it but it sets it up and imposes it to itself. Hoene-Wroński deduced this law from two elements, namely from the knowledge and the being. The development consists of associating and diametrical splitting of those elements. Taking successive positions: neutral, diametrical or dominating one upon the other they take various forms and exert various influences and in this way form the whole richness of the world. The knowledge is subjective, active, spontaneous, it acts in a free and intentional way. The being is objective, passive, inert, unable to create any feature. It is a material on which a variety of phenomena is being put. Knowledge is universal, it applies to an unbounded number of individual objects. It gives the meaning and the variety to the world. In the world of things both elements, i.e., the knowledge and the being, interact and determine each other.

The law of creation is derived by Wroński from the very nature of the Absolute. The latter grants it to itself. In the first creation act the Absolute is being divided into knowledge (subject) and being (subject). The knowledge is the auto-creation (*autogenia*), the being is the auto-establishing (*autotezja*). Further the autocreation goes in accordance with the law of creation and leads to the full autocreation of

---

5 One can see here connections with the Husserl's transcendental reduction.

the Creator. The law of creation guides this autocreation and similtaneously is being derived from it as its component.

In the posthumously published work *Messianisme, philosophie absolue* (1876) presenting the functioning of the law of creation Wroński constructs an architectonics of the world. The whole reality is captured in "systems" (*systematy*). A system of a smaller degree of generality follows from a system of a higher degree. At the top of this hierarchy stands the system of auto-creation of God. Next are:

(1) the formation of God by itself,
(2) the creation of the reality,
(3) the creation of the world,
(4) the creation of a human being,
(5) the formation of a human being by himself/herself,
(6) the formation of the absolute religion,
(7) the progressive development of the mankind.

In this way Wroński derived the whole universe from the single law of creation. He was convinced that he succeeded at the place where Descartes failed – namely in deducing the whole human knowledge from a single principle and according to one single method. This implies the need of a reform of the hitherto existing human knowledge, the need to reform the science. To do this one should apply the law of creation – this would make possible to order and to steer them towards the ultimate goals. The reform of the science should begin – according to Wroński – from mathematics. The reason of this was his conviction that the value of the law of creation can be checked just in the case of mathematics. Hence he aimed to deduce all theorems and methods of mathematics from the general principle.

In *Introduction to a Course of Mathematics* (London 1821) Wroński explained in a straightforward way the plans to reform mathematics – this work should present his plans to the general public. He begins by stating that any positive knowledge is based on mathematics or at least uses mathematics. He distinguishes four periods in the development of mathematics. The first one is the period when mathematics was carried on *in concreto*, i.e., there was no abstraction from the material reality, mathematics had a practical character (so was the case of mathematics in ancient Egypt and Babilon). The second period is the period of Greek mathematics. It can be characterized by the fact that abstraction was used but – according to Hoene-Wroński – mathematical truths were "only particular facts [cases] and have still not reached the general truths". The third period is the time from Cardan and Fermat till Kepler and Wallis. Some general truths did appear in mathematics, but they were isolated, they were "individual mathematical products". In particular the formulas for solutions of equations of degree 3 and 4 have been found but there was no idea about the general setting of the problem. The last fourth period has begun with Newton and Leibniz . Then methods have been developed which can be applied to "all the appearances of the nature". This period is characterized by the usage of sums of sequences, the only common tool so far.

The characteristic feature of all periods in the hitherto development of mathematics was – according to Hoene-Wroński – that they were based on some relative principles. This means that there were no absolute principles and – on the other hand – science should be based just on such ideas. Hence the prediction of the new higher stage in the development of mathematics. Its basis should be the reform proposed by Wroński. It consists of the division of mathematics into theory and technie (*technia*). All mathematical truths should be deduced from the unique highest law and in this way they should receive the absolute certainty.

Let's come back to the philosophical system of Hoene-Wroński. Note that he did not explain what is in fact the Absolute, what are its essence and nature. Hence various interpretations are possible: it can be understood as God, as the mind, as *Ding an sich*. Wroński himself understood the Absolute still in a more abstract way than all those interpretations. He did not define its essence being convinced that a source of any being and any knowledge is in fact beyond being and knowledge and consequently cannot have any definition, neither ontological nor epistemological. Any definitions of the Absolute are always definitions not in its nature but only in relation to other beings and objects. So is, in particular, the case when one says that the Absolute possesses the highest reality, highest certainty, highest stability, that in it there is the infinite truth, the infinite right, the universal necessity, the complete harmony, the perfect identity, the maximal independence and the eternal source of creativeness. In order to get to know the Absolute "one should overcome the worldly conditions of a rational being" and to go up to the absolute reason. This can be done in a so called "pure inspection", hence by intuition which is simultaneously a cognitive act as well as an existential act.

3. The Philosophy of History of Hoene-Wroński

Wroński's aim was not only to describe and to define the Absolute but rather to derive from it the whole diversity of the being according to a single law – the law of creation.

This law was for Wroński not only a principle according to which the being has been formed but also a principle that should be respected by any human being who wants to fulfil his/her destiny. Under this condition the mankind can move from the present political system which is full of antagonisms and inconsistencies to a completely intelligent system. In this way we come to the next feature of Hoene-Wroński's system, namely to his messianism.

Wroński distinguished four periods in the hitherto existing history. Each of them put different aims, in particular: materialistic aims (Orient), moral ones (Greece, Rome), religious ones (the Middle Ages), intellectual ones (modern times till the 18th century). The 19th century is a transitional period. There are two trends struggling with each other: the conservative one that puts as its aim the right, and the liberal one aiming at the truth. The future of the mankind is in the unity of the right and the truth, of religion and science, in creating the epoch of "the absolute right and truth". And here the significant rôle should be played by Slav nations – they will overcome the contrasts and antagonisms of the Romance and Teutonic peoples and lead the mankind to the period of absolute aims, to the fulfilment of the human destiny, to the discovery of the truth and to the immortality.

Since the philosophy leads to the cognition of the Absolute, it identifies itself with religion, it transforms the revealed religion into the intellectual religion. Consequently a discoverer of the absolute religion becomes in a certain sense a Messiah who having come to the boundaries of any cognition proclaims the unity of science and religion. Hence Wroński distinguishes two fundamental domains of the philosophy: the domain providing the knowledge about the reality (he called it theory (*teoria*) or *autotezja*) and the domain directing the reality to enable it to reach its aims (he called it *technia* or *autogenia*). Just the second component (which joins the philosophy and politics) makes the philosophy of Hoene-Wroński richer in comparison with, e.g., the system of Hegel or the whole German idealism.

By the application of the law of creation, statements of the revealed religion should received the certainty of the knowledge. This idea united the absolute philosophy and the absolute religion developed on the basis of Christianity and it should be realized in the history of the mankind tending towards the ultimate aims. Messianism should fulfil the promises given to the mankind by Messiah-Christ.

The progress of the mankind consists of the constant and permanent ascent to the higher principles of knowledge. In the last period of the development the reality of the Absolute will be scientifically established. Then the dogmas and mysteries of the revealed religion will be explained, sehelian (from the Hebrew *sehel* – mind) church will be established. In this church the religious worship will consists of practising the fine arts. It will be a complement and the fulfilment of the Christianity. Wroński developed those ideas in particular in a letter to the pope Leo XII (1827).

A man will successively regain freedom from the inert conditions of his/her corporal nature and will still more and more use spontaneous conditions of his/her spiritual freedom. This progress will be accompanied by the development of the consciousness till the appearance of the absolute consciousness. At the final step a man will subordinate his/her freedom to the law of creation and will consciously join the creative activity of the absolute mind.

The big reform of the mankind can be realized by a synthesis of act and thought, hence by uniting the activity of two nations leading in those domains, i.e., France and Germany. After 1830 Wroński digressed this conception and began to proclaim the messianism of Slav nations guided by the Russian tsar. According to him Slavs have a special historical mission – the mission of synthesizing the German thought (i.e., the philosophy) and the French act (i.e., social programs) and of reaching the ultimate aims. Hence Slavs should close the epoch of antinomies and open the epoch of truth. In the latter the idea of "the God's Kingdom on the earth" should be realized.

Hoene-Wroński advertised the messianism till the end of his life, developed it in numerous works and popularized it. He attempted to arouse the interest of his contemporaries in his ideas, addressed monarchs, printed proclamations to nations, wrote prospectuses, founded associations, published journals. He deeply believed in the infinite power of the human mind. He was convinced that only the mind regaining its freedom from the earthly limitations will lead the mankind to its culminant pre-destinations.

The philosophical doctrine of Hoene-Wroński is in fact idealistic. Mystical elements are intertwined in it with the radically rationalistic ones. Hence it belongs both to the rationalistic trend as well as to the romantic one. It is a specific combination of metaphysics, the philosophy of history, religion, ethics and politics. This explains Wroński's interests not only in the philosophy as such but also his engagement in the policy.

4. The Reception of the Philosophical Ideas of Hoene-Wroński and His Followers

The philosophical conceptions of Hoene-Wroński as well as his philosophy of history have found various and differentiated evaluations and have caused various reactions. Honoré de Balzac saw in him one of the most intelligent persons (cf. his letter to Madame Hańska from 1st August 1834 – it can be found in *Pisma* of T. Boy-Żeleński, vol. XII, Warszawa 1958, p. 231). Add that Balzac devoted to Hoene-Wroński a novel *Balthazar Claes ou la Recherche de l'absolu* published in 1834.

The contemporaries criticized Wroński or did not understand him. The reason was on the one hand the ambiguity of his considerations, on the other his attitude full of haughtiness and predominance having their source in his conviction of the absolute value of his conceptions. Karol Libelt[6] wrote that the world has lost in him a thinking writer of exceptional talent because his mind has chosen a wrong direction. Józef Ujejski[7] reproached Wroński with charlatanry and even

---
6   Karol Libelt (1807–1875), Polish Messianic philosopher, political and social worker.
7   Józef Ujejski (1883–1937), historian of Polish literature, professor of Warsaw University.

paranoia. Wincenty Lutosławski[8], one of the last neomessianists, taunted him with the lack of the university education and with the self-exaltation. He wrote also that from the writings of Wroński one can learn in a very complicated way very simple truths.

Wroński's ideas awoke interest of occultists (independently of the fact that he separated himself from them). Papus (Gerard Encausse) wrote in his *Traité méthodique de science occulte* (1891) that Hoene-Wroński owes his deepest ideas to cabbala. É. Lévi mentioned Wroński a couple of times in his *Histoire de la magie* (1860).

The first notorious continuator of the philosophical ideas of Hoene-Wroński was Antoni Bukaty (1808–1876), a philosopher, historian and engineer. He was active in France where he moved after the decline of the November uprising (1830).

Another continuator of Wroński was Samuel Dickstein (1851–1939), a mathematician, historian of mathematics and an educator. He catalogued and described scientifically the Kórnik collection of Wroński's writings – he made it in the book *Katalog dzieł i rękopisów Hoene-Wrońskiego* [*The Catalogue of Works and Manuscripts of Hoene-Wroński*] published in Cracow in 1896. He is also the author of the book *Hoene-Wroński. Jego życie i praca* [Hoene-Wroński. His Life and Work] (1896). A catalogue of Wroński's works was also prepared by Bolesław J. Gawecki – cf. his *Wroński i o Wrońskim* [Wroński and about Wroński] (1958).

In Poland the interest for Wroński's ideas and conceptions was shown mainly by people being not professional philosophers. In the interwar period the Messianistic Institute was established in Warsaw – it was active between 1919 and 1933. Its leader was J. Jankowski. Among its members were Paulin Chomicz, Czesław Jastrzębiec-Kozłowski and Jerzy Braun. In the institute many translations of Wroński's works have been prepared. In 1933 Association of Hoene-Wroński has been established in Warsaw – Paulin Chomicz became its president. In the board there were Cz. Jastrzębiec-Kozłowski and J. Braun. The organ of the association was the journal *Zet*. After the Second World War the main initiator of the revival of Wroński's thought in Poland was J. Braun. In 1962 he founded in the Warsaw Club of the Catholic Intelligentsia [Klub Inteligencji Katolickiej] a Section of Polish Philosophy. The philosophical conceptions of Wroński have been studied there. In 1965 Braun left Poland. After his departure the studies have been continued in two sections: Section of the Philosophy of Religion (directed by Czesław Domaradzki) and Section of Polish Religious Rationalism (led by Jan Łuszczewski). In those sections were active among others: Jastrzębiec-Kozłowski, A. Madej and L. Łukomski as well as J. Niementowski. Łukomski is the author of the book *Twórca filozofii absolutnej. Rzecz o Hoene-Wrońskim* [The Founder

---

8   Wincenty Lutosławski (1863–1954), professor of philosophy.

of the Absolute Philosophy. About Hoene-Wroński] published in 1982. Both sections existed tillthe mid 80s.

5. Conclusions

Summing up one should say that Wroński belonged to the best European metaphysicians of the beginning of the 19th century. He distinguished himself by the scientific background and the extensiveness of aims. He was in fact an outstanding person and one of the most original thinkers. He knew many (a dozen or so) languages, had good knowledge in various domains, he was a scholar [erudite]. On the other hand his considerations were too abstract, his thought too ambiguous, he had problems to express clearly his ideas. Simultaneously he was absolutely self-confident (what sometimes was close to arrogance) and has ignored works of others and formulated arbitrary opinions about them. He was rather a difficult person in contact and this has certainly influenced the fate of his ideas.

The philosophical ideas of Hoene-Wroński did not find any echo. He had no big circle of supporters, successors or students. He was alone. Hoene-Wroński belongs to scientists who are rather forgotten today.

Philosophical Reflection on Mathematics in Poland in the Interwar Period

The aim of this essay is to indicate main trends and tendencies, main standpoints and views in the philosophical reflection on mathematics in Poland between the wars, i.e., between 1918 and 1939.

Why just this period? Because it was the time of intensive development of mathematics (Polish Mathematical School) and of logic (Warsaw Logical School, Lvov-Warsaw Philosophical School) in Poland. Hence a natural question arises whether this development of mathematics and logic was accompanied by philosophical reflection on those disciplines, whether the researches were founded on and stimulated by certain fixed philosophical presuppositions. On the other hand philosophy of mathematics and logic is based on and uses certain results of metamathematics, of the foundations of mathematics and of logic. Did logical achievements influence the philosophical reflection?

1. Before 1918

To discuss the philosophy of mathematics between the wars one should start earlier and consider predecessors.

In the 19th century there was in Poland no significant work on logic or on philosophy of mathematics. One can mention only Józef Maria Hoene-Wroński (1776–1853) but his ideas did not find any resonance. The reason was the ambiguity of his ideas and the unclear language in which they were formulated.

The situation began to change at the end of the 19th century. The main centuries which played important rôle here were Lvov and Cracow. In Lvov the main figure was of course Kazimierz Twardowski (1866–1938), the founder of the Lvov-Warsaw Philosophical School. His rôle in the development of the analytic tradition of philosophy and of logic in Poland is well known. The atmosphere in Lvov stimulated also mathematicians. Here in 1908 Wacław Sierpiński (1882–1969) obtained his *Habilitation* and started his lectures, in particular lectures in set theory which were one of the first lectures in set theory in the world. What is important from our point of view is the fact that Sierpiński has chosen as the subject of his *Habilitation lecture* a problem from the philosophy of mathematics. His lecture was devoted to the problem of the rôle and meaning of the concept of a correspondence in mathematics (cf. Sierpiński 1909). He described it as one of the most important notions of mathematics, as "a source of all best ideas". The reason of that he saw in the fact that, as H. Poincaré wrote in *La Science et l'Hypothèse*: "Mathematicians do not study objects but relations between them; hence it is not important for them when some objects are replaced by another provided the relations between them remain unchanged". (One can see here structuralistic ideas.) He added also that the source and base of all applications of mathematics was the existence of an ideal correspondence between the domain of abstract mathematical ideas and the domain of the reality.

Also another mathematician who played a crucial rôle in the rise of Polish Mathematical School, i.e., Zygmunt Janiszewski (1888–1920) showed interest in the philosophy of mathematics. He received his habilitation in Lvov in 1913. His habilitation thesis was in topology but the title of his *Habilitation lecture* was „On realism and idealism in mathematics" (cf. Janiszewski 1916), hence it was devoted to the philosophy of mathematics. Janiszewski considered there the discussion between realists and idealists on the problem of existence in mathematics. This problem appeared with great power in connection with set theory and especially with Zermelo's well-ordering theorem (1904). Janiszewski presented in his lecture several opinions in this dispute and showed in the conclusion a scepticism towards perspective of finding a final solution to this problem because it is a part of the old dispute between nominalism and platonism.

Why did Sierpiński and Janiszewski choose subjects from the philosophy of mathematics for their habilitation lectures? One of the reasons could be the fact that the habilitation procedures were at the Philosophical Faculty and most of the members of the Faculty Council were non-mathematicians. Therefore Sierpiński and Janiszewski have chosen general subjects. But on the other hand they could choose also popular subjects from mathematics itself. The fact that they have chosen subjects just from the philosophy of mathematics can indicate that there was a good atmosphere in Lvov for foundations and philosophy of mathematics and that they both were interested not only in mathematics but also in its philosophy. Both were convinced that one needed a conception of mathematics on which the development of this discipline in Poland could be founded and suggested that set theory could be a basis of it.

The interests of Polish mathematicians in logic and philosophy of mathematics as well as their conviction of the importance of those domains for mathematics can be seen also in the interesting book published in 1915 *Poradnik dla samouków* [Handbook (Guide) for Autodidacts] devoted to mathematics. One finds there chapters by Jan Łukasiewicz (introductory chapter „About science"), W. Sierpiński (about set theory) and some chapters by Z. Janiszewski who was the main contributor and the soul of the whole enterprise. Janiszewski wrote chapters on the foundations of geometry, on mathematical logic and on philosophical problems of mathematics. The last one was an extensive chapter presenting main problems and views of the philosophy of mathematics with an extensive bibliography in which the current positions of the literature were well represented. In the volume III of the guide (published in 1923) there was also a chapter „On the meaning of mathematical logic for mathematics" by Jan Sleszyński. It is worth mentioning here that Sleszyński wrote there that mathematical logic cannot be reduced to the methodology of mathematics, because it is an autonomous discipline. He defended also logic against various objections formulated by mathematicians and philosophers (among others by Poincaré).

The above remarks indicate that connections between mathematicians and logicians in Warsaw were in fact very good and that they collaborated closely. We return to that later.

The second center which should be considered here was Cracow. Main figures there were Stanisław Zaremba (1863–1942) and Jan Sleszyński (1854–1931) (whom we mentioned already). Zaremba was working mainly in analysis and applications of mathematics but was interested also in the philosophy and methodology of mathematics. He published several papers in the last domains, in particular „Pogląd na te kierunki w badaniach matematycznych, które mają znaczenie teoretyczno-poznawcze" [Remarks on those trends in mathematical investigations which have epistemological meaning] and „Uwagi o metodzie w matematyce i fizyce" [Remarks on the methods of mathematics and physics]. In the first one he considered the meaning of the studies in the foundations of geometry and in set theory for the philosophy, especially for the epistemology. In the second one he claimed that the investigations of the nature are the most important source of new mathematical discoveries and that the rôle of mathematics in physics consists in providing a tool to deduce corollaries from hypotheses obtained by observation and experience.

Talking about Zaremba one should also mention the controversy about the concept of a magnitude. In 1916 Jan Łukasiewicz (1878–1956) devoted one of his courses at Warsaw University to the methodology of deductive sciences. During his lectures he discussed the book by Zaremba *Arytmetyka teoretyczna* [Theoretical Arithmetic] (1912) and analyzed it from the methodological point of view challenging some principles adopted by Zaremba as well as his definition of a magnitude (in particular he criticized the usage of sentences with no contents). Łukasiewicz published his remarks in a paper (1916). This was the beginning of a dispute in which several persons took part, among others Kazimierz Kuratowski, Tadeusz Czeżowski, Leon Chwistek and of course Zaremba. The essence of the dispute concerned in fact not the concept of a magnitude but the rôle of logic in mathematics. Zaremba represented the view that logic should be "in mathematics", should be *ancilla mathematicae* (cf. the title of his work *La logique en mathématique*, 1926) whereas Łukasiewicz saw the (mathematical) logic as an autonomous discipline providing the foundations and methodology of mathematics. The latter idea was also accepted by leaders and founders of Polish Mathematical School in Warsaw who stressed the rôle of set theory, of the foundations of mathematics and of mathematical logic and saw the logic in the center of mathematics (by Zaremba its place was in the periphery of mathematics).

Zaremba and Sleszyński influenced the interest in logic and the philosophy of mathematics of some young mathematicians in Cracow, among others of Witold

Wilkosz (1891–1941). He wrote some papers in which he discussed the meaning of mathematical logic to mathematics and the process of abstraction (trying to base it on the abstraction principle) (cf. Wilkosz 1938; 1939).

With Cracow was connected also Edward Stamm (1886–1940) who studied in Switzerland and Austria. He was a teacher of mathematics in a small town and the author of some papers devoted to the algebra of logic and to the philosophy of mathematics. Especially interesting is the paper „Czem jest i czem będzie matematyka?" [What is and what will be mathematics?] (1910) in which analyzing the development of mathematics he comes to the conclusion that „mathematics is not a science but a method, it is the ideal deductive-symbolic stage of a science in general".

2. The Period 1918–1939

We should start by stressing the fact of close collaboration and mutual influences of logicians and mathematicians in Warsaw in the considered period. Both groups saw the mathematical logic and the methodology of mathematics as disciplines which are autonomous with respect to both mathematics and philosophy on the one hand and on the other they were convinced that those disciplines play a fundamental rôle in developing mathematics. According to them mathematics and mathematical logic should be neutral towards various philosophical controversies, they should be developed independently of any philosophical presuppositions. This attitude can be illustrated for example by the conviction of Polish mathematicians that the philosophy of the axiom of choice must be separated from its rôle in mathematics. Sierpiński wrote in (1965, p. 95):

> Still, apart from our personal inclination to accept the axiom of choice, we must take into consideration, in any case, its rôle in the set theory and in the calculus. On the other hand, since the axiom of choice has been questioned by some mathematicians, it is important to know which theorems are proved with its aid and to realize the exact point at which the proof has been based on the axiom of choice; for it has frequently happened that various authors have made use of the axiom of choice in their proofs without being aware of it. And after all, even no one questioned the axiom of choice, it would not be without interest to investigate which proofs are based on it and which theorems are proved without its aid – this, as we know, is also done with regard to other axioms.

This means simply that one should disregard philosophical controversies (and treat them as a "private" matter) and investigate (controversial) axioms as purely mathematical constructions using any fruitful methods.

One of the consequences of the described attitude of Polish logicians and mathematicians was the fact that they did not attempt to develop a comprehensive philosophy of mathematics and logic (Stanisław Leśniewski and Leon Chwistek were here the exceptions!). They formulated their philosophical opinions concerning mathematics or logic only occasionally and only on problems which were just interested for them or on which they actually worked. Consequently there were in Poland no genuine philosophers of mathematics. Philosophical remarks were formulated by logicians and mathematicians only on the margin of their proper mathematical or logical works (and had no meaning for the results themselves).

The current trends and views in the philosophy of mathematics, i.e., logicism, intuitionism and formalism, were of course well known (and there appeared papers discussing those tendencies, their meaning and development). But none of them was represented in Warsaw School. Moreover, it did not represent any other trend, it had no official philosophy of logic and mathematics. This followed from the belief of the autonomy of logic and mathematics with respect to philosophy. Opinions in the field of the philosophy of logic and mathematics were treated as "private" problems and philosophical declarations were made reluctantly and seldom. If they were made then it was stressed, directly or indirectly, that these were personal opinions.

Though some of logical investigations were motivated by philosophical problems (e.g., the many-valued logics by Łukasiewicz) but the formal, logical constructions were always separated from their philosophical interpretations. This attitude was still strengthened by Alfred Tarski (1901–1983) and Andrzej Mostowski (1913–1975) who claimed that a logician or a mathematician can have philosophical views or sympathies quite different from those which could be suggested by the scope of problems he is working on. They provided also good examples of this attitude by their own work. Mostowski wrote about Tarski (cf. Mostowski 1967c, p. 81):

> Tarski, in oral discussions, has often indicated his sympathies with nominalism. While he never accepted the 'reism' of Tadeusz Kotarbiński, he was certainly attracted to it in the early phase of his work. However, the set-theoretical methods that form the basis of his logical and mathematical studies compel him constantly to use the abstract and general notions that a nominalist seeks to avoid. In the absence of more extensive publications by Tarski on philosophical subjects, this conflict appears to have remained unresolved.

Mostowski on the other hand was a sympathizer of constructivism but in his logical and foundational investigations did not take into account the methodological limitations put by it.

Another example are the investigations on intuitionistic logic carried out among others by Tarski without accepting intuitionism as the philosophy of mathematics. Program of Janiszewski and Polish Mathematical School created set-theoretical foundations of mathematics in the methodological and not philosophical sense.

What were the separated philosophical opinions formulated by Polish logicians, philosophers and mathematicians? Let us start by the problem of psychologism. Psychologism was popular in the philosophy of logic and mathematics in the late 19th century. According to it the objects studied by logic and mathematics exist as psychic entities and come to be known just like other psychic facts. Already Twardowski took a step towards antipsychologism. Next step was a paper by Łukasiewicz „Logika a psychologia" [Logic and psychology] (1907) where he declared himself firmly for antipsychologism in logic. His arguments were as follows: (1) logical laws are certain and psychological ones (being empirical in fact) only probable, (2) laws of logic and laws of psychology differ in content because the former concern the connections between the truth and falsehood of judgements and the latter state relationships between psychic phenomena, (3) the terms 'thinking' and 'judgement' have different meaning in psychology and in logic. Łukasiewicz stated finally in (1907):

> The clarification of the relationship between logic and psychology may prove to the advantage of both disciplines. Logic will be purified of the weeds of psychologism and empiricism, which hamper its true development, and the psychology of cognition will rid itself of elements of apriorism, behind which the genuine light of its truth could not fully show itself. It must be borne in mind that logic is an *a priori* science, like mathematics, while psychology, like any natural science, is, and must be, based on experience.

Łukasiewicz's arguments against psychologism were similar to those of Husserl and Meinong. They were universally accepted in Poland. Their consequence was the conviction that the certainty of theorems of logic cannot be explained by psychological arguments. This was in fact a negative solution to the problem of certainty of logic. Since almost all Polish logicians were sympathizers of genetic empiricism hence any aprioristic solution of this problem could not be accepted.

Polish logicians did not accept the concept of logic as pure syntax. This view was popular at that time, it was developed under the influence of Hilbert's metamathematics and the philosophy of language of Vienna Circle. An exception was here Chwistek who treated his semantical systems as formal systems of expressions. Warsaw School represented the semantical point of view. In this context one should see the semantical foundations of logic founded by Tarski in the thirties. An original approach to the problem of the nature of logic was represented by Leśniewski. One can call it intuitive formalism. He attempted the complete formalization of logical systems but claimed that formal expressions always code a fixed intuitive contents.

The semantical point of view implied the rejection of analytical concept of logic, i.e., the rejection of the thesis that logic is a collection of tautologies that are contentually empty (this is a thesis on logic and mathematics *versus* reality). Leśniewski (and Kotarbiński) claimed that logic describes the most general features of being, logic plays a rôle of a general theory of the real world.

Tarski described the concept of tautology as vague and did not see any objective basis for the division of terms into logical and extra-logical. Consequently he did not treat the borderline between formal and empirical disciplines as sharply marked.

Andrzej Mostowski wrote in (1955b, p. 42) that various metamathematical results:

> obtained by the mathematical method confirm therefore the assertion of materialistic philosophy that mathematics is in the last resort a natural science, that its notions and methods are rooted in experience and that attempts at establishing the foundations of mathematics without taking into account its originating in natural sciences are bound to fail.

And added:

> An explanation of the nature of mathematics does not belong to mathematics but to philosophy, and is possible only within the limits of a broadly conceived philosophical view treating mathematics not as detached from other sciences but taking into account its being rooted in natural sciences, its applications, its associations with other sciences and, finally, its history.

Łukasiewicz's views concerning the considered problem of relations between logic and mathematics on the one hand and reality on the other were changing. In (1912) he claimed that logical and mathematical judgements are *a priori* truths about the world of ideal entities. Hence he treated both disciplines as unrelated to experience. The discovery of many-valued logics implied that Łukasiewicz maintained that logic systems can be given an ontological interpretation and that experience will help to decide which of systems of logic is fulfilled in the reality (cf. 1936). Later he tended to the conventionalism and relativism. In (1952) he wrote:

> We have no means to decide which of the n-valued systems of logic [...] is true. Logic is not a science of the laws of thought or of any real object; it is, in my opinion, only an instrument which enables us to draw asserted conclusions from asserted premises. [...] The more useful and richer a logical system is, the more valuable it is.

Kazimierz Ajdukiewicz (1890–1963) was also a sympathizer of conventionalism (in a radical version). He claimed that logic is something implied by the meaning rules (rules of sense), both axiomatic and deductive ones. Later he abandon the radical conventionalism and claimed that laws of logic refer indirectly to the experience and that they should be treated as rules of inference, hence they belong to the metascience and are mainly of a methodological character.

Next problem which was discussed and commented was the problem of nominalism. A declared nominalist was Leśniewski. Hence he denied the existence of any general objects. Consequently the systems he created consisted of a finite number of individual inscriptions. Tarski had nominalistic leanings (inherited from Leśniewski) but the needs of metamathematics made him to abandon these sympathies. In particular he referred to formula types, that is classes of equiform formulas but these classes of formulas were treated by him as consisting of formulas interpreted as physical bodies.

Łukasiewicz's view towards nominalism was different. He maintained that arguments of Tarski defending nominalism were not sufficient. He thought that logicians merely use nominalistic terminology but in fact they are not nominalists. He inclined to interpret logic in an outright neo-platonic spirit. In (1937) he wrote:

> In concluding these remarks I should like to outline an image which is connected with the most profound intuitions which I always experience in the face of logistic. That image will perhaps shed more light on the true background of that discipline, at least in my case, than all discursive description could. Now, whenever I work on even the least significant logistic problem – for instance, when I search for the shortest axiom of the implicational propositional calculus – I always have the impression that I am facing a powerful, most coherent and most resistant structure. I sense that structure as if it were a concrete, tangible object, made of the hardest metal, a hundred times stronger than steel and concrete. I cannot change anything in it; I do not create anything of my own will, but by strenuous work I discover in it ever new details and arrive at unshakable and eternal truths. Where is and what is that ideal structure? A believer would say that it is in God and His thought.

Note that Łukasiewicz stressed that logic itself cannot solve the philosophical controversy over universals. Hence any claims that logic is nominalistic or not are groundless. Similar opinions towards the neutrality of logic with respect to the problem of universals held Ajdukiewicz and Czeżowski. Add that Kotarbiński in the early and radical version of his reism held views similar to those of Leśniewski.

As remarked above Leon Chwistek (1884–1944) and Stanisław Leśniewski (1886–1939) were two exceptions in the described attitude towards philosophical problems and issues, more exactly they were interested only in those logical problems which were implied by their own philosophical views in the foundations of mathematics, their philosophical views generated their interest in particular problems, their logical investigations were motivated by their philosophical views.

Leśniewski had the chair of the philosophy of mathematics at the University of Warsaw (since 1919 till his death in 1939). He represented a philosophical approach to logic though he was convinced that philosophical investigations are hopeless and lead to no definite solutions. His aim was to construct a system of logic that would satisfy two general requirements: it should serve as the foundations for mathematics and should be constructed in a manner free of any ambiguities. He did it by constructing three systems: protothetic, ontology and mereology. Leśniewski meant logical systems in a nominalistic way. Language was for him a collection of concrete individual inscriptions. There existed only those expressions which have been actually written, he admitted no 'potential' existence. This was called by him 'constructive nominalism'. It was connected with intuitive formalism. Consequently he rejected the interpretation of logic and mathematics as games using symbols devoid of meaning. According to him every language system says 'something' 'about something', is a way to express what is intuitively true, is an indispensable way of encoding and transmitting logical intuitions. Add also that Leśniewski maintained that logic should be meant as extensional and bivalence.

Chwistek is known mainly for his logical works, i.e., for his simplification of the theory of types of Russell and Whitehead (he did it in a nominalistic spirit). His aim in logical investigations was to create a comprehensive system of logic and mathematics based on a theory of expressions (called by him rational metamathematics). His results and ideas had rather limited influence. The reason was the complicated and nonstandard notation used by him as well as his way of presenting the results.

He represented rationalism (called by him critical rationalism) and rejected irrationalism. According to him there are two sources of knowledge: experience and deduction. Methods used in science and in the philosophy should be constructive. The aim of science is to describe by mathematical expressions objects given in an experience. Mathematical formulas are only descriptions of an experience and cannot be treated as laws concerning objects which are not given by experience.

One of the most known philosophical conceptions of Chwistek was his theory of the plurality of the reality. It was published for the first time in his paper from 1917 "Trzy odczyty odnoszące się do pojęcia istnienia" [Three lectures concerning the concept of existence] and found its final presentation in his book *Granice nauki* [Limits of Science] (1935). He postulated, according to various types of experience, four types of reality: reality of impressions, reality of images, reality of things and physical reality (constructed in science). He attempted also to characterize properties of those types of realities by suitable sets of axioms.

Chwistek represented in logic and mathematics nominalism and was against formalism. He claimed that objects of deductive systems are expressions and one cannot accept any other objects. According to him geometry is an experimental discipline. The development of non-Euclidean geometries was considered by him as the most important achievement in science. It rejected the Kantian idealism and the view that geometry is given *a priori*. Geometry and other mathematical theories as well as theories of the science should be developed constructively, i.e., one should base them on such axioms and definitions that the theorems deduced from them should be in accordance with experience. This would suggest that he would accept conventionalism. But in *Granice nauki* he rejected it. Moreover he claimed that conventionalism was incorrect not only in science but it was also a source of wrong views in social problems (it reduced truth to the usefulness and efficiency and in this way led to the reinforcement of the ruling class).

3. Alfred Tarski

In previous sections we mentioned some views of Tarski concerning the philosophy of logic and mathematics. Since Tarski played on important rôle in the development of logic let us say more about his philosophy.

Start by noting that Tarski was interested in philosophical problems and very actively participated in the philosophical life of his time. He was convinced of philosophical significance of his works, in particular of his work on truth. In (1933a) he wrote:

> I shall be satisfied if this paper convince the reader that the method used above already now is an indispensable apparatus which may be helpful in considerations of purely philosophical problems. [...] The central problem of this paper – construction of a definition of a true sentence and founding a scientific basis of a theory of truth – belongs to epistemology and is one of the main problems in this domain of philosophy. Hence I expect that just specialists in epistemology will take an interest in it, that – not becoming discouraged by difficult notions and methods, which so far have not been applied in this field – will analyze in a critical way results contained in it and will be able to use them in their further studies.

He described himself as (cf. Tarski 1944):

> Being a mathematician (as well as a logician, perhaps a philosopher of a sort) [...].

Tarski's philosophical attitude was anti-metaphysical, he supported the idea of scientific philosophy. He accepted a program of "small philosophy" which aims at detailed and systematic analysis of the concepts used in philosophy. Such a philosophy is minimalistic, anti-speculative and sceptical towards many fundamental problems of traditional philosophy. This attitude was inherited by Tarski from the Lvov-Warsaw School and strengthened by contacts with the Vienna Circle. He maintained also empiricism and abandoned the analytic/synthetic distinction and stressed that logical and empirical truths belong to the same generic category. Influenced by Leśniewski and Kotarbiński he was inclined to rather a strongly nominalistic understanding of expressions. According to this sentences are treated as concrete physical objects and languages as consisting of token-expressions. Needs of metalogical studies forced the understanding of them as expressions-types. Tarski sharply contrasted colloquial, natural language and formalized language.

Tarski was inclined to identify mathematics with the deductive method. He maintained that there is no hard borderline between formal and empirical sciences. He admitted the rejection of logical and mathematical theories on empirical grounds. He claimed also that there is no sharp demarcation between logical and factual truth and that the concept of tautology is unclear.

One must stress that all those were his "private" philosophical views which did not influence his logical and mathematical researches, in other words, his researches were independent of any philosophical presuppositions. In the paper "Über einige fundamentale Begriffe der Methodologie der deduktiven Wissenschaften" (1930) he explicitly wrote:

> [...] it should be noted that no particular philosophical standpoint regarding the foundations of mathematics is presupposed in the present work.

This was typical for him and for the whole Warsaw School in logic. This independence of logical and mathematical studies and philosophical views explains the cognitive conflict and discrepancy between Tarski's nominalistic and empiricistic sympathies and his "platonic" mathematical and logical practice. Note that his attitude enabled him to contribute to various important foundational streams without the necessity of accepting their philosophical assumptions and attempting to reconcile the philosophy and the research practice. His program of metamathematics can be summarized by his words from the paper (1954) where he wrote:

> As an essential contribution of the Polish school to the development of metamathematics one can regard the fact that from the very beginning it admitted into metamathematical research all fruitful methods, whether finitary or not.

Note that this attitude was in full accordance with the attitude of Polish mathematicians indicated above. According to it one should study the problems using any fruitful methods and making no philosophical presuppositions. There is no need to announce one's philosophical views concerning the investigated problems because this does not belong to scientific duties, this is a "private" affair.

4. Conclusions

As we showed above Polish logicians and mathematicians believed that philosophical problems of logic and mathematics are important. They knew quite well the current views and trends in this field, commented upon them and formulated several own opinions concerning the philosophy of mathematics. But on the other hand they treated logic and mathematics as autonomous disciplines independent of the philosophical reflection on them, independent of any philosophical presuppositions. They sharply separated mathematical and logical research practice and philosophical discussions concerning logic and mathematics. Philosophical views and opinions were treated as "private" matter that should not influence the mathematical and metamathematical investigations where all correct methods can and should be used.

Philosophy of Mathematics in the Warsaw Mathematical School

When speaking about the philosophy of mathematics in the Warsaw Mathematical School one should have in mind three persons: Wacław Sierpiński, Zygmunt Janiszewski and Stefan Mazurkiewicz. They formulated and expressed their philosophical views towards mathematics mostly in connection with set theory.

Let us start by the *Habilitation* procedure of W. Sierpiński. It took place in 1908 in Lvov. As a subject of the *Habilitation* lecture Sierpiński had chosen a problem from the philosophy of mathematics. The title of the lecture (held before the Council of the Philosophical Faculty of Jan Kazimierz University on 6th July 1908) was: „Pojęcie odpowiedniości w matematyce" [The concept of correspondence in mathematics]. It was published under the same title in the journal *Przegląd Filozoficzny* in 1909 (cf. Sierpiński 1909).

The aim of Sierpiński was to consider the rôle and meaning of the concept of correspondence in mathematics. He considered various domains in which this concept occurs, in particular he spoke about the equipollent sets and cardinal numbers, about operations, analytical geometry, complex numbers, geometry (cartography, projective geometry, descriptive geometry), analysis and at the end about functions. He comes to the conclusion that the concept of correspondence belongs to the most important concepts of mathematics. In (1909, p. 8) he writes:

> It infiltrates all domains of human thought; it is a foundation on which we build our basic concepts; it is a source of all best ideas.[1]

A justification of this fact he finds in Poincaré who in *La Science et l'hypothèse* (1902, p. 32) wrote:

> Mathematicians do not investigate objects but relations between them: hence they can freely replace the investigated objects by any other insofar the interrelations are not changed.[2]

And Sierpiński finishes his considerations stating that:

---

1 „Przenika ono wszystkie dziedziny myśli matematycznej; jest podstawą, na której budujemy inne zasadnicze pojęcia; jest źródłem wszystkich najwspanialszych pomysłów."
2 „Les mathématiciens n'étudient pas des objets, mais des relations entre les objets; il leur est donc indifférent de remplacer ces objets par d'autres, pourvu que les relations ne changent pas."

[...] the fact that mathematics – a domain that is so abstract, finds so many real applications can be explained by the existence of the correspondence between the domain of abstract objects and the reality.[3]

This is a strong thesis concerning one of the fundamental problems of the philosophy of mathematics – the problem of connections between the pure mathematics and the applied mathematics as well as the problem of the mathematical character and structure of the physical world. Sierpiński does not solve the problem and does not justify his claim. But it is not important from our point of view. More important is the very fact that he has chosen just a subject from the philosophy of mathematics as a subject of his *Habilitation* lecture.

A similar choice was made some years later by Zygmunt Janiszewski who as a subject of his *Habilitation* lecture (held on 11th July 1913 at the meeting of the Council of the Philosophical Faculty of Jan Kazimierz University in Lvov) has chosen (though his *Habilitationsschrift* was devoted to topology) a problem of the controversy between realists and idealists in the philosophy of mathematics. The title of his lecture was: „O realizmie i idealizmie w matematyce" [On realism and idealism in mathematics] – it was published in 1916 under the same title in the journal *Przegląd Filozoficzny* (cf. Janiszewski 1916).

The controversy between realism and idealism was (and still is) present in the philosophy of mathematics from the very beginning (compare the ontological conceptions concerning the objects of mathematics by Plato who can be treated as a founder of the idealism and by Aristotle who can be seen as the founder of realism). The controversy occurred with great vividness on the turn of the nineteenth and the twentieth centuries in connection with Cantor's set theory and especially after 1904 Zermelo's proof of the well-ordering theorem which turned the attention of mathematicians at the controversial axiom of choice.[4] The essence of this controversy can be reduced to the question: "what does it mean 'to exist' in mathematics?". Note that both the axiom of choice as well as Zermelo's well-ordering theorem state the existence of certain objects – the selector in the case of the axiom of choice and the well-ordering in case of Zermelo's theorem – in a non-constructive way, i.e., they provide no information on the postulated objects or on their possible construction.

Janiszewski analyzes in his paper possible answers and solutions of realists and idealists and indicates difficulties they meet. He considers also sufficient and necessary conditions of the existence in mathematics. The necessary condition of the existence is of course the consistency. But does it also suffice? Idealists

---

3 „[...] fakt, że nauka, tak oderwana, jaką jest matematyka, znajduje tyle zastosowań realnych, wytłumaczyć daje się istnieniem doskonałej odpowiedniości między dziedziną abstrakcji a dziedziną realnej rzeczywistości."

4 The well-ordering theorem is in fact equivalent – on the basis of an appropriate system of set theory – to the axiom of choice.

answer this questions positively. According to them 'to exist' means exactly the same as 'to be consistent'. Realists on the other hand maintain that the answer is negative, i.e., they claim that in mathematics only those objects do exist that have "a (good) definition" (Janiszewski 1916, p. 163). In this way another question arises: what is a *good* definition?

According to realists a set is determined and defined when – if not all of its elements can be individually defined then at least a rule of construction of an arbitrary element of the set is provided (cf. Janiszewski 1916, p. 168). Idealists claim that a set can be determined and defined without determining its particular individual elements. A set is given when a criterion of being its element is known (such a principle was adopted already by Cantor).

Janiszewski comes to the conclusion that the controversy between idealists and realists in the philosophy of mathematics shows that (cf. 1916, p. 169):

> [...] contrary to the common opinion that mathematical considerations are certain and clear one finds here controversies.[5]

Such situation occurred several times in mathematics. But every time solutions have been found. Will it be so also in the philosophy of mathematics? Janiszewski is here a pessimist and writes (cf. 1916, p. 170):

> One should doubt it. Differences between philosophical views which occur in the case of this controversy and which are its source are in fact the same as it happened in the quarrel between nominalists and platonists in the Middle Ages, a quarrel that lasts also today between positivism and idealism.[6]

Note that Janiszewski does not declare on which side of the controversy his sympathies are. He restricts himself to the presentation of various opinions and arguments. And this is typical – as will be seen later – for the whole group of Warsaw mathematicians.

Let us come back to the problem mentioned already above in connection with the *Habilitation* lecture of Sierpiński, namely let us ask why did Sierpiński and Janiszewski choose as subjects of their *Habilitation* lectures just problems from the philosophy of mathematics though they were in fact genuine mathematicians and not philosophers? Was their decision a consequence of the fact that their *Habilitation* procedures took place at the Philosophical Faculty and that most of the members of the Faculty Council were just classical scholars and not mathema-

---

5 „[...] w przeciwieństwie do rozpowszechnionego mniemania o bezwzględnej oczywistości i pewności rozumowań matematycznych i tu spotykamy kwestie sporne."
6 „O tym należy wątpić. Różnica bowiem filozoficznych poglądów, która się objawia w tym sporze, która jest jego źródłem – jest ta odwieczną różnicą, która powodowała przez średniowiecze ciągnący się spór między nominalistami a platończykami, który ciągnie się i dziś między pozytywizmem a idealizmem."

ticians? In this situation a strictly mathematical (more technical) subject could not arouse their interest. But in any case both could choose for example a subject from mathematics on a more popular level. The fact that they have chosen just subjects from the philosophy of mathematics indicates that at the university in Lvov there was a good intellectual atmosphere towards the foundations and philosophy of mathematics and that they both were interested not only in mathematics itself but also in the philosophical problems of this discipline. Both were also convinced that a definite conception of developing mathematics in Poland is needed in order to cultivate it and to develop it, moreover, methodological foundations of it should be defined – as we shall see they considered set theory to be such a foundation.

The interests of Janiszewski in the philosophy of mathematics can be seen already earlier on the example of the book *Poradnik dla samouków* [Handbook (Guide) for Autodidacts] published in 1915. Janiszewski was *spiritus movens* of the whole project and the author of the greatest number of chapters – besides the general introduction and the concluding remarks as well as the information chapter he wrote about differential equations, functional, difference and integral equations, about series, foundations of geometry and about logic and the philosophy of mathematics.[7] Just the last two papers, i.e., „Logistyka" [Logistics] (Janiszewski 1915a) and „Zagadnienia filozoficzne matematyki" [Philosophical problems of mathematics] (Janiszewski 1915b) are the most important from our point of view.

The chapter "Logistics" is devoted to the presentation of mathematical logic (called also symbolic logic or – especially at that time – logistics). Janiszewski begins by explaining why in this book devoted in fact to mathematics there is a chapter on logic and he gives fours reasons (cf. 1915a, p. 449):

> a) logistics is formulated in a form of a calculus (algebra of logic) and mathematics is thought as a science about any calculus,
> b) it is the unique discipline that can be applied in mathematics,
> c) in some parts (e.g., in the theory of relations) it considers the same objects as mathematics but treated a bit more generally,
> d) the logistic *calculus* has not only a logical interpretation but also a mathematical one, hence it belongs without any doubts also to mathematics (namely to set theory).

And he adds in the footnote that also an interpretation of logistics in number theory is possible.

---

[7] Besides Janiszewski particular chapters of *Poradnik* were written by: S. Kwietniewski – chapters about analytical, synthetic, descriptive and differential geometry as well as about the history of mathematics, W. Sierpiński – chapters about arithmetic, number theory, algebra, set theory, real functions, differential and integral calculus, S. Zaremba – chapters about analytic functions, partial differential equations, group theory and variation calculus as well as S. Mazurkiewicz – a chapter about probability. The introductory chapter "About science" was written by J. Łukasiewicz.

Janiszewski characterizes logistics saying that it is *"formal logic* (i.e., a science about forms of pure thought) *using the mathematical method*; more exactly: a method which so far was applied in a broader extent only by mathematics" (p. 449).

As one of the most important features which distinguishes logistics and simultaneously differentiates it from other forms and domains of logic is – according to him – the fact that one uses in it symbolism.[8]

In the discussed chapter Janiszewski presents the main facts from the history of logistics and says about its main achievements. He stresses that the attacks on mathematical logic and the depreciating of its meaning do not provide in fact any serious arguments and "witty and full of deeper thoughts as well as evil-minded chapters of Poincaré's book *Science et méthode* about logistics are rather satire than critique" (1915a, p. 456).[9]

It is interesting to see his comments on the relations between logic and mathematics as well as on the status of logic. Janiszewski is conscious of the fact that mathematical logic can be a convenient and useful tool of the analysis of language and reasonings and that "sometimes the logistical calculus can provide good methods and help to draw conclusions" (1915a, footnote 1 on page 456). He explicitly declares (cf. 1915a, p. 455):

---

8   And he adds that this feature "became the reason why it is so unpopular among philosophers" (1915a, p. 450).

9   Poincaré wrote in *Science et méthode* (1908, Livre II, Chapitre III: Les Mathématiques et la Logique, VII. La pasigraphie): "The essential part of this language consists of some algebraic symbols denoting connectives: if, and, or. Maybe they are useful, but if they will help to renew the whole philosophy is another question. It is hard to suppose that the word *if* as written in the form $\supset$ gains some new power.

This invention of Peano was called in former times a *pasigraphy*, i.e., the art of writing a mathematical treatise using no word of the colloquial language. This name indicates very well the applicability of this art. Later on it became more dignified by being called *logistic*. This word is used in Military Schools to denote the art of guiding and placing apart the army in the camp; it is clear that the new logistic had nothing to do with that, that the new name claims to do a revolution in logic."

(L'élément essentiel de ce langage, ce sont certains signes algébriques qui représentent les différentes conjonctions: si, et, ou, donc. Que ces signes soient commodes, c'est possible; mais qu'ils soient destinés à renouveler toute la philosophie, c'est une autre affaire. Il est difficile d'admettre que le mot *si* acquiert, quand on l'écrit $\supset$, une vertu qu'il n'avait pas quand on l'écrivait si.

Cette invention de M. Peano s'est appelée d'abord la *pasigraphie*, c'est-à-dire l'art d'écrire un traité de mathématiques sans employer un seul mot de la langue usuelle. Ce nom en définissait très exactement la portée. Depuis, on l'a élevée à une dignité plus éminente, en lui conférant le titre de *logistique*. Ce mot est, paraît-il, employé à l'École de Guerre, pour désigner l'art du maréchal des logis, l'art de fair marcher et de cantonner les troupes; mais ici aucune confusion n'est à craindre et on voit tout de suite que ce nom nouveau implique le dessein de révolutionner la logique.)

> Some knowledge of logistic should be recommended to everybody who wants to know something about today's state of logic, hence in particular to philosophers and in some sense also to mathematicians [...]. It becomes for them indispensable when they want to deal with the philosophy of mathematics.[10]

Knaster writes in (1960, p. 2) that Janiszewski attempted "to get the deep knowledge of mathematical logic called at that time logistics and began to use it practically". He applied mathematical logic "first of all to solve methodologically mathematical problems by using broadly a specific set-theoretical symbolism" as well as "to unclose gaps and unclear places in the structure of mathematical concepts even so fundamental as a line or a plane".

It is worth noting that in the discussed chapter "Logistics" Janiszewski stresses that (mathematical) logic is an independent and autonomous mathematical discipline and not only a method or tool of mathematics (cf. 1915a, p. 456) as well as that "its aim is not the (at least immediate) practical benefit" (1915a, footnote 1 on page 454). This remark is really important when one takes into account the fact that Janiszewski had studied in France! Such a "pro logical" attitude and the emphasizing of the rôle and meaning of mathematical logic for mathematics itself together with the firm stressing of its autonomy and independence is very important and characteristic for the Warsaw School and certainly has contributed to the development of the Warsaw School of Logic.

The second of Janiszewski's papers mentioned above, i.e., the paper „Zagadnienia filozoficzne matematyki" (1915b) is devoted to philosophical problems of mathematics. The author presents problems of a philosophical nature that are connected with mathematics, in particular the problem of the deductive vs. inductive character of mathematics, of the character of mathematical induction, of the correctness of definitions, of the nature of mathematical objects and of the art of their existence, he presents the controversy between idealists and realists, describes the rôle and meaning of antinomies, considers philosophical problems connected with the concept of a space and the problem of the nature and character of geometrical theories as well as the sensibleness of the question about their truthfulness. References to each of those problems are provided and at the end a list (with comments) of general literature in the philosophy of mathematics has been attached. Those lists show that Janiszewski had a good knowledge of the actual philosophical literature concerning mathematics. He formulated problems in a refined way. What is characteristic is the fact that – as in his other publications – he never expressed his own views and restricted himself to the (skilled and competent) presentation of views of others showing in this way the full spectrum of opinions and the complexity of problems. He

---

10 „Pewne zaznajomienie się z logistyką należy polecić każdemu, kto chce mieć pojęcie o dzisiejszym stanie logiki, szczególniej więc fachowym filozofom, a poniekąd i matematykom [...]. Staje się zaś ona dla nich niezbędna, jeśli zechcą się zająć filozofią matematyki."

stressed the independence of the research work of a mathematician of some of the philosophical controversies though he remarked that there are controversial philosophical problems that can influence the research in mathematics. He wrote in (1915b, p. 470):

> Problems mentioned in previous sections are located – so to say – beyond the scope of the activity of a mathematician: views he can have, if any, will have no influence – at least directly – on his work in the mathematics and will not hinder the communication with other mathematicians. Independently of how do they understand natural numbers or what do they think about mathematical induction, all mathematicians will use them in the same way. However there are such controversial problems that have a direct influence on actual mathematical investigations. They concern the *validity* and *soundness* of certain mathematical reasonings as well as the *objectivity* of certain mathematical concepts.[11]

Among the latter Janiszewski mentions the controversy concerning the imaginary quantities, infinitesimal calculus, the summation of series or the Poncelet's principle of continuity which have now only historical character as well as – still important, as he writes – the problem of the correctness of definitions (should impredicative definitions be allowed in mathematics?) or problems connected with set theory.

Set theory just mentioned played an important rôle in the Warsaw School. The story began with the discovery made by Sierpiński. In 1907 he discovered the amazing fact that the plane and the line have the same number of points. Soon he learned[12] that this fact had been discovered already thirty years before by Georg Cantor and that it belongs to the fundamental results of the new mathematical discipline, namely to set theory. This was the beginning of Sierpiński's interests in this domain. Being since 1910 a professor of Jan Kazimierz University in Lvov[13]

---

11 „Zagadnienia, poruszone w poprzednich paragrafach, znajdują się, że tak powiemy, poza obrębem działalności matematyka: jakiekolwiek będzie on miał poglądy na nie, czy też nie będzie ich mieć wcale, to nie wywrze [to] – przynajmniej bezpośrednio – wpływu na jego pracę w obrębie matematyki i w tym obrębie nie utrudni porozumienia z innymi matematykami. Bez względu na to, za co uważają liczby naturalne albo indukcję matematyczną, wszyscy matematycy będą się nimi posługiwać w jednakowy sposób. Istnieją jednak i takie kwestie sporne, które mają wpływ bezpośredni na aktualną pracę matematyczną. Dotyczą one *ważności* pewnych rozumowań matematycznych i *przedmiotowości* niektórych pojęć matematycznych."
12 Mostowski writes in (1975, p. 9) that when Sierpiński discovered this fact, he wrote to his colleague T. Banachiewicz, the future professor of astronomy of the Jagiellonian University, who at that time studied in Göttingen, asking him whether this result is known. Banachiewicz answered the question sending a telegram containg the unique word: "Cantor". In this way he called Sierpński's attention to Cantor's works – and the latter began to study them.
13 He held one of two chairs in mathematics, the other was held by Józef Puzyna.

he gave there lectures, among others, in set theory.[14] He wrote also a handbook *Zarys teorii mnogości* [An Outline of Set Theory] (1912).

At the beginning of the First World War Sierpiński was sent by Russian authorities to an internment camp in Wiatka.[15] Thanks to the help of his Russian colleagues he was able to move to Moscow where he collaborated with N. Lusin and learned about the theory of analytic sets which was developed there. In the future he appeared to be one of the most important persons who developed this new domain of set theory called descriptive set theory.

In Lvov Sierpiński aroused young mathematicians' interest in set theory. Among them were Zygmunt Janiszewski, Stefan Mazurkiewicz and Stanisław Ruziewicz. When in 1915 the Russian authorities evacuated his university from Warsaw to Rostov upon Don and when some months later Polish university has been opened in Warsaw, among its first professors were just Z. Janiszewski and S. Mazurkiewicz. At the end of 1918 W. Sierpiński joint them and received the chair of mathematics. In this way there were at one place three scholars interested in set theory.[16]

In 1917 Janiszewski wrote a paper „O potrzebach matematyki w Polsce" [On the needs of mathematics in Poland] (1917). This small (6 pages) paper became a program for the whole generation of Polish mathematicians. Janiszewski postulated there that the future scientific activity should be concentrated on one domain of mathematics[17] and that a new mathematical journal should be erected. In (1917, pp. 15 and 18) he wrote:

> According to the above described project a strictly scientific journal should be erected, a journal that would be devoted exclusively to one of those branches of mathematics in which we have outstanding, really creative and numerous scientific workers. In this journal [...] only papers written in one of the four languages recognized in mathematics as international would be accepted [...]. This journal would contain, beside original papers, also bibliographies of this branch, summaries and even reprints of papers published elsewhere, in particular translations of valuable papers printed in non "international" languages, hence first of all papers written in Polish which are wasted being unknown; finally correspondence: answers to questions [...].

---

14 The opinion – proclaimed sometimes – that Sierpiński's lectures were the first in the world in this new domain of mathematics is erroneous. Earlier lectures in set theory were given by Ernst Zermelo (Göttingen, 1900–1901), Felix Hausdorff (Leipzig, 1901) and Edmund Landau (Berlin, 1902–1903, 1904–1905).
15 Sierpiński was on vacations when the war began.
16 Ruziewicz became professor of the Technical University and of Jan Kazimierz University in Lvov as well as a rector of the Academy of Foreign Trade.
17 How important this was can be seen from the following anecdote told by E. Marczewski in (1948, pp. 17–18) where he wrote: "when [...] in 1911 Puzyna, Sierpiński, Zaremba and Żorawski met in the section of mathematics at the Conference of Scientists and Physicians in Cracow, they found no common subject to discuss: their scientific interests were extremely different."

[...] let us come back to the problem of mathematical creativity. Here an appropriate atmosphere can be created only when people will work on common problems. Collaborators are indispensable for a scientist. A person working alone does mostly decay. The reasons are not only of a psychic nature (lack of a stimulant): an isolated scientist *knows* much less than those working together. Only the results of researches reach him – they are then already polished and published and this happens usually some years after they have been obtained. An isolated scholar did not see the way in which they have been obtained, he did not experience together with their creators the process of discovering. "We are far from those hotbeds and kettles in which mathematics is being made, we come delayed and – there's no remedy for it – we must remain behind" as one of Russian mathematicians once said to me in Göttingen about his compatriots. Even more this can be applied to us!

Consequently if we do not want to "remain always behind" we should use radical means and to reach the basis of the evil. We should create such a "hotbed" by us! And we can reach this aim only by concentrating most of our mathematicians on the work in one branch of mathematics. It happens now by itself – one should only come to its aid. It is sure that the erection by us of a special journal devoted to one branch of mathematics will attract many [scholars] to the work just in this branch.

The journal would help to create by us such a "hotbed" also in another way: we would become then a technical center of mathematical publications in this branch. Manuscripts of new papers would be sent to us and others would be in contact with us.[18]

A natural candidate for such a branch of mathematics on which the research activity and efforts should be concentrated was just set theory and related domains like topology, theory of real functions etc. – it was just the domain of scientific interests of the group of Warsaw mathematicians who moved to Warsaw from Lvov as well as of a part of Lvov mathematicians. In order to make possible the

---

18 „W myśl powyższego projektu należałoby założyć u nas czasopismo ściśle naukowe, poświęcone wyłącznie jednej z tych gałęzi matematyki, w których mamy pracowników wybitnych, prawdziwie twórczych i licznych. Czasopismo to [...] przyjmowałoby artykuły w każdym z czterech językach uznanych w matematyce za międzynarodowe [...]. Pismo to zawierałoby, obok artykułów oryginalnych, bibliografie tej gałęzi, streszczenia, a nawet przedruki ważniejszych artykułów, drukowanych gdzie indziej, szczególnie zaś tłumaczenia artykułów wartościowych, drukowanych w językach nie „międzynarodowych", a więc przede wszystkim prac polskich, które marnują się nieznane; wreszcie korespondencje: odpowiedzi na zapytania [...].

[...] powróćmy do sprawy twórczości matematycznej. Tu atmosferę odpowiednią może wytworzyć dopiero zajmowanie się wspólnymi tematami. Konieczni prawie dla badacza są współpracownicy. Odosobniony najczęściej zamiera. Przyczyny tego są nie tylko psychiczne, brak pobudki: odosobniony *wie* o wiele mniej od tych, co pracują wspólnie. Do niego dochodzą tylko wyniki badań, idee już dojrzałe, wykończone, często w kilka lat po swym powstaniu, gdy ukażą się w druku. Odosobniony nie widział, jak i z czego one powstawały, nie przeżywał tego procesu razem z ich twórcami. „Jesteśmy z daleka od tych kuźni czy kotłów, w których wytwarza się matematyka, przychodzimy spóźnieni i, nie ma rady, musimy pozostać w tyle" mówił mi w Getyndze o swoich rodakach pewien uczony matematyk rosyjski. O ileż bardziej stosuje się to do nas!"

second of Janiszewski's proposals, a new journal under the title *Fundamenta Mathematicae* was founded. On the cover of the first issue of *Fundamenta* it was written[19] that it is a journal devoted to "set theory and related problems (direct applications of set theory), Analysis Situs[20], mathematical logic, axiomatic studies". The first volume appeared in 1920.[21]

Janiszewski and others saw and accepted the connections of set theory with other (both classical as well as being just developed) domains of mathematics and looked at it not as an isolated theory. H. Lebesgue in a paper „Á propos d'une nouvelle revue mathématique: *Fundamenta Mathematicae*" (1922) written on the occasion of the publication of the second volume of *Fundamenta* wrote that "set theory has been removed beyond the mathematics by great priests of the theory of analytic functions", and if "now this ostracism towards set theory disappears" this happens thanks to the fact that "set theory which arose from the theory of analytic functions could become useful for its elder sister and could show to the people of goodwill its advantages and its richness".

The conviction of the founders of the Polish Mathematical School concerning the rôle and meaning of set theory in mathematics found its emphatic expression in the book *Poradnik dla samouków* [Handbook (Guide) for Autodidacts] mentioned above. In the chapter „Teoria mnogości w stosunku do innych działów matematyki" [Set Theory in Relation to Other Domains of Mathematics] written by Stefan Mazurkiewicz and published in the third volume of *Poradnik* (it was a supplement to the volume I) he wrote (1923, pp. 89–90):

> Considering Janiszewski's table of "domains of mathematics" (Handbook, vol. I, pp. 22/23) we see the special place of set theory in this classification. This table has two wings, what agrees with the traditional division of mathematics into two branches: on the left one has analysis (together with arithmetic and algebra), on the right – geometry. In the middle one finds only two theories: set theory and group theory. – Notice that moving in the table from the top to the bottom one moves from simpler, more primitive and self-contained domains to domains that are more complex and that need some auxiliary means from outside. In this way one obtains a sort of a pyramid of mathe-

---

Otóż, jeśli nie chcemy zawsze „pozostawać w tyle", musimy chwycić się środków radykalnych, sięgnąć do podstaw złego. Musimy stworzyć taką „kuźnię" u siebie! Osiągnąć zaś to możemy tylko przez skupienie większości naszych matematyków w pracy nad jedną gałęzią matematyki. Dokonywa się to obecnie samo przez się, trzeba tylko temu prądowi dopomóc. Otóż niewątpliwie utworzenie u nas specjalnego pisma dla jednej gałęzi matematyki pociągnie wielu do pracy w tej gałęzi.

Lecz jeszcze w inny sposób pismo dopomogło by do wytworzenia się u nas tej „kuźni": bylibyśmy wtedy ośrodkiem technicznym publikacji matematycznych w tej gałęzi. Do nas przysyłano by rękopisy nowych prac i utrzymywano by z nami stosunki."

19 This phrase was repeated in each of the following volumes.
20 Called today topology – my remark, R.M.
21 Unfortunately Janiszewski did not live to see the publication of this volume – he died at the age of just 31 years on 3rd January 1920 during the influenza pandemic.

matical skills based of course on the top. This top is just set theory having in the table the top position. Directly under it there are foundations of arithmetic, foundations of geometry and topology. At last one sees numerous "connection lines" that go (usually centrifugally) from set theory in all directions. Summing up one can say that the table grants set theory almost the dominating (since both fundamental and central) position in mathematics, moreover it stresses the influence of set theory on other domains.[22]

Further Mazurkiewicz considers the meaning and rôle of set theory in the theory of real functions, in analysis, in geometry and in the foundations of mathematics. He stresses that the theory of real functions "gave the first impulse to the development of set theory and today it is in the most part a direct application of the latter" (cf. 1923, p. 90). And he adds that "set theory leads in the theory of real functions first of all to the systematization of problems and to the introduction of a structure into the shapeless mass of small results" (1923, p. 92). He shows also that the investigations in the functional calculus are essentially dependent of set theory, in particular they depend on the generalization of the very concept of a function. In geometry – according to Mazurkiewicz – set theory did not find so far "a broader application and it will probably not find it" (1923, p. 97). Nevertheless he notices that we owe to set theory "the immense enrichment of our knowledge of spacial forms" (1923, p. 97).

The above quotations show that in the Warsaw School of Mathematics one treated set theory as the foundation of mathematics in the methodological and not in the philosophical (i.e., ontological and epistemological) sense. One treated it rather as an auxiliary (though having the fundamental meaning) theory than as a separate, lonely and self-contained theory. One was aware of the fact that set theory (like topology) starts to be developed and is – as Mazurkiewicz put it in the paper published in the third volume of *Poradnik* mentioned above – "in embryo"

---

22 „Rozważając ułożoną przez Janiszewskiego tablicę „podziału matematyki" (Poradnik, t. I, str. 22/23), dostrzegamy, że stanowisko teorii mnogości zostało w tablicy tej wyznaczone w sposób bardzo szczególny. Tablica jest dwuskrzydłowa, co jest zgodne z tradycyjnym podziałem matematyki na dwie gałęzie: po lewej stronie mamy analizę (łącznie z arytmetyką i algebrą), po prawej geometrię. Na linii środkowej znajdujemy dwie tylko teorie: teorię mnogości i teorię grup. – Zauważmy nadto, że przesuwając się w tablicy omawianej od góry ku dołowi, przechodzimy na ogół od działów prostszych, bardziej pierwotnych i samowystarczalnych – do bardziej złożonych i wymagających z zewnątrz czerpanych środków pomocniczych, tym sposobem mamy tu rodzaj piramidy umiejętności matematycznych, opartej oczywiście na wierzchołku. Otóż tym wierzchołkiem jest teoria mnogości, która zajmuje w tablicy miejsce szczytowe, mając pod sobą bezpośrednio podstawy arytmetyki, podstawy geometrii i topologię. – Wreszcie widzimy liczne „linie związku", rozchodzące się (przeważnie odśrodkowo) od teorii mnogości we wszystkich kierunkach. – Reasumując, powiedzieć można, że tablica nadaje teorii mnogości stanowisko niemal dominujące w matematyce (gdyż zarazem podstawowe i centralne), ponadto zaś uwydatnia jej oddziaływanie na inne działy."

(1923, p. 98). This fact "strongly disturbs the possibility of a broader application of them [i.e., of set theory and topology – my remark, R.M.] in mathematics" (1923, p. 98). However "along with the development of set theory itself its meaning will certainly grow"(*ibidem*).

The treatment of set theory as the foundation of mathematics in the methodological sense found its expression in the stress put on its applications in other domains of mathematics. An expression of it is the fact that in papers published in *Fundamenta Mathematicae* rather not much place was taken by "inner" problems of set theory. Much more attention was paid to its applications in other theories like topology, function theory or analysis.

One should stress here that the members of the Warsaw School were conscious of the connections between set theory on the one hand and logic and the foundations of mathematics as well as the philosophy of mathematics on the other. In the paper „Teoria mnogości w stosunku do innych działów matematyki" (1923) from the third volume of *Poradnik* mentioned above Mazurkiewicz refers to the paper by Janiszewski „Zagadnienia filozoficzne matematyki" from volume I (cf. Janiszewski 1915b) and notices that (cf. 1923, p. 98)

> [...] the discovery of some contradictions, i.e., of antinomies, in set theory became one of the motives of the revision of principles of the formal logic[23]

and that

> [...] on the base of the concept of a set there was undertaken (by the school of Peano and later by Russell and Whitehead) an attempt to pack the whole of mathematics into the frames of a uniform hypothetico-deductive system, an attempt that appeared to be imperfect but on the other hand extremely interesting thanks to tendencies to a synthesis connected with it.[24]

In the paper quoted by Mazurkiewicz, Janiszewski considers philosophical problems of set theory from the point of view of the controversy between realists and idealists and comes to the conclusion that set theory is indispensable in considering questions of the philosophy of mathematics. He writes (cf. 1915b, p. 486):

> To study the philosophy of mathematics one should *well* know set theory, arithmetic, foundations of geometry and basic concepts of the infinitesimal analysis; further

---

23 „[...] ujawnienie w łonie teorii mnogości pewnych sprzeczności, tj. antynomij, stało się jednym z motywów rewizji zasad logiki formalnej"
24 „[...] na gruncie pojęcia zbioru podjęta została (przez szkołę Peany, a następnie przez Russella i Whiteheada) próba wtłoczenia całej matematyki w ramy jednolitego systemu hipotetyczno-dedukcyjnego, próba wprawdzie ułomna, jednak niezwykle interesująca z uwagi na tkwiące w niej tendencje do syntezy."

the knowledge of the logistic is necessary; one needs at last a general philosophical education.²⁵

It is worth to quote here also the sequel of Janiszewski's opinion. He writes (1915b, p. 486):

> This does not suffice to work actively in this domain [i.e., in the philosophy of mathematics – my remark, R,M.]; it is necessary to understand mathematics in a deeper way and this can be expected only from persons who worked creatively in this domain. Let the example of so many philosophers who – being even well mathematically educated – made in their papers devoted to the philosophy of mathematics [various] mathematical mistakes and showed in this way the lack of understanding of mathematics (though not unacquaintance with it!) be a discouraging example. The lack of the philosophical education often causes by mathematicians dealing with those problems a misunderstanding of their philosophical aspect and the omission of many problems.²⁶

The consciousness of the rôle and meaning of logic in mathematics among members of the Warsaw School found its expression also in the fact that in the Editorial Board of *Fundamenta Mathematicae* beside three mathematicians (i.e., Z. Janiszewski, S. Mazurkiewicz and W. Sierpiński) there were also two logicians: Stanisław Leśniewski (1886–1939) and Jan Łukasiewicz (1878–1956).²⁷ Both were members of the Lvov-Warsaw Philosophical School founded by Kazimierz Twardowski and held chairs of philosophy at the Faculty of Mathematics and Natural Sciences of Warsaw University. Their duty in the board was to take care of the development of mathematical logic and the foundations of mathematics – it was planed that issues of *Fundamenta* will be alternately devoted to set theory and its applications and mathematical logic and the foundations of mathematics.²⁸ The importance attached by the founders of *Fundamenta* to logic and the foundations is already stressed by the very name of the journal. This collaboration of mathematicians and logicians (who had a philosophical and

---

25 „Do studiowania filozofii matematyki należy znać *dobrze* teorię mnogości, arytmetykę, podstawy geometrii i podstawowe pojęcia analizy nieskończonościowej; następnie konieczna jest znajomość logistyki; wreszcie potrzebne jest ogólne wykształcenie filozoficzne."
26 „Do czynnej jednak pracy na tym polu [tzn. w zakresie filozofii matematyki – uwaga moja, R.M.] to nie wystarczy; koniecznym jest głębsze zrozumienie matematyki, czego można oczekiwać tylko od tych, którzy sami w tej dziedzinie pracowali w sposób twórczy. Niech przykład tylu filozofów, którzy, mając duże nawet wykształcenie matematyczne, popełnili w swych pracach nad filozofią matematyki błędy matematyczne i wykazali niezrozumienie (choć nie nieznajomość!) matematyki, działa tu odstraszająco. Brak znowu filozoficznego wykształcenia powoduje często u matematyków, zajmujących się tymi zagadnieniami, niezrozumienie filozoficznej ich strony, przeoczenie po prostu całej masy zagadnień."
27 Leśniewski and Łukasiewicz were in the Editorial Board till 1928.
28 This plan was not fulfilled. The reason was that the number of papers in logic and the foundations submitted to the journal was too small.

not a mathematical background[29]) is really typical for the Warsaw School and distinguishes it from other schools.[30] It brought many interesting fruits. Thanks to the union of two different traditions and approaches both set theory as well as mathematical logic and the foundations of mathematics were developed in Warsaw in the interwar period in a remarkable way – Warsaw became in fact the first center of set-theoretical and logical investigations in the world. The collaboration of logicians and mathematicians contributed also to the broadening of the perspective in investigations concerning set theory and the foundations of mathematics. Whereas works by Sierpiński and Mazurkiewicz were devoted to problems connected with paradoxes of the infinity, to the axiom of choice and the continuum hypothesis or the descriptive set theory (hence to mathematical problems connected with set theory), works by Leśniewski and Łukasiewicz and their students (A. Lindenbaum, A. Mostowski, A. Tarski and others) were devoted to metamathematical problems of this theory, in particular to the problem of the independence of the axiom of choice, the dependence of this axiom and the continuum hypothesis on the hypothesis of large cardinals, interrelations between various definitions of the concept of finiteness, etc.

To finish our considerations let us mention one more essential feature of the Warsaw School of Mathematics. This school was an adherent of no definite tendency in the philosophy of mathematics – though one had good knowledge of actual trends and theories.[31] Important was only the correctness and fruitfulness of the applied methods. Important were the results and not the particular methods used. It found its expression first of all in the case of investigations of the axiom of choice. The latter was the source of controversies, being rejected by ones and accepted and applied by others. The members of the Warsaw School represented the position that one should investigate directly not the very axiom of choice and discuss its status from the point of view of the philosophy of mathematics but rather study mathematical consequences of this axiom and in this way to replace the philosophical considerations by precise mathematical ones. Sierpiński put it clearly in the following way (1965, p. 95):

> Still, apart from our personal inclination to accept the axiom of choice, we must take into consideration, in any case, its rôle in the set theory and in the calculus. On the other hand, since the axiom of choice has been questioned by some mathematicians, it

---

29 It should be stressed that they – in particular Łukasiewicz – had extremely good mathematical intuition. Their lectures found very good reception among students of mathematics and were appreciated by them.
30 More about this collaboration writes Duda in (2004).
31 Add that Janiszewski used to say about himself that he is not a mathematician but a philosopher and that he "[...] is doing mathematics in order to state how far can reach the human mind by the logical reasoning alone" ([...] zajmuje się matematyką dlatego, aby przekonać się, jak daleko może umysł ludzki dojść samym logicznym rozumowaniem) (Steinhaus 1921).

is important to know which theorems are proved with its aid and to realize the exact point at which the proof has been based on the axiom of choice; for it has frequently happened that various authors have made use of the axiom of choice in their proofs without being aware of it. And after all, even no one questioned the axiom of choice, it would not be without interest to investigate which proofs are based on it and which theorems are proved without its aid – this, as we know, is also done with regard to other axioms.

Andrzej Mostowski on the Foundations and Philosophy of Mathematics

Co-authored by Jan Woleński

The relations between mathematics, its foundations and philosophy are fairly subtle and complicated. One can speak about mathematical foundations of mathematics and philosophical foundations of mathematics. In fact, logicism, formalism and intuitionism can be perceived as foundational projects of investigating mathematics by mathematical means as well as different philosophies answering such questions as, for example, mathematical existence or the epistemological status of deduction. The expression "fundamental problems of mathematic" which comprise both approaches is a convenient label. Mathematicians working in technical problems of the foundations of mathematics have different attitudes related to the question how fundamental problems are mutually related. Some of them entirely disregard philosophical problems of mathematics, other, for example, Brouwer or Hilbert, are inclined to base their metamathematical research on explicitly accepted philosophical assumptions, which essentially influence the further course of technical work, but still other see mathematical and philosophical foundations of mathematics as independent although connected in a way and indispensable for understanding mathematical activity.

The third attitude is characteristic for Polish tradition in logic and the foundations of mathematics. Chwistek and Leśniewski were the only exceptions in this respect. The former proposed a certain version of logicism, but the latter, after the early stage of working on an improvement of *Principia Mathematica*, began to develop a radical version of finitism and constructivism based on extremely nominalistic presumptions. The considerable rest of Polish logicians and mathematicians represented the view guided by two following principles:

(P1) all commonly accepted mathematical methods should be applied in metamathematical investigations;

(P2) metamathematical research cannot be limited by any a priori accepted philosophical standpoint.

These two principles do not imply that mathematics is free of own genuine philosophical problems or that mathematicians should neglect these problems as exceeding their professional activities. According to Polish school, although formal metamathematical results do not solve philosophical controversies about mathematics, yet the former illuminate the latter.

The tradition of Polish analytic philosophy, originated with Twardowski in Lvov and continued by the Lvov-Warsaw School (see Woleński 1989 for a comprehensive presentation of this) supplemented (P1) and (P2). According to Twardowski and his students, we must clearly and sharply distinguish world-views and the scientific philosophical work. This idea was particularly stressed by Łukasiewicz, the main architect of the Warsaw school of logic. He regarded various philosophical problems arising in formal sciences as belonging to world-views of mathematicians and logicians, but the work consisting in constructing logical and mathematical systems together with metalogical (metamathematical) investigations constituted for him the subject of logic and mathematics as special sciences. Hence, philosophical views cannot be a stance for measuring the correctness of formal results. Yet philosophy may serve as a source of logical constructions; Łukasiewicz's many-valued logic is a good example in this respect. Since all members of the Warsaw school of logic were Łukasiewicz's students, his opinion about the relation of logic to philosophy became influential.

Due to the above characterized attitude, Polish mathematicians did not accept any of the "big three" in the foundations, that is, logicism, formalism or intuitionism. The Polish view, as we can call it, has a particular and surprising consequence consisting in a freedom of accepting philosophical opinions sometimes being at odds with applied methods. Perhaps Tarski was an extreme example of this practice. He used all admissible mathematical methods in his logical works, in particular infinitary ones, usually associated with Platonism in the philosophy of mathematics, but he contributed to all mentioned grand projects by the idea of logical concepts as invariants (related to logicism), the theory of consequence operations (a component of formalism) and the topological semantics for intuitionistic logic. Tarski himself stressed that this methodological attitude, sometime labelled as "methodological Platonism", became a characteristic feature of Polish school and its essential contribution to metamathematics (Tarski 1954, p. 713; page-reference to the reprint):

> As an essential contribution of the Polish school to the development of metamathematics one can regard the fact that from the very beginning it admitted into metamathematical research all fruitful methods, whether finitary or not.

On the other hand, he had explicit sympathies to empiricism, nominalism, reism and finitism (he even called himself "a tortured nominalist"). Due to a strict departure of mathematics and philosophy (philosophical opinions of mathematicians were considered as somehow private in Poland) as well as locating them on different levels, no contradiction occurs in Tarski's position, although we certainly encounter here an example of a cognitive dissonance to some extent. Probably Tarski saw the situation in such a way and perhaps it explains why he usually abstained from a wider elaboration of his philosophical views, at least in his writings. Tarski was more involved in philosophy in oral debates concerning philosophy. It is well documented by the records of discussions between Tarski, Carnap, Quine and Russell in 1940–1941; these protocols were written by Carnap and are deposited in the University of Pittsburgh (see Mancosu 2005).

Let us turn now to Mostowski's views and opinions concerning the foundations and philosophy of mathematics. The strongest expression of his philosophical position were perhaps these ((a) Mostowski 1953a, p. 231, (b) Mostowski 1955, p. 16; (c) Mostowski 1955, p. 42):

> (a) There is no doubt that all mathematical concepts have been developed by abstraction from concepts formed on the basis of a direct experience. But he adds that this statement is not sufficient and the process of abstraction should be deeper analyzed.

> (b) Materialistic philosophy has since long been opposed to such attempts [i.e., attempts of treating mathematics solely as a collection of formal axiomatic systems – R. M., J. W.] and has shown the idealistic character both of Hilbert's program which consists in defining the content of mathematics by its axioms and of the neopositivistic program consisting in the explanation of the content of mathematics by an analysis of the language.

> (c) Results obtained by mathematical method confirm therefore the assertion of materialistic philosophy that mathematics is in the last resort a natural science, that its notions and methods are rooted in experience and that attempts at establishing the foundations of mathematics without taking into account its originating in natural sciences are bound to fail.

Unfortunately there is a problem of interpretation, because these papers were written in the first half of the fifties and the ideological atmosphere of that time could have had an influence on it (look at a typical slang of this time "confirm therefore the assertion of materialistic philosophy" or "materialistic philosophy shown", required in philosophical remarks at the time). It is not possible now to decide to what extent outside factors influenced the paper. On the other hand the author could restrict himself to purely mathematical issues and avoid entirely any philosophical remarks and declaration. If he did not do it we can treat his remarks as genuine, also because Mostowski's declared sympathies were consistent with materialism. He inherited his general philosophical attitude from Tarski, perhaps also some inclination to empiricism and a respect for nominalism (see Murawski 2004b for remarks about nominalistic tendencies among Polish logicians). We have also certain evidence that Mostowski sympathized with reism (the view that there are solely individual corporeal things) to some extent (quoted after Kotarbińska 1984, p. 73):

> Please imagine yourself that I heaved a sigh for reism there [that is, at the school on the foundations of mathematics – R. M., J. W.]. Presented ideas were results of speculations, so breakneck, so far elusive to intuition and incomprehensible, that reism appeared as an oasis in which one could rest and to breathe of fresh air.

Mostowski stressed that constructivistic trends in the foundations of mathematics are nearer to the nominalistic philosophy than to the idealistic one (in the platonic sense). This nominalistic character implies that constructivism does not accept the general notions of mathematics as given but try to construct them and implies that mathematical concepts can be identified with their definitions. According to Mostowski, an obvious advantage of nominalism is the fact that several important mathematical theories have been reconstructed in a satisfactory way on a nominalistic basis and those reconstructions have turned out to be equivalent to the classical theories.

On the other hand, Mostowski, respected the principles (P1) and (P2) and freely used infinitary methods and strongly insisted that formal work in mathematics and logic should not be bounded by philosophical assumptions. This position is clearly expressed in Mostowski (1949–1950), a paper devoted to the problem of the classification of logical systems. Mostowski stressed that though the investigations of this problem are purely formal, they nevertheless assume a definite philosophical point of view with respect to logical systems. In particular, he declares that the logical systems are to be considered not as empty schemes devoid of any interpretation and explicitly accepts the objective existence of the mathematical reality populated, e. g., by the set of all integers or the set of all real numbers. The objective existence means here being independent of all linguistic constructions. The aim of logic and mathematics is stated as follows (p. 164; page-reference to the reprint):

> The role of logical and mathematical systems is to describe this reality. Every logical sentence has thus a meaning: it says that the mathematical reality has this or some other property. If the mathematical reality does in fact possess the given property, then the sentence expressing this property is true, otherwise it is false.
> [...]
> The point of view characterized above allows us to make plausible the existence of undecidable sentences in almost all logical systems. Such sentences evidently exist if it is not possi9ble to prove all the true sentences in the considered system. Now every proof is in a logical system consists in a succession of some operations (called the rules of proof) which can be mechanically performed on one or two expressions. The unprovability of certain true sentences can be explained by the fact that the properties of "mathematical reality" are more complicated than the properties which can be established by successive applications of the rules of proof to the axioms.

These rather rich philosophical remarks are immediately frozen by the next remark:

> We do not intend to defend the philosophical correctness or even the philosophical acceptability of the point of view here described. It is evident that it is entirely opposite to the point of view of nominalism and related trends.

Taking into account Mostowski's sympathies, even fairly moderate, to nominalism and empiricism we find in his views the same tension which he attributed to Tarski (Mostowski 1967c, p. 81):

Tarski in oral discussions, has often indicated his sympathies with nominalism. While he never accepted the "reism" of Tadeusz Kotarbiński, he was certainly attracted to it in the early phase of his work. However, the set-theoretical methods that form the basis of his logical and mathematical studies compel him constantly to use the abstract and general notions that a nominalist seeks to avoid. In the absence of more extensive publications by Tarski on philosophical subjects, this conflict appears to have remained unresolved.

It seems, nevertheless, that Mostowski felt himself obliged to a more extensive and systematic treatment of his views in the philosophy of mathematics than it occurred in Tarski's case. It was probably a result of the following view (Mostowski 1948a, p. IV; unfortunately this book was published only in Polish, although English translation was planned – compare, e.g., the back cover of Kuratowski–Mostowski 1952 where Mostowski's *Mathematical Logic* is announced as a book in preparation):

> We are not able to deprive logic (independently of how it would be formal) of some, even sub-conscious, philosophical background. A conscious choice in this respect is more difficult, because, in the light of the contemporary discussion on the foundations of mathematics, it is impossible to say with certainty which view, of many competing, is the best or even good.

Mostowski considered this situation as one of the main difficulties encountered by the authors willing to write a book on mathematical logic. Yet he thought that the choice in question exceeds formal logic as such. After 17 years, he said (Mostowski 1965, p. 149):

> We see that the issue between Platonists, formalists and intuitionists is as undecided to-day as it was fifty years ago.

Guided by such evaluations Mostowski tried to intentionally avoid in his textbook any discussion of philosophical problems since they go beyond the limits of formal logic. For working purposes he treated a logical system as a language in which one speaks about sets and relations. An extensionality axiom has been adopted and it has been assumed that elements of the language satisfy principles of the simple theory of types. In Mostowski's opinion such a standpoint is convenient for formal considerations and coincides with views which are more or less consciously adopted by most of mathematicians (what does not necessarily mean that it must be accepted without any doubts by philosophers). However, one can observe that any logical system is interpreted (it speaks about set-theoretical objects). This corresponds with Leśniewski's view (also adopted by Tarski) known as "intuitionistic (better 'intuitive') formalism", that mathematics does not consist in purely formal games, because it uses languages equipped with comprehensible meanings, although formalized. This is perhaps the main reason why Mostowski was never especially attracted by formalism.

Mostowski, although sceptical toward prospects of ultimate solutions of principal philosophical controversies, was convinced that mathematical treatment of philosophical problems is illuminating. The following quotation is perhaps characteristic (Mostowski 1955, p. 3; similar remarks occurs in many other Mostowski's books and articles):

> The present stage of investigations on the foundations of mathematics opened at the time when the theory of sets was introduced. The abstractness of that theory and its departure from the traditional stock of notions which are accessible to experience, as well as the possibility of applying many of its results to concrete classical problems, made it necessary to analyze its epistemological foundations. This necessity became all urgent at the moment when antinomies were discovered, However, there is no doubt that the problem of establishing the foundations of the theory of sets would have been formulated and discussed even if no antinomy had appeared in set theory.

The general philosophical problems stemming from the discussions on the foundations of set theory cover:

(a) the question of the nature of mathematical concepts and its relation to the world;

(b) the nature of mathematical proofs and their correctness.

Mostowski had no illusions that (a) and (b) might be mathematically decided (Mostowski 1955, p. 3; see also Mostowski 1955a):

> These problems are of a philosophical nature and we can hardly expect to solve them within the limits of mathematics alone and by applying only mathematical methods.

However, he stressed that these general questions led to more specific ones which are subjected to a formal treatment, namely

(a') the rôle and limits of axiomatic method,
(a") the constructive tendencies,
(b') the axiomatization of logic,
(b") the decision problem.

And he ended his report with the following remark (Mostowski 1955, p. 42):

> Thus, as we see, the investigations of the foundations of mathematics are not without importance although they do not stand for a full investigation on the foundations of mathematics. Their results are to use for mathematics as well as for philosophy. In this sense they fulfil the tasks assigned to them.

The passing from (a) and (b) to (a')–(a") and (b')–(b") was related to Mostowski's historical perspective (see Mostowski 1965) in which the big three in the philosophy of mathematics, that is, logicism, constructivism and formalism, formed at the beginning of the 20th century, have been replaced in the 1930s by three new schools, namely set-theoretical, constructivistic and metamathematical. They are represented by Tarski's semantic theory of truth, Gödel's incompleteness theorems and Heyting's axiomatization of intuitionistic logic, respectively. In general, in the set-theoretical school the semantical properties of expressions of formalized languages are studied, constructivistic investigations are concentrated on various formal (mathematical) conceptual constructions and logical connections between them, and metamathematical investigations are devoted mainly to logical connections "inside" axiomatic systems and to logical properties of them. Clearly, these new paradigms of the foundations of mathematics prefer mathematical methods over philosophical analyzes.

Investigating constructivism played a special place in Mostowski's reflection about the foundations in mathematics. For a while he even believed that this direction will give the ultimate base for mathematics in the future (Mostowski 1948a, p. VI):

> I am inclined to think that the satisfactory solution of the problem of the foundations of mathematics will follow the line pointed out by constructivism or a view close to it. Yet it is not a sufficient base for writing a textbook of logic at the present moment.

Mostowski abandoned later the idea that constructivism is generally superior over other views, although he saw its advantages in particular cases – for instance, in arithmetic – because it dispenses us with assuming the existence of actual infinity or allows solutions for which nominalism is enough (remember that constructivism and nominalism are closely related to Mostowski's view). In general, Mostowski maintained that finitary, predicative and constructive (these qualifications are related, but not equivalent; we will use "constructive" and "constructivism" as general labels) methods are not sufficient for mathematics (see also Mostowski 1972b, pp. 29–32). However, contrary to many logicians and mathematicians who shared this critical view, Mostowski did not stop at pointing out limitations of constructivism, but very seriously took and examined claims of this program. According to him, constructivism is sometimes more philosophically satisfactory, for example, in the already mentioned case of arithmetic. Applied mathematics is another domain in which constructive approach is very promising. Thus, drawing the exact scope of constructive methods in the framework of classical mathematics constitutes an important task for mathematics as well as for philosophy. This was the background of Mostowski's idea of the degrees of constructivity (see Mostowski 1953a, Mostowski 1959) as attached to various mathematical theories (arithmetic of natural numbers, arithmetic of real numbers, theory of real functions or set theory; the axiom that every set is constructible expresses an amount of constructivity, although it is rather weak). A fruitful approach to this question requires a deeper analysis of how classical and constructivist formal schemes are intertranslatable.

Mostowski's works on constructivism entered more deeply into the relation between classical and constructive mathematics than any earlier comparisons of both foundational projects. Mostowski's position can be well-described by a certain distinction introduced by Heyting, namely that of theories of the constructible and constructive theories (in fact, this distinction was motivated by papers of Mostowski and Grzegorczyk delivered at the the Symposium on constructivity in Amsterdam in 1957; see Grzegorczyk 1959, Heyting 1959, Mostowski 1959). A theory of the constructible has three features: (i) a mathematical theory is presupposed in order to define the class of constructible objects; (ii) the notion of constructibility is defined, it is not a primitive concept; (iii) a liberty in the choice of a definition of the constructible, although a sufficient correspondence to our intuitive notion of a mathematical construction is required. On the other hand, a constructive theory takes the concept of constructibility as primitive and it is obliged to provide its precise characterization, for example, by the postulates generated by intuitionistic logic. Doubtless, Mostowski, strongly influenced by computable analysis of Banach and Mazur (see Mazur 1963), developed a theory of the constructible with various degree of constructivity.

Mostowski's understanding of constructivism is well expressed in his following words (Mostowski 1959, p. 180):

> My conception of constructivism will be as naive as possible and will consists in the following. I shall consider theories of real numbers and real functions in which not arbitrary real numbers or real functions are considered by only numbers of functions which belong to a certain class specified in advance. According to the choice of this class, we shall obtain different theories of arithmetic and analysis. Our choice of the initial class will not be arbitrary: we shall try to make the choice so that the elements of the chosen class satisfy certain conditions of calculability of effectiveness. We shall start with stringent conditions and then loosen them gradually and we shall see that it is possible in this way to systematize a good deal of older and also of more recent work of constructivists. I shall pay no attention to the way in which the classes just mentioned are defined and shall impose no limitations on methods of proof acceptable in dealing with numbers or functions belonging to these classes. This naive approach to constructivism is certainly objectionable from the constructivist point of view. It does not represent a constructivist development of a branch of mathematics but gives merely a glance of constructivism, so to say, from outside. The value (if any) of such an approach I see in the possibility of reviewing on a common background several of the simplest constructivistic conceptions; but more refined ones an especially those which, like intuitionism, impose restrictions on methods of proof must necessarily be excluded from such a review.

Summarizing Mostowski's attitude toward constructivism, we can say that he looked at this view from the classical point of view. Now it is clear that this understanding of constructivism had to irritate Heyting and provoke him to introducing the mentioned distinction between theories of the constructible and constructive theories. Heyting himself considered constructivism proper (intuitionism) as the position, which accepts only constructive theories. The difference is, then, that whereas Heyting's position was purely constructivistic, Mostowski represented a combination of constructvism and set-theoretical program, which constituted for him a workable basis for mathematically tractable foundations of mathematics. Perhaps a clear distinction between mathematical and philosophical foundations of mathematics is the most important general Mostowski's idea in his approach to foundational problems of formal science.

Another question elaborated by Mostowski concerned the interplay between syntax and semantics, the problem suggested by Gödel's incompleteness theorems and Tarski's theorem on the undefinability of truth; Mostowski was a great expert in these fundamental metamathematical results and refined general semantic method of proofs of them (see Mostowski 1946, Mostowski 1957a). From a general and philosophical point of view, he was the first who clearly and mathematically pointed out that semantics requires infinitistic methods, but finitary ones are suitable for syntax. The precise specification of differences between syntactic and semantic formulation of the incompleteness theorems became a by-product of this simple observation, which leads to an important conclusions (Mostowski 1965, p. 42, p. 50):

> The interpretation of a language is defined by means of set-theoretical concepts, which gives rise to the close relations between semantics and the set-theoretical, infinistic philosophy of mathematics; whereas the theory of computability leans toward a more finitistic philosophy. [...]
>
> By way of conclusion, let us try to evaluate Herbrand's and Gentzen's theorems from a more general point of view. There are undoubtedly two opposing trends in the study of the foundations of mathematics: the infinistic or set-theoretical and the finitistic or arithmetical. Herbrand's and Genzten's original discoveries belong of course to the second of these trends but the subsequent which has been based on these results has borrowed many ideas from the first. This influence of the set-theoretical approach is clearly visible in Bernays' consistency theorem [if all the axioms are effectively true in a model **M**, the same holds for their logical consequences – R. M, J. W.] in which semantic notions are consciously imitated in finitistic terms. We may say that Herbrand's and Gentzen's methods allow us to make finitistic certain particular cases of set-theoretical constructions.

The phrase "certain particular cases of set theoretical constructions" is here extremely important, because it alludes to limitations of purely syntactic methods as compared with richer semantic procedures.

Mostowski considered also other philosophical aspects of Gödel's incompleteness theorems although he was mainly concerned with their formal shape (Mostowski 1946 and Mostowski 1957a are among the first expositions of this celebrated results). In particular, he touched the concept of a proof and the idea of a formalization of mathematics. Mostowski declares that he is not going to enter into the discussion of philosophical problems whether questions which are unsolvable today are in fact "essentially undecidable" or not. He sees the source of difficulties here in the fact that we do not have a precise notion of a correct mathematical proof. A notion of a formal proof developed in mathematical logic made it possible to construct and investigate formal systems. It has been believed that such systems encompass the whole of mathematics, i.e., that any intuitively correct mathematical reasoning can be formalized in such systems. But since it is essentially impossible to prove that a given formal system coincides with the intuitive mathematics, hence (Mostowski 1957a, pp. 3–4)

> there is no immediate connection between the problem of completeness of any proposed formal system and the problem of existence of essentially unsolvable mathematical problems [but] [...] the problem of completeness of formalized systems is [...] important because it makes explicit the degree of difficulty of formalization of intuitive mathematics even if we restrict ourselves to that portion of mathematics which deals with integers. [...] in spite of all efforts of the logicians we are still very far from an exact understanding in what consists the notion of truth in mathematics.

Thus, there is a tension between the research practice of mathematicians and the idea that mathematics can captured be by formalized systems (see Mostowski 1972a). Mostowski says (Mostowski 1972a, pp. 82, 83, 84) that

> a full formalization of mathematics seems to be nowadays an out-of-date idea. Antinomies in set theory do not frighten any more. Mathematics [...] is being developed not paying any attention to what is happening in its foundations. [...] A mathematical proof is something much more complicated than a simple succession of elementary rules contained in the so called inference rules. [...] Therefore one must necessarily show moderation in stressing the role of logical rules in [mathematical] proofs. [...] The tendency to mechanize mathematical reasonings seems to me to be a highly dehumanized activity: as E. L. Post once wrote, the essence of mathematics consists in concepts of truth and meaning.

However, despite these reservations and doubts concerning formalization and its future prospects, Mostowski was definitively optimistic as far as the matter concerns relations between logic and mathematics (Mostowski 1972a, p. 83):

> the collaboration of logic and mathematics was fruitful and probably will still bring important results.

In fact it would be rather difficult to expect another attitude by logicians coming from Poland.

Set theory became a favorite field of Mostowski's studies from a philosophical as well as from a mathematical point of view (see Mostowski 1969 as his main contribution to formal set theory). Hence, Mostowski's philosophy of mathematics and his view how the foundations of mathematics should be done, can be additionally illustrated by his remarks about various problems of set theory. He was fully aware that philosophy must enter into set theory and its foundations (see Kuratowski–Mostowski 1952, pp. V–VI). Moreover, studies of the philosophy of set theory are important, because (p. V; these remarks are omitted in English translation of this book, although a general link between philosophy and set theory is indicated):

> There exists so far no comprehensive philosophical discussion of basic assumptions of set theory. The problem whether and to what extent abstract concepts of set theory (and in particular of those parts of it in which sets of very high cardinality are considered) are connected with the basic notions of mathematics being directly connected with the practice has not been clarified so far. Such an analysis is needed because by Cantor, the inventor of set theory, basic notions of this theory were encompassed by a certain mysticism.

One of the most important questions of set theory is the problem on which axioms should set theory be based. There is no absolute freedom in the choice of axioms, but one should choose axioms that guarantee that the theory based on them will have (Kuratowski–Mostowski 1952, p. V):

> an essential scientific value, i.e., will be able to serve in the process of getting known the material world, either directly or indirectly via other domains of mathematics for which it will be a tool.

Explaining the foundations of set theory is a challenging task due to its importance for the foundations and the philosophy of mathematics (Kuratowski–Mostowski 1952, p. vii):

> The great importance of mathematics as a *tool* for other mathematical fields, including branches of mathematics connected directly with applications, presently prevails, as it seems to be, the importance of investigations in set theory itself.
> Investigations on the foundations of set theory play also a great role in the general foundations of mathematics. The analysis of such concepts as consistency of axioms, their independence, categoricity, completeness, effectiveness of proofs – are closely connected with set theory.
> In this domain the influence of set theory can be especially strongly seen. In particular, thanks to the definition of a finite set and to the introduction of cardinals, the arithmetic of natural numbers could be founded on a firm basis. Simultaneously new problems connected with the general concept of an infinite set has been established and precisely formulated. This concept has no mystical character any more as it was the case through ages.

As it is to be expected a special attention was paid by Mostowski to controversial axioms and results as the axiom of choice and the continuum hypothesis. He (Kuratowski also shared this attitude) proceeded in a way characteristic for Polish mathematicians, that is, by separating the philosophy behind the axiom of choice on the one hand and its mathematical content and its rôle in mathematics on the other. This approach was initiated in Sierpiński (1918) and summarized in the following manner (Sierpiński 1965, p. 95; this remark was constantly repeated by Sierpiński since the 1920s, see Sierpiński 1923, p. 102–103, for instance):

> Still, apart from our personal inclination to accept the axiom of choice, we must take into consideration, in any case, its role in the set theory and in the calculus. On the other hand, since the axiom of choice has been questioned by some mathematicians, it is important to know which theorems are proved with its aid and to realize the exact point at which the proof has been based on the axiom of choice; for it has frequently happened that various authors have made use of the axiom of choice in their proofs without being aware of it. And after all, even if no-one questioned the axiom of choice, it would not be without interest to investigate which proofs are based on it and which theorems are proved without its aid – this, as we know, is also done with regards to other axioms.

Mostowski came back to philosophical problems of set theory also later, in particular, when he summarized the place of set theory in mathematics and science (see Mostowski 1972b) or commented Cohen's results on the independence of the axiom of choice and the continuum hypothesis (see Mostowski 1964, Mostowski 1967a, Mostowski 1967b, Mostowski 1968). In this context, he considered the following questions: What are sets and how their laws can be discovered? What are in particular sets of reals? Can every set be determined by defining the property of its elements (and consequently, is it identical with this property) or is it a certain abstract object which exists independently of our mental constructions? He concludes (Mostowski 1968, p. 177):

> Unfortunately the problem of truth in mathematics is not simple. Repeat: If sets existed in the same sense as physical objects then we could expect that the truth or falsity of the continuum hypothesis would be ultimately discovered. But if sets are only our own mental construction then the answer to the question whether the continuum hypothesis is true or false can depend on what constructions we will accept as allowed.

Since (see also earlier remarks on the concept of truth in the light of Gödel's results) – according to Mostowski – nothing can be said about the admissibility of platonism in set theory, one does not know whether the question about the truth or falsity of the continuum hypothesis has any sense. On the other hand formal problems concerning its consistency and independence are reasonable and interesting to high degree. The independence results of Cohen (supplementing the earlier results on consistency due to Gödel) do not solve the problem of truth in set theory. Moreover, since the continuum hypothesis and the axiom of choice cannot be decided (on the basis of accepted axioms of set theory), this suggests (see Mostowski 1968, p. 176) one of the most important arguments against mathematical Platonism. The situation is analogous to the situation in geometry: axiomatic set theory is now in the same situation as axiomatic geometry was after works of Klein and Poincaré which indicated the real meaning of the problem of truth of the parallel axiom. After the results of Cohen various possible but mutually inconsistent axiomatic set theories can be constructed. If this will be the case, then (Mostowski 1968, p. 182)

> we shall be forced to admit that in the match between platonism and formalism the latter has again scored one point.

In fact, Mostowski was inclined to a stronger conclusion (see Mostowski 1967a, Mostowski 1967b). He says that the complicated and not fully clear nature of the concept of set and consequently the possibility of various axiomatizations of set theory imply that – despite of great mathematical and philosophical importance of set theory – there are no chances that it will become a central mathematical discipline. On the one hand most (if not all) mathematical notions can be interpreted within set theory (this remarkable fact requires an explanation), but, on the other hand, we encounter essentially different concepts of set, which are equally suitable as basic for intuitive set theory. Mostowski concludes (Mostowski 1967a, pp. 94–95):

> Of course if there are a multitude of set theories then none of them can claim the central place in mathematics. Only their common part could claim such a position; but it is debatable whether this common part will contain all the axioms needed for a reduction of mathematics to set theory.

This concurs with Mostowski's doubts towards the axiomatization of set theory formulated in one of his earlier papers ((a) Mostowski 1955, p. 19; (b) Mostowski 1972b, p. 28–29):

> (a) A particularly perturbing fact which calls for explanation is that recently various new axioms have been added to the system of axioms of the theory of sets or the formulations of axioms have been altered; in consequence we have at present to choose between a great many essentially different systems of axioms of the set theory, yet there are no criteria indicating the proper choice among all these numerous systems. He was convinced that the ultimate formulation of axioms of set theory should be proceeded by a discussion of the fundamental assumptions of this theory.
>
> (b) [...] the mere incompleteness of Z–F is not an alarming symptom by itself. What is disturbing is our ignorance of where to look for additional information which would permit us to solve problems which seem very simple and natural but which are nevertheless left open by the axioms of Z–F. We come here very close to fundamental problems of the philosophy of mathematics whose basic question is: what is mathematics about? A formalist would say that it is about nothing; that it is just a game played with arbitrarily selected axioms and rules of proof. The incompleteness of Z–F is thus of no concern for a formalist. Platonists on the contrary believe in the 'objective existence'

of mathematical objects. A set-theoretical Platonist believe therefore that we should continue to think more about sets and experiment with them until we finally discover new axioms which, added to Z–F, will permit us to solve all outstanding problems. [...] Whatever the final outcome of the fight between these two opposing trends will be, it is obvious that we should concentrate on the study of concepts which seem perfectly clear and perspicuous to us. In Cantor's time the concept of an arbitrary set seemed to be a very clear concept, but the antinomies proved that this was not so. Today, this concept has been replaced by that of an arbitrary subset of a given set. In addition, the belief that all subsets of a given set form a set is almost universally accepted. However, it is by no means true that these views are shared by all mathematicians. Even Gödel himself, who [...] should be counted among the Platonists, has once expressed the view that the concept of an arbitrary subset of a given set is in need of clarification. [...]. The present writer believes (although he cannot present convincing evidence to support this view) that it is in this direction where the future of set theory lies.

The last quotation with typical reservations and caution expressed by Mostowski suggests that one should return once again to his picture of relations between philosophy and mathematics. Considering the general question whether mathematical objects can be treated as fully defined by appropriate systems of axioms (arithmetic of natural numbers is a good example), Mostowski states that the decision belongs not to mathematics but to philosophy and concludes (Mostowski 1955, pp. 15–16, 41–42):

The only consistent standpoint, confirming to common sense as well as to mathematical usage, is that according to which the source and ultimate "raison d'etre" of the notion of number, both natural and real, is experience and practical applicability. The same refers to notions of the theory of sets, provided we consider them within rather narrow limits, sufficient for the requirements of the classical branches of mathematics.

If we adopt this point of view, we are bound to draw the conclusion that there exist only one arithmetic of natural numbers, one arithmetic of real numbers and one theory of sets; therefore it is not possible to define these branches of mathematics by systems of axioms which are supposed to establish once and for all their scope and their content.

Systems of axioms play an important role in those theories: they systematize a certain fragment of these theories, namely that which includes out present knowledge; they often facilitate the exposition of a theory and are therefore of didactic value.

Incompleteness results of Gödel showing that natural numbers cannot be fully characterized by a system of axioms and that there exist non-isomorphic models of arithmetic should not lead to pessimistic conclusions because they provide a tool to obtain several independence results.

Similar, and even more difficult problems, arise in the foundations of set theory. Here the main difficulty is the indefiniteness of the notion of an arbitrary set as well as the status of such axioms as the axiom of choice.

[...]

The problem of the foundations of mathematics is not a single concrete mathematical problem which, once solved, may be forgotten. The considerations regarding the foundations of science are just as old as science itself and mathematics is no exception to this rule. For many centuries the essence and content of mathematics have been, and probably will remain also in future, an object of considerations for philosophers. In the course of time mathematics changes and this also necessitates a change of views on its foundations.

[...] An explanation of the nature of mathematics does not belong to mathematics, but to philosophy, and it is possible only within the limits of a broadly conceived philosophical view treating mathematics not as detached from other sciences but taking into account its being rooted in natural sciences, its applications, its associations with other sciences and, finally, its history.

* * *

What conclusions can be drawn from the above analysis of Mostowski's philosophical remarks? First of all one should stress that he was aware of philosophical problems connected with mathematics and its foundations and of their importance and meaning. On the other hand he tried to avoid (with few exceptions) any definite philosophical declarations concentrating instead on strongly mathematical and technical side of issues. If it was necessary then some general philosophical declarations have been made (but, in fact, unwillingly). He was aware of the meaning of results obtained in the foundations of mathematics by mathematical methods for the philosophy of mathematics but simultaneously was convinced that those results cannot give definite solutions to problems of the philosophical nature. Therefore he rather presented various possible solutions instead of making any concrete declarations. Philosophical problems and possible solutions to them were discussed by him on the margin of proper metamathematical and foundational studies, in introductory remarks only and – what is very important – did not influenced the latter. Mostowski strongly avoided philosophical comments and remarks in technical papers. Philosophical perspective on the one hand and metamathematical and foundational one on the other were strictly separated by him. Though some of his results were – as one can suppose – inspired and motivated by philosophical considerations (e.g., independence of definitions of finiteness, constructions that led to the so called today Kleene–Mostowski hierarchy, constructions of models with automorphisms) but he never wrote about that and never formulated them explicitly concentrating on mathematical and metamathemtaical studies. Hence one can only formulate here some hypotheses and conjectures which can be neither confirmed nor rejected. Although Mostowski did not develop a new "ism" in the philosophy of mathematics, his works essentially contributed to this field and can be regarded as paradigmatic cases of a very reasonable interplay of mathematical and philosophical ideas. And he was a perfect (perhaps even the most perfect) example of the Polish attitude to the foundations and philosophy of mathematics.

Part IV
Mathematical Logic in Poland

Stanisław Piątkiewicz and the Beginnings of Mathematical Logic in Poland

Co-authored by Tadeusz Batóg

The paper presents information on the life and work of Stanisław Piątkiewicz (1849– ?). His *Algebra w logice* [*Algebra in Logic*] of 1888 contains an exposition of the algebra of logic and its use in representing syllogisms. This was the first original Polish publication on symbolic logic. It appeared 20 years before analogous works by Łukasiewicz and Stamm.

1. Traditional Views on the Origins of Mathematical Logic in Poland

The successes of Polish logic after the First World War stimulated several authors to investigate its origins. All of them credited the beginnings of logic in Poland to Kazimierz Twardowski (1866–1938), who founded the Lvov-Warsaw school of philosophy while a professor at Lvov University, and to Jan Łukasiewicz (1878–1956), a logician who was a professor of philosophy at Warsaw University. Thus Adjukiewicz, for example, wrote in (1934, p 401): "Łukasiewicz, well-educated not only in philosophy but also in mathematics, became a discoverer of logistic for Poland. Perhaps this resulted from the influence of Twardowski, who was the first in Poland to speak about the algebra of logic, doing so in his university lectures in [...] 1899–1900". Likewise, Ingarden remarked in (1938, p. 28): "Twardowski was the first in Poland [...] to lecture about new attempts to reform logic".

Other authors writing on this subject have generally accepted the claims of Ajdukiewicz and Ingarden. Such was the case, for example, with Jordan (1945, p. 9) and Skolimowski (1967, p. 54). Kotarbiński presented Łukasiewicz as the first representative of mathematical logic in Poland, and Twardowski as his distinguished teacher (cf. 1959 and 1967). Recent publications, such as that of Woleński (1989, p. 82) essentially repeat these views.

What is wrong with the claims of Ajdukiewicz, Ingarden, and their followers is this: they insinuate that, before Twardowski and Łukasiewicz, there was no response at all in Poland to mathematical logic. In fact, mathematical logic was known in Poland and was discussed there in print long before Łukasiewicz's book on the subject appeared in 1910. Clearly, the first publications to inform Polish readers of the algebra of logic were translations of works by foreign authors. One of those translations stimulated the first original Polish contribution to this subject. This was not by Łukasiewicz, but by the mathematician Stanisław Piątkiewicz.

2. The Life and Work of S. Piątkiewicz

Works dealing with the algebra of logic have not generally been translated into Polish. Thus, for example, those of George Boole and Augustus De Morgan do not exist in Polish even today. Nevertheless, three books treating the algebra of logic, at least in part, were translated into Polish during the 1870s and 1880s.

The first of these was *Logic*, a textbook by the Scottish philosopher Alexander Bain, published in Polish translation in 1878. This book devoted a chapter to Boole and De Morgan. The second was *Logique*, an elementary handbook by Louis Liard, translated in 1886, which mentioned Boole's algebra of logic. The third, *Elementary Lessons in Logic* by William Stanley Jevons, appeared in translation the same year. Results in the algebra of logic were mentioned there only in passing.

The first translation, that of Bain's book, is historically the most important, since it directly influenced Piątkiewicz. This Polish mathematician, who also had philosophical interests, was born on September 21, 1849 in the village of Dębowiec (Dembowiec) within the Austro-Hungarian Empire. Although he was from a poor family, he graduated from Lvov University in 1871 with a degree in mathematics and physics. The following year he began teaching in a grammar school in Przemyśl, and in 1879 was assigned to one in Lvov, where he remained until 1890. Then he returned to Przemyśl as director of the local grammar school, where he taught logic and psychology. He remained there until his retirement in 1906.[1]

While he taught in Lvov, Piątkiewicz took an interest in mathematical logic. His teaching duties may have been what led him to study logic more thoroughly and to read the translation of Bain. The result was Piątkiewicz's fifty-page paper, *Algebra w logice* [*Algebra in Logic*], published in 1888 in the reports of the Royal Imperial Grammar School No. IV in Lvov. These reports were very similar to what in Germany are called *Programmschriften*.

3. The Contents of *Algebra w logice*

In *Algebra w logice*, Piątkiewicz referred to a wide range of authors on mathematical logic, from Leibniz and Boole to Gottfried Ploucquet and William Hamilton, De Morgan and Jevons, Robert Grassmann and Ernst Schröder. Piątkiewicz was

---

[1] Most of the biographical data were taken from the printed reports of the directors of the First Grammar School in Przemyśl and of the Fourth Grammar School in Lvov. The place and the year of birth of Piątkiewicz are given by S. Goliński in (1894, p. 100). The exact data of the birth was appointed on the base of the book of baptisms of the parish Dębowiec. Dates of the beginning and the end of the university studies are given according to the information obtained from the District State Archive in Lvov. Some of the biographical details as well as some sources were accessible to us by the courtesy of Dr. S. Kostrzewska-Kratochwilowa, Rev. J. Wójtowicz and two former pupils of S. Piątkiewicz and then grammar school professors S. Jurek and J. Kolankowski.

also aware of an extensive secondary literature, and referred explicitly to Bain's book as the source of his own contribution (cf. 1888, p. 5). The only work of consequence that Piątkiewicz did not mention was Gottlob Frege's *Begriffsschrift*, which had not yet attracted wide interest in any country, although it would do so later.

The first section of Piątkiewicz's work defended the algebra of logic against various objections to it, for example, that natural language can adequately express the formal laws of thought without the use of symbols. He emphasized the vagueness of natural language and wished to show "that the presentation of logic in an algebraic way is possible, that it is not superfluous and that it will contribute to broadening the scope of formal logic" (1888, p. 6).

In his second section, entitled "Identity Law. Association (Addition) of Concepts", Piątkiewicz revealed his psychologistic approach to logic, which had been influenced by Herbart. This psychologism did not reduce the value of Piątkiewicz's formal theory, since, as a mathematician, he presented the theory independently of any psychological interpretation.

Piątkiewicz's psychologism was the traditional view. He was similarly traditional in not distinguishing between unproved propositions and derived theorems – in contrast with Schröder's *Operationskreis* of 1877. Thus Piątkiewicz was guided by traditional logic in his choice of elementary laws. Among the laws governing identity, for example, he mentioned only the so-called law of identity ($a = a$), which was much valued by earlier logicians and philosophers. He did not mention the important law stating that identity is transitive, and mistakenly treated the symmetry of identity as contained in the law of identity. Symmetry and transitivity were used tacitly.

The main subject of this second section was the addition (union) of classes. Here he followed Jevons, instead of Boole, in defining addition for all classes rather than for merely disjoint ones. Piątkiewicz discussed some relations between classes, such as subordination, overlapping, exclusion, and being different. His definitions of these relations were not completely precise, resulting in some errors of reasoning. (As an aside, we note that Piątkiewicz, like Bolzano, argued that the intension and extension of a concept are not reciprocal.)

The third section of Piątkiewicz's report was called "Specification (Multiplication) of Concepts", i.e., intersection of classes. To guarantee that multiplication of classes is defined in every case, he accepted the empty class 0. Some formal proofs were presented, e.g., for the theorem that if $ac = bc$ and $a+c = b+c$, then $a = b$. Yet he did not grasp the difference between the formal proof of a law and its geometric interpretation, as when he considered the distributive law.

Section four, entitled "Negation of Concepts, Negative Concepts. The Law of Contradiction and the Law of the Excluded Middle", treated those two laws, which, together with the law of identity, were the traditional laws of logic before Boole. The two laws were stated as $aa_1 = 0$ and $a+a_1 = 1$, where a_1 denoted (as in Schröder's *Operationskreis*) the complement of a class *a*. Piątkiewicz stated De Morgan's laws and proved the uniqueness of negation. The proofs given in this section were fairly precise.

"Logical Equations", the fifth section, is the heart of Piątkiewicz's work. It contained a detailed presentation of Schröder's method of solving logical equations. In this section Piątkiewicz showed his competence in using the logical calculus.

The sixth and final section ("Applications of Previous Sections") was the longest one. Most of it dealt with the deduction of particular syllogistic moods by means of the theory of logical equations and with the formulation of rules of a generalized syllogistic in which sentences may contain negated subjects and predicates. Here there were various imprecise statements, but they merely continued the imprecisions already present in Boole.

4. Conclusion

As this synopsis of *Algebra w logice* shows, Piątkiewicz informed his readers about Boolean logic in a competent way. Although his work contained no new results, it showed more knowledge of symbolic logic than was then possessed by any Polish university teacher of logic. At the time he was the only person in Poland who was fully conscious of the progress being made in logic.

Unfortunately Piątkiewicz did not succeed in awakening in Poland an interest in the new logic. The reasons are clear. He did not hold an academic position. After completing *Algebra w logice*, he moved to Przemyśl, far from any center of research. Moreover, during those years mathematicians, whether in Poland or elsewhere, seldom paid attention to logic. Logicians, who were philosophers, were not capable of reading papers in mathematical logic. Hence Piątkiewicz's work remained unknown for a long time.

Only twenty years later did there again arise in Poland a level of logical knowledge comparable to that of Piątkiewicz – thanks to the work of Łukasiewicz and to the article (1911a, 1912) by the Polish mathematician Edward Stamm (1886–1940), who went on to write numerous works on the algebra of logic. But at first Łukasiewicz and Stamm had no original results either. The well-known appendix to Łukasiewicz's work on Aristotle's logic (1910), which Jordan (1945, p. 11) called the first presentation of modern logic in Poland, was in fact only a summary of Louis Couturat's elementary exposition *L'algebre de la logique* (1905). It is true that, methodologically, this appendix was at a higher level than Piątkiewicz's work. But Łukasiewicz's appendix already belonged to another epoch in the development of logic. For in 1910 had appeared the first volume of *Principia Mathematica* by Whitehead and Russell, which was first discussed in Poland by Chwistek (1912).

In any case, mathematical logic did not begin in Poland in 1910 or in 1899. Piątkiewicz's *Algebra w logice* of 1888 marks the real beginning of mathematical logic there. Between 1888 and 1899 one can note another significant date. In 1891 the logic of Giuseppe Peano was briefly mentioned by Samuel Dickstein (1851–1939), a Polish mathematician at Warsaw University who was also a historian of mathematics (cf. 1891, p. 39). Thus knowledge of mathematical logic came to Poland several years earlier than is usually claimed.

Contribution of Polish Logicians to Recursion Theory[1]

1. Introduction

The first need of a systematic study of functions whose values can be calculated by a finite process (they are usually called computable) can be seen in the Hilbert school. It was connected with the decision problem for first-order logic (and generally, for first-order theories) considered by Hilbert and his students in connection with the program of Hilbert. The aim of this program was to justify classical mathematics by finitistic means.

One usually shows that a given decision problem has a positive solution simply by providing an appropriate algorithm. And no precise definition of a finitistic method is needed here. On the other hand if we want to prove that a given specific problem does not admit an algorithmic solution then a precise general definition of such a method is necessary.

Historically the first precise definition of a class of computable functions was given by Kurt Gödel in his famous paper "Über formal unentscheidbare Sätze der 'Principia Mathematica' und verwandter Systeme. I" (1931a) in which he proved the essential incompleteness of the arithmetic of natural numbers. The class of functions defined by Gödel is called today the class of primitive recursive functions. It consists of those functions on the set of natural numbers $\mathbb{N}$ which can be obtained from some specific initial functions by the operations of substitution and recursion.

Since then several other definitions were proposed. It is significant that all of them proved to be equivalent. Let us mention here definitions using the notion of representability (Church, Gödel), arithmetical definitions (Kleene), definitions using canonical systems (Post), abstract machines (Turing) or algorithms (Markov). They led to the notion of a general recursive function and to the establishing of a new domain of mathematical logic called recursion theory.

Investigations were continued in two directions. On the one hand subclasses of the class of recursive functions were studied. Those subclasses were obtained by putting various restrictions on operations on functions admitted in the process of constructing new functions from given ones. This led for example to the notion of elementary functions of L. Kalmár or to the study of various forms of multiple recursion (R. Péter). On the other hand one tried to generalize the notions of computability and recursiveness to objects of higher types, in particular to functionals, i.e., to operations defined on n-tuples of natural

---

1 The upper bound of the period covered in the paper is fixed as 1963, i.e., the year in which A. Mostowski's paper "Thirty Years of Foundational Studies", *Acta Philosophica Fennica* 17 (1965), 1–180 was published.

functions (functions from $\mathbb{N}^n$ to $\mathbb{N}$) and k-tuples of elements of $\mathbb{N}$ and assuming values from $\mathbb{N}$.

Precise definitions of various types of computable functions provided a basis for further logical classifications of mathematical and logical notions. Using recursive functions and quantifiers one can classify non-recursive arithmetical notions. This is the idea which lies behind the hierarchies of Kleene and Mostowski and their extensions.

Another approach to the classification of notions (which became an independent domain of study) was provided by the theory of degrees of unsolvability. The subject of this theory are degrees defined as equivalence classes of various reducibility relations.

Observe that the study of computable functions and relations is strictly connected with constructivistic trends in mathematics. The tendency to restrict mathematics to constructive objects and methods was given a precise tool by recursion theory – the vague notion of constructivity could be precisely defined. It gave new impulses to the study of constructive foundations of mathematics, in particular of analysis, leading to various systems of computable analysis.

One should also add that computable functions (and consequently the recursion theory) are important for the philosophy of mathematics too. They provide namely a basis for a nominalistic philosophy of mathematics which rejects such abstract objects as sets, functions etc. and accepts only those objects which can be named. Recursive functions and relations (and various variants of them) can be considered as such objects and therefore evidently accepted by the nominalists. On the other hand one should note that there are obviously nominalistic theories of mathematics which admit a broader class of objects than the class of computable functions. We mean here the descriptive set theory, i.e., the theory of Borel sets and of analytical sets initiated by the French semi-intuitionists (Borel, Baire, Lebesgue) and developed in the thirties by set-theoretically minded mathematicians grouped around Luzin in Moscow and around Sierpiński in Warsaw. Semi-intuitionists started from the notion of a real number and insisted that only nameable sets and functions defined on the set of all reals are admissible. It has turned out that the descriptive set theory shows many analogies to the theory of computable (recursive) functions (cf., e.g., Addison 1954 and 1958/1959).

The present paper is devoted to the contribution of Polish logicians to the researches described above. Section 2 says about studies on subclasses of the class of recursive functions – first of all about the Grzegorczyk's hierarchy of primitive recursive functions. Section 3 is devoted to a hierarchy of notions introduced independently by Kleene and Mostowski and called today the Kleene––Mostowski hierarchy or the arithmetical hierarchy. Examples of the classification of particular mathematical notions in this hierarchy are also given. Further (Section 4) studies on the generalizations of the notion of recursiveness to the case of functionals are discussed. Section 5 says about Banach–Mazur's and Grzegorczyk's work on the constructive foundations of mathematics, in particular of analysis. The final Section 6 is devoted to model-theoretical applications of recursion theory, in particular to the discussion of results concerning the class (in the Kleene–Mostowski hierarchy) of possible models of theories.

2. Grzegorczyk's Hierarchy

The hierarchy of the primitive recursive functions, called today the Grzegorczyk's hierarchy, was introduced in Andrzej Grzegorczyk's paper (1953). Grzegorczyk defined there an increasing sequence $\mathscr{E}^0, \mathscr{E}^1, \mathscr{E}^2, \mathscr{E}^3, \ldots$ of classes of recursive functions and proved several properties of them. Let us recall basic notions and theorems.

Consider the following sequence of functions:

$$f_0(x,y) = y+1,$$
$$f_1(x,y) = x+y,$$
$$f_2(x,y) = (x+1)(y+1),$$

and for $n \geqslant 2$:

$$f_{n+1}(0,y) = f_n(y+1, y+1),$$
$$f_{n+1}(x+1, y+1) = f_{n+1}(x, f_{n+1}(x,y)).$$

One can show that the functions f_n for $n > 0$ are strictly increasing with respect to both arguments. Using those functions Grzegorczyk defines now the classes $\mathscr{E}^n$ ($n \in \mathbb{N}$) of functions in the following way: $\mathscr{E}^n$ is the smallest class of functions including the functions $x+1, U_1(x,y) = x, U_2(x,y) = y, f_n(x,y)$ as initial functions and closed under the operations of substitution and limited recursion.

It is proved that the class $\mathscr{E}^3$ coincides with the class of elementary (computable) functions introduced by L. Kalmár in (1943) (cf. also Péter 1951, p. 60) and characterized as the smallest class of functions including as initial functions $x+1, x+y, x \dot{-} y$ and closed under the operations of substitution, limited summation and limited multiplication (it can be shown that this class is also closed under the operation of limited minimum). Main properties of the functions belonging to the class $\mathscr{E}^0$ are the following: (1) for each function $f \in \mathscr{E}^0$ there exists a number k_0 such that for every n, $f(n) < n + k_0$ and (2) each recursively enumerable set is enumerated by a certain function of the class $\mathscr{E}^0$.

The classes $\mathscr{E}^n$ ($n \in \mathbb{N}$) form a hierarchy of functions, i.e., in particular for every n: (1) $\mathscr{E}^n \subseteq \mathscr{E}^{n+1}$ and (2) the function $f_{n+1}(x,x)$ increases faster than any function of the class $\mathscr{E}^n$ and consequently $\mathscr{E}^n \neq \mathscr{E}^{n+1}$. One can also prove that for $n > 2$ the class $\mathscr{E}^{n+1}$ contains a universal function for the class $\mathscr{E}_1^n$, i.e., for the class of unary functions of the class $\mathscr{E}^n$.

Grzegorczyk's hierarchy is a hierarchy of primitive recursive functions. In fact one shows that the class of primitive recursive functions $\mathscr{P}$ is equal to the sum of all the classes $\mathscr{E}^n$ ($n \in \mathbb{N}$). This leads to the corollary that in the definition of the class $\mathscr{P}$ the operation of recursion cannot be eliminated or exchanged into the operation of limited recursion.

Grzegorczyk's hierarchy was extended to transfinite ordinals by M. H. Löb and S. S. Wainer (cf. Löb–Wainer 1970a and 1970b as well as Wainer 1970). They introduced a general procedure for generating hierarchies which can be applied to a wide variety of classes of number-theoretic functions (even to classes containing non-recursive functions) and proved that the hierarchies may be extended through the ordinals of Cantor's second number class without collapsing. In this way in particular a proper extension of Grzegorczyk's hierarchy can be obtained – in fact the extension and the original Grzegorczyk's hierarchy coincide at the level ω. The most interesting part of Löb–Wainer's hierarchies is that below the ordinal ε_0.

When discussing the contribution of Polish logicians to the general recursion theory one should also mention the paper by Antoni Janiczak (1955) (edited by A. Grzegorczyk from the notes left by the author who died prematurely in 1951). Some interesting properties of the class of partially recursive functions are studied there. The class **PR** of partially recursive functions is defined as follows: $f \in$ **PR** if and only if there exist two recursive functions g and h such that

$$\forall z \ [z \in D^*(f) \equiv \exists x (h(z,x) = 0)],$$
$$\forall z \ [z \in D^*(f) \longrightarrow f(z) = g((\min x)[h(z,x) = 0])],$$

where $D^*(f)$ is the set of arguments of f. The author proves that: (a) there exists a function $f \in$ **PR** which assumes only two values 0 and 1 and which cannot be extended to any recursive function, (b) there exists a function $f \in$ **PR** which cannot be majorized by any recursive function g, (c) if the set $D^*(f)$ is the complement of a recursively enumerable set and f can be extended to a function $g \in$ **PR** then f can be extended to a recursive function h, hence, in particular, if $D^*(f)$ is a recursive set and $f \in$ **PR** then f can be extended to a recursive function.

3. Kleene–Mostowski Hierarchy

In Section 1 we mentioned the group of set-theoretically minded mathematicians around W. Sierpiński in Warsaw. Their work contributed to the development of the descriptive set theory. In their studies considerable space was devoted to the estimation of the Borel class or projection class of given sets.

Those researches were an inspiration for Andrzej Mostowski. Using an analogy between the operation of projection (applied in the descriptive set theory) and the strictly logical operation of the existential quantifier he constructed a hierarchy of arithmetical notions called today the Kleene–Mostowski hierarchy or the arithmetical hierarchy.

Mostowski's work was done during the war. His results could be published only in 1947 – they appeared in the paper "On definable sets of positive integers" in *Fundamenta Mathematicae*. Earlier, in 1943, a paper "Recursive predicates and quantifiers" by S. C. Kleene appeared – it contained results similar to those of Mostowski. This paper was of course unknown in Poland at that time – it became available only when Mostowski's paper was under press (cf. Mostowski 1947, p. 112). Mostowski admits also in (1947) that A. Tarski informed him that he had found in 1942 results very similar to his results.

Though the papers by Kleene and by Mostowski contained similar results obtained independently (for that reason the hierarchy introduced in them is called today the Kleene–Mostowski hierarchy), it should be stressed here that both authors had quite different inspirations. Kleene was motivated by some recursion-theoretic considerations rooted in the incompleteness theorem of Gödel. But he saw also some analogies between logical operations of existential and universal quantifiers and geometrical operations of projection and intersection, resp., (cf. Kleene 1943, p. 50). Mostowski on the other hand was motivated by the results of the descriptive set theory. The analogy between the properties of classes of the hierarchy defined by him and the properties of classes of projective sets is mentioned at several places in his paper (1947).

Mostowski develops his theory starting not from the notion of general recursiveness (as it was the case by Kleene) but, for the convenience of the reader, from a notion of a decidable propositional function. He introduces classes $P_n^{(k)}$ and $Q_n^{(k)}$ (for $k, n \in \mathbb{N}$) forming a hierarchy. The index k denotes here the arity of classified notions (relations) and the index n – the level of the hierarchy. Hence $P_n^{(k)}$ and $Q_n^{(k)}$ are classes of k-ary relations of the nth level (in the notation used today the upper index is omitted and the classes $P_n^{(k)}$ and $Q_n^{(k)}$ are denoted as Σ_n and Π_n, resp., or as Σ_n^0 and Π_n^0, if one wants to distinguish this hierarchy from the analytical hierarchy – see below).

The initial classes $P_0^{(k)}$ and $Q_0^{(k)}$ are equal and are defined as the class of k-ary decidable propositional functions. Next classes are introduced inductively. A relation R is said to belong to the class $P_{n+1}^{(k)}$ if and only if there exists a relation R_1 of the class $Q_n^{(k+1)}$ such that for any k-tuple $a_1, \ldots, a_k$ it holds:

$$R(a_1, \ldots, a_k) \equiv \exists x R_1(a_1, \ldots, a_k, x).$$

A relation R is of the class $Q_{n+1}^{(k)}$ if and only if its complement is of the class $P_{n+1}^{(k)}$.

Classes defined in this way possess several interesting properties. Mostowski proved in (1947) appropriate theorems on the sum, product and cartesian product of relations of a given class. Further it is shown that for $n \geqslant 0$ the common part of the classes $P_n^{(k)}$ and $Q_n^{(k)}$ is a field of sets and that

$$P_n^{(k)} \subseteq P_{n+1}^{(k)} \cap Q_{n+1}^{(k)},$$
$$Q_n^{(k)} \subseteq P_{n+1}^{(k)} \cap Q_{n+1}^{(k)}.$$

Using Cantor's diagonal theorem one proves that

$$P_n^{(k)} \neq P_{n+1}^{(k)} \quad \text{and} \quad Q_n^{(k)} \neq Q_{n+1}^{(k)}$$

as well as

$$P_n^{(k)} \text{ non } \subseteq Q_n^{(k)} \quad \text{and} \quad Q_n^{(k)} \text{ non } \subseteq P_n^{(k)},$$

hence the classes $P_n^{(k)}$ and $Q_n^{(k)}$ for $n \in \mathbb{N}$ form a hierarchy.

Mostowski's paper (1947) also contains various applications. First of all it is shown that certain generalizations of Gödel's incompleteness theorem for systems based on infinitary rules hold. In particular it is proved (Theorem 4.21) that if the system S is ω-consistent and fulfils certain condition C_s then there exists a sentence ϑ undecidable in S. The condition C_s is the conjunction of the following conditions: the (arithmetized counterpart of the) function of substitution is of the class $P_s^{(k+2)}$, the (arithmetized counterpart of the) function of negation is of the class $P_s^{(2)}$, the (arithmetized counterpart of the) relation of provability is of the class $P_s^{(2)}$ and the (arithmetized counterpart of the) relation of being a theorem is of the class $P_s^{(1)}$. It is obvious that systems with finitary inference rules (and based on a recursive set of axioms) fulfil the condition C_0. Conditions C_s ($s > 0$) can be fulfilled by systems with infinitary rules (an example of such a system is Rosser's system introduced in his paper 1937).

It is also proved in Mostowski (1947) that systems S which are ω-consistent and fulfil the condition C_s are not closed under the rule of infinite induction. Even more, it is shown that if S fulfils the condition C_s ($s \geqslant 0$) then the set of sentences which can be obtained from the axioms of S by n^{th} iterations of the rule of infinite induction is not closed under this rule and is incomplete if it is ω-consistent.

The hierarchy of relations $P_n^{(k)}$ and $Q_n^{(k)}$ ($n \in \mathbb{N}$) is extended in Mostowski (1947) to the case of functions. A function f is said to be of the given class if and only if its graph belongs to this class. Mostowski proves also Post's theorem and says that "it is an exact analogue of the well-known Souslin's theorem concerning sets which are analytical together with their complements" (p. 106). In the footnote (p. 82) it is explained that after having finished the first draft of this paper the author became acquainted with Post's paper (1944) where the discussed result is published.

Results of (1947) were extended by Mostowski in the sequel paper (1948b). Assuming that primitive recursive k-ary relations belong to the class $P_0^{(k)}$ and that the relation $qB_\varphi n$ (which means: q is the Gödel number of a formal proof of $\varphi(n)$) is primitive recursive, it is shown that $P_0^{(k)}$ is the class of general recursive n-ary relations. The main result concerns the class $R_1^{(k)}$ defined as the smallest finitely additive field of sets containing the classes $P_1^{(k)}$ and $Q_1^{(k)}$. It is proved that $R_1^{(k)} \neq P_2^{(k)} \cap Q_2^{(k)}$. And again the analogy with projective sets is stressed. The author says: "These theorems are quite analogous to the following well known results concerning projective sets: if a set B is an **A**-set as well as a **CA**-set, it must be Borelian; but a set which is a **PCA**-set as well as a **CPCA**-set does not necessarily belong to the smallest (denumerably additive) field of sets over **A** + **CA**" (p. 115).

Mostowski applied the hierarchy introduced in the paper (1947) to the classification of various mathematical and logical notions. One should mention here his papers (1949; 1949–1950; 1955c).

In Mostowski (1949–1950) the hierarchy is extended up to the first nonconstructive ordinal ω_1 (denoted today as ω_1^{CK}) and it is used to the classification of certain theories defined in a semantical way. A logical system S is said to be of the class P_α or Q_α if and only if the set of (numbers which correspond to) true sentences of S belongs to the class $P_\alpha^{(1)}$ or $Q_\alpha^{(1)}$, resp.. It is known that the propositional calculus is a system of the class P_0 and the first-order predicate calculus is a system of the class P_1. Mostowski shows also in (1949–1950) that the system S_e of elementary arithmetic introduced by Carnap in (1934) is not recursively enumerable and that the class of this system is exactly Q_1. The theory S_e is defined as a first-order system with infinitely many predicates and function symbols (for every primitive recursive relation and function there is an appropriate symbol) in which one has one connective (binegation) and bounded quantifiers only. It is also shown that the usual first-order system S_a of general arithmetic (here arbitrary quantifications are allowed) is exactly of the class P_ω and that the class of the system of arithmetic of real numbers S_r (in the language of S_a enriched by set variables and the membership relation $\in$) cannot be determined unless we extend the definition of classes $P_\alpha^{(k)}$ and $Q_\alpha^{(k)}$ beyond ω_1.

It is worth mentioning here that in (1949–1950) Mostowski formulates explicitly his philosophical views. He says (on p. 247):

> We shall not consider logical systems as void schemata deprived of any interpretation. On the contrary we shall assume the objective existence of a kind of "mathematical reality" (e.g., of the set of all integers or of the set of all real numbers). By objective existence we mean existence independently of all linguistic constructions. The rôle of logical and mathematical systems is to describe this reality.

Mostowski's most known and important result on classification of notions is contained in his paper (1955c). He considers there fractions of the form $10^{-x}\alpha(x,y)$ where α is a primitive recursive function and investigates the set of those integers y for which $\lim_x 10^{-x}\alpha(x,y)$ exists and belongs to a preassigned class of real numbers. A typical result is stated for example in Theorem 7. It says that the set $Z_\alpha^{(5)}$ of those y's for which $\lim_x 10^{-x}\alpha(x,y)$ exists and is integral is the most general set of the class $Q_3^{(1)}$, i.e., if α ranges over the set of primitive recursive functions then the set $Z_\alpha^{(5)}$ ranges over the whole class $Q_3^{(1)}$. Results obtained in the paper allow to construct examples of sets definable by means of two or three quantifiers but not definable by a smaller number of them. In particular from Theorem 7 quoted above it follows that if $U(n,x,y)$ is a (general recursive) function universal for the class of primitive recursive functions with 2 arguments then the set of pairs (n,y) such that $\lim_x 10^{-x}U(n,x,y)$ exists and has an integral value belongs to the class $Q_3^{(2)}$ but not to the class $P_3^{(2)}$, hence the notion of the limit of a sequence is the notion of the class $P_3^{(2)}$ and cannot be simplified. This means that all three quantifiers used in its definition are indispensable and cannot be dropped or replaced by dual ones.

The arithmetical hierarchy of Kleene and Mostowski was extended in two ways. First it was expanded into the constructive transfinite – cf. Mostowski (1949––1950) and Kleene (1955). In this way the hyperarithmetic hierarchy was obtained. Again there are several analogies between this hierarchy and the descriptive set theory. They were explicitly formulated by Addison in (1958/1959).

The second extension was made by Kleene in (1955) and led to the analytical hierarchy. It is in fact an extension of the hyperarithmetical hierarchy and is defined in a similar way as the arithmetical hierarchy. One starts here from arithmetical relations (i.e., from the class $\bigcup_{n\in\mathbb{N}}(\Sigma_n \cup \Pi_n)$) and divides relations into classes according to their definitions taking into account the number and type of quantifiers – but now we allow formulae containing not only quantifiers over natural numbers but also quantifiers whose range consists of sets of integers. The particular classes are denoted by Σ_n^1 and Π_n^1 (in this notation the classes of the arithmetical hierarchy are denoted as Σ_n^0 and Π_n^0 instead of Σ_n and Π_n). It can be proved that the class of hyperarithmetical relations is exactly the intersection of the classes Σ_1^1 and Π_1^1 (therefore it is usually denoted by Δ_1^1).

Using the arithmetical and analytical hierarchies Mostowski distinguished in (1959) various degrees of constructivism. Let K be an arbitrary class of number-theoretic functions. One defines a class **K** of real numbers associated with the class K in the following way: a real number a belongs to the class **K** if and only if for every n:

$$\left| |a - [a]| - \frac{f(n)}{n} \right| < \frac{1}{n}$$

where f is a fixed function from K. Choosing now appropriately the initial class K one obtains classes **K** satisfying certain general theses of constructivism. In particular for $K = \Sigma_0^0 = \Pi_0^0$ we obtain the class $\mathbf{K}_0$ of recursive real numbers. It is the largest class of reals whose approximations can be calculated by means of algorithms. It should be noted here that they are accepted by most constructivists though the totality $\mathbf{K}_0$ of them is not generally accepted. The reason is that for its definition a universal quantification over the set of natural numbers is needed and this presupposes the existence of an actually infinite totality. Hence a radical constructivist will accept every particular element of the class $\mathbf{K}_0$ but not the whole set $\mathbf{K}_0$.

In a similar way we define the class $\mathbf{K}_\alpha$ of reals which can be approximated by functions of the class $\Pi_\alpha^0 \cap \Sigma_\alpha^0$. It can be shown that $\mathbf{K}_\alpha$ is a real closed field and that $\mathbf{K}_\alpha \neq \mathbf{K}_\beta$ for $\alpha \neq \beta$.

Special attention should be payed to the class $\mathbf{K}_\omega$ obtained from the class of arithmetical functions $\bigcup_{n \in \mathbb{N}} \Sigma_n^0 \, (= \bigcup_{n \in \mathbb{N}} \Pi_n^0)$. It consists of the reals which are elementarily definable, i.e., it coincides with the universe of Hermann Weyl's constructive analysis (cf. Weyl 1918). Elements of $\mathbf{K}_\omega$ are exactly those reals which are accepted by a constructivist who does not reject any infinite sets but who claims that all such sets should be reducible to the set of natural numbers.

The advantage of Mostowski's approach is that it enables us to give precise definitions of various notions of constructivism and in this way make possible comparison of various senses of constructivity.

4. Computable Functionals

As mentioned in the introduction the notions of recursiveness and computability were extended in various ways. One of them consisted in generalization of those notions to the case of objects of higher types. The first step in this direction was the paper (1958) of K. Gödel where primitive recursive functionals were considered. Another one was the Kleene's paper (1959) in which computable functionals were introduced and investigated.

This trend in the recursion-theoretical researches was also present in the works of Polish logicians, first of all in the works of Andrzej Grzegorczyk. Studies on primitive recursive and computable functionals are connected with the study of constructive foundations of mathematics and in particular of analysis which we will discuss in the next section (and therefore we will return, in a specific context, to problems considered here).

The notion of a primitive recursive functional was considered by Grzegorczyk in the paper (1964) where he proposed a very elegant definition of it. The starting point is the definition of a very general class $\mathscr{R}$ of recursive objects of finite types. It may be considered as identical with the class of recursive functionals mentioned by Gödel in (1958) and by G. Kreisel in (1959). Grzegorczyk compares his definition of the notion of a primitive recursive functional with that of Kleene (1959) and proves that the class defined by Kleene is a proper subclass of the class of recursive functionals assuming only natural numbers as values. It is also shown how one can apply the new notion to Gödel's interpretation of intuitionistic arithmetic (cf. Gödel 1958). This notion makes the interpretation easier.

Grzegorczyk wrote also some papers devoted to the notion of a computable functional. In the paper (1955a) an inductive definition of it is given. The class of computable functionals is defined there as the smallest class containing some initial functionals and closed under the operations of substitution and of effective minimum. If one cancels in this definition the condition of effectiveness of the minimum operation then the definition of the class of elementarily definable functionals is obtained. It was considered by Grzegorczyk in the paper (1955b). The latter is connected with some ideas of H. Weyl expressed in (1918). Weyl proposed there a restriction of the logical methods of analysis to the elementarily definable ones. This vague notion was given a precise definition in the paper Grzegorczyk (1955b). It is stressed in it that many theorems of the classical analysis can be obtained by means of elementary methods. In particular it is shown that the classical analysis of continuous functions can be reproduced in an elementary manner.

The mathematical definition of the computable functionals given in Grzegorczyk (1955a) was shown to be equivalent to a metamathematical one (cf. Grzegorczyk 1955c). It was proved namely that a functional is computable if and only if it is representable in a certain system of arithmetic with no quantifiers but with the μ-operation and with appropriate axioms defining the function-variables (this system is similar to the system (S) considered by A. Mostowski in his book 1952).

5. Constructive Foundations of Mathematics

One of the reactions to the difficulties in the foundations of mathematics disclosed by antinomies was the proposal to exclude all general set-theoretical notions from mathematics and to limit it to the study of those objects that can be effectively defined or constructed. An extreme example of this tendency is provided by intuitionism of L. E. J. Brouwer. Less extreme are the trends which do not challenge the classical rules of proof (as intuitionism does) and demand only the restriction of the class of admissible mathematical objects to constructive ones. Since there is no unique notion of constructivity there are various (mutually conflicting) constructivistic programs. One of them is based on the recursion theory and in particular on the notion of a computable (recursive) function. It was applied first of all to the reconstruction of the classical analysis. Its characteristic feature is the full acceptance of the notion of integer taken from the classical arithmetic (this notion is not analyzed any further). On the other hand several limitations are put on the notion of a real number and all other mathematical notions. Their aim is to eliminate all non-constructive notions (cf. Grzegorczyk 1959).

Various proposals in this direction have been formulated. Two main approaches can be distinguished here. The first one assumes as being known from classical analysis the general notion of functions and among them one singles out a narrower class of computable real functions. The second one, which seems to answer better the aims of computable analysis, defines the class of computable real functions by means of certain operations performed on simple primitive functions (a more general notion of a functional, i.e., of a function which assigns numerical values to every system of arguments composed of numbers and functions, is often used here).

The investigations by Stefan Banach and Stanisław Mazur belong to the first type. They were begun in 1936 (cf. Banach–Mazur 1937). Their aim was to investigate a fragment of analysis admitting only numbers whose expansions into decimal fractions are represented by primitive recursive functions. The studies were continued by S. Mazur – the most extensive presentation of the obtained results can be found in Mazur's paper *Computable analysis* (1963). The paper (edited by A. Grzegorczyk and H. Rasiowa) is based on notes of the lectures of S. Mazur in the academic year 1949/50 at the Institute of Mathematics of the Polish Academy of Sciences in Warsaw. The lectures gave a systematic exposition of the results obtained by Banach and Mazur in the period 1936–1939 (their earlier publication was made impossible because of the Second World War) and some of results obtained by Mazur after the war (in particular results concerning general recursive mathematical objects).

The Banach–Mazur's computable analysis begins with the study of the notion of a recursive or computable real number. A positive real number a is said to be recursive if and only if there exists a primitive recursive function f such that for any $n = 0, 1, \ldots$:

$$\left| a - \frac{f(n)}{n+1} \right| < \frac{1}{n+1}.$$

The number a is called computable if and only if there exists a (general) recursive function f with the indicated property. It can be proved that:

(1) all algebraic numbers are recursive,
(2) the set of recursive numbers and the set of computable numbers constitute fields of numbers,
(3) these two fields are algebraically closed in the field of real numbers.

Further the notions of a recursive and computable sequence is introduced. A sequence $\{a_k\}$ is called recursive (computable) if and only if there exists a primitive recursive (general recursive) function f such that for every $k, n \in \mathbb{N}$:

$$\left| a_k - \frac{f(k,n)}{n+1} \right| < \frac{1}{n+1}.$$

It can be shown that the set of computable sequences is closed under the four arithmetical operations: $+, -, \cdot, :$; on the other hand the set of recursive sequences is not closed under the operation of division. The sequence $\{a_k\}$ is said to be recursively (computably) convergent to a if and only if there exists a primitive recursive (general recursive) function h such that for $k > h(m)$ the following holds:

$$|a_k - a| < \frac{1}{m+1}.$$

The notion of a convergent sequence has the following properties:

(1) the four arithmetical operations performed on sequences recursively (computably) convergent give the sequences recursively (computably) convergent to the limits obtained by the four arithmetical operations,

(2) the limit of a sequence recursively (computably) convergent is recursive (computable) provided that the sequence is recursive (computable),

(3) there exists a recursive sequence which converges to 0 in a non-computable manner,

(3) for each computable sequence convergent to a computable number there exists a computable subsequence computably convergent,

(4) there exists a recursive monotonic convergent sequence which does not converge to a computable number – hence the theorem of Bolzano–Weierstrass is not satisfied in the domain of computable sequences and numbers.

There are also other anomalies. For example the functions "signum" and "the integral part" lead out of the class of computable sequences.

The next main notion considered by Banach and Mazur is the notion of recursive and computable real function. It is defined in the following way: a real function defined on a set of recursive (computable) numbers is said to be recursive (computable) if and only if it transforms any recursive (computable) sequence into a recursive (computable) one. We can prove that those functions possess the following properties:

(1) the four arithmetical operations performed on computable functions do not lead out of the class of computable functions,

(2) the operations of addition, multiplication and substraction do not lead out of the class of recursive functions, however the function $1/x$ is not recursive on the set of all positive recursive numbers,

(3) the functions "signum x" and "integer-part of x" considered even on the set of all recursive numbers of the closed interval $[0, 1]$ are not computable,

(4) the function "signum x" is computable (but not recursive) on the set of all computable (recursive) numbers different from 0,

(5) the function "integral-part of x" is computable (but not recursive) on the set of all computable (recursive) non-integers,

(6) a function defined by a power-series with a recursive (computable) sequence of coefficients and with a recursive (computable) center, considered on

the set of all recursive (computable) numbers from the interval of convergence of this series, is not necessarily recursive (computable) on a set of all recursive (computable) numbers from an arbitrary closed interval contained in the interval of convergence.

It is also shown that a computable function on a set of all computable numbers of an interval is continuous and has the property of Darboux. On the contrary, recursive functions do not necessarily have the property of Darboux.

The above mentioned property that a computable function is continuous lead to the conclusion that in the computable analysis in which only computable numbers, sequences and functions are admitted one can take into consideration only those parts of classical analysis which treat exclusively of continuous functions. On the other hand it is not clear whether the entire classical theory of continuous functions can be obtained in computable analysis. The reason is that it is not known whether a computable function defined in a computable closed interval assumes the maximum of its values at a computable point. However it has been shown that numerous theorems of the theory of continuous functions are transferable to computable analysis.

Banach and Mazur's approach to constructive analysis is similar to that of E. Specker (1949). It can be proved that a function is recursive in the sense of Specker if and only if it is recursive and recursively uniformly continuous in the sense of Mazur.

The second approach to the constructive analysis described at the beginning of this section is represented by A. Grzegorczyk (1955a). It consists of using the notion of a functional. One says that a functional $\Phi(f_1,\ldots,f_k,x_1,\ldots,x_n)$ is computable if and only if it can be derived from the primitive functionals:

$$U_1(f,g,x) = f(x),$$
$$U_2(f,g,x) = g(x),$$
$$S(f,x) = x+1,$$
$$M(f,x,y) = x-y,$$
$$P(f,x,y) = x^y$$

by a finite number of the following operations: the substitution of the functional for the numerical variable, the identification of variables and the effective minimum.

The real function $\varphi(x)$ defined in the interval $[a,b]$, where $0 < a < b$, is called computable if and only if there exists a computable functional $\Phi(f,n)$ such that for each x ($a < x < b$) and each function f with natural values satisfying the condition

$$\left| x - \frac{f(n)}{n} \right| < \frac{1}{n}$$

for $n = 1, 2, \ldots$, the following inequality is satisfied:

$$\left| \varphi(x) - \frac{\Phi(f,n)}{n} \right| < \frac{1}{n}$$

for $n = 1, 2, \ldots$ In other words, the functional Φ transforms the function $f(n)/n$ approximating the argument x into the function $\Phi(f,n)/n$ approximating in the same degree the value $\varphi(x)$.

It can be shown that a function computable in the sense of Grzegorczyk is also computable in the sense of Banach–Mazur, hence it is continuous. One can also prove that a function computable in the sense of Grzegorczyk defined in the interval $[a,b]$ with computable endpoints, assumes its maximum value at the computable point of the interval $[a,b]$.

The comparison of various notions of a computable real function can be found in Grzegorczyk (1957).

Let us add at the end of this section that one can also build a system of constructive mathematics (and in particular of constructive analysis) by replacing in the definitions accepted by Banach and Mazur the class of computable functions by other broader classes. For example, one can admit elementarily definable functions, i.e., functions definable by means of the quantifiers bounding the integral variables only. This approach was considered by Grzegorczyk in (1955b). It was shown there that the classical analysis of continuous functions can be reproduced in an elementary manner. However it is not clear in what exactly the definable analysis deviates from the classical one.

6. Complexity of Models

In Section 3 we mentioned that mathematical and logical notions can be classified in the Kleene–Mostowski hierarchy. In this section an application of this hierarchy to model theory will be indicated.

From the theorems of Löwenheim–Skolem and of Gödel it follows that every consistent first-order theory T has a countable model. Hence it has a model with the universe being the set $\mathbb{N}$ of natural numbers. Thus predicates and function symbols of the language $L(T)$ of T can be interpreted as relations on the set $\mathbb{N}$ and as functions from $\mathbb{N}^n$ to $\mathbb{N}$, resp.. So we can classify them in the Kleene––Mostowski hierarchy. It will be said that a model $\mathfrak{M} = \langle \mathbb{N}, R_1, \ldots, R_k, f_1, \ldots, f_l \rangle$ is of the class Π_n^0 (Σ_n^0) if and only if all the relations R_i and functions f_j are of the class Π_n^0 (Σ_n^0).

Using Lindenbaum's method of completing a theory one can show that every consistent first-order theory has a model of the type $\Sigma_2^0 \cap \Pi_2^0$. Hence there arises the problem if a given theory can possess a simpler model.

A. Mostowski in (1953b) gave an example of a finitely axiomatizable theory which has no model of the class Σ_1^0, i.e., no recursively enumerable model. This theory is a modification of the axiomatic system of set theory proposed by P. Bernays in (1937). The modification consists in allowing more primitive notions – there are in fact twenty such notions in the system. The axioms of the theory are those given by Bernays with obvious changes which are the consequence of the choice of primitive notions.

This result was strengthened by Mostowski in (1955d) where an example of a finitely axiomatizable theory (so in fact of a single formula) which has no recursively enumerable model was given. What is important here is the fact that this example is simpler than that from the paper (1953b). In fact it does not refer to the axiomatic set theory and uses only tools known from the theory of recursive functions. The description of it is very long and complicated – therefore it cannot be given here. Note only that modifying it one can obtain a finitely axiomatizable theory (a formula) which possesses no model belonging to the class $\Sigma_1^0 \cap \Pi_1^0$.

Mostowski admits in (1955d) that he attempted to find an example of a theory with no model belonging to the smallest field of sets generated by the classes Σ_1^0 and Π_1^0. This problem was formulated by G. Kreisel in (1953) (p. 47; cf. also Mostowski in collaboration with A. Grzegorczyk et al. (1955b), p. 30). It was solved by G. Hensel and H. Putnam. Putnam showed in (1965) that every axiomatizable, consistent, first-order theory based on a finite number of predicates has a model in the field Σ_1^* generated by the recursively enumerable relations. This results was extended in the paper by Hensel–Putnam (1969) to the case of theories of the described type with identity. More precisely Hensel and Putnam showed that such a theory, if it possesses an infinite normal model, then it has a model in the class Σ_1^*. The exhibited model is the simplest possible in the sense that it contains Ramsey indiscernibles and only those extra elements needed for completion.

The problem of the lowest possible class of models of theories was studied also by A. Grzegorczyk. In the paper (1962a) he pointed out four first-order theories formalized in a language with two individual constants 0 and S and one binary function $|$. They are all finitely axiomatizable and form an increasing sequence: $F \subseteq F^{(1)} \subseteq F^{(2)} \subseteq F^{(3)}$. It is proved that: (1) the theory F is essentially undecidable (cf. Grzegorczyk's paper 1962) and has no primitive recursive model, (2) the theory $F^{(1)}$ has no primitive recursive model but has a general recursive model, (3) $F^{(2)}$ has no general recursive model but has an arithmetical ω-standard model and (4) the theory $F^{(3)}$ has no arithmetical ω-standard model.

To the list of the results described above one should add one more due to Mostowski. In the paper (1957b) he considered recursive models for axiomatic systems of arithmetic and proved that under certain assumptions every such model is isomorphic to the classical model. Hence the result says that, up to isomorphism, the only recursive model of arithmetic is the standard one.

Logical Investigations at the University of Poznań in 1945–1955

Co-authored by Jerzy Pogonowski

1. Some Introductory Information

The history of Warsaw Logical School (being an important part of Lvov-Warsaw Philosophical School) is well known.[1] But logical investigations have been carried out also at other scientific and academic centers of Poland, not only in Warsaw. Hence the question: how did they look like? What were the connections with Warsaw School? Were they connected with the tradition of Lvov-Warsaw School?

We want to investigate those questions with respect to Poznań University. We are interested in the period 1945–1955.[2] To understand it better we shall start by describing briefly logical investigations in Poznań before the Second World War.

The university in Poznań was founded in 1919. This happened 400 years after the foundation of the Lubrański Academy and 308 years after the formal creation of the university in Poznań by king Zygmunt III on the base of a Jesuit college (the university was not created then as a consequence of objections by the Cracow Academy). The initiators of the foundation of a university after the First World War were scholars gathered at Poznań Friends of Scholarship (Poznańskie Towarzystwo Przyjaciół Nauk; shortly: PTPN) existing since 1857. The university was founded as Piast University (Wszechnica Piastowska). In 1920 the name was changed to that of the University of Poznań.

At the beginning there were two faculties: Faculty of Philosophy and Faculty of Law. Since 1925 there existed five faculties: Law and Economy, Medicine, Humanities, Mathematics and Natural Sciences, Agriculture and Forestry.

Already in 1919 at the Faculty of Philosophy a Chair of Theory and Methodology of Natural Sciences and Humanities was founded. Its head was Władysław Mieczysław Kozłowski (1858–1935). He was scientifically active in many domains: botany, history, theory of literature and others but he is known first of all as a philosopher working also in logic. His main achievements in the latter are: an interesting and original classification of sciences and his works in the

---

1 Cf. Woleński (1989).
2 Our study is based on materials from the archive of Adam Mickiewicz University, the archive of Poznań Friends of Scholarscip [Poznańskie Towarzystwo Przyjaciół Nauk] and on some studies on the history of the University (cf. Grot 1971, 1972). Remembrances of some persons (former students of Kazimierz Ajdukiewicz), whose scientific career has started in the considered period as well as (rather few) existing papers on the activity of particular scholars were also used.

methodology of sociology. He was also the author of a university handbook in logic *Podstawy logiki, czyli zasady nauk* (Warszawa 1916). It was already a bit out of date when Kozłowski taught logic at the University of Poznań, nonetheless it was the first Polish handbook of logic in which Boolean algebras and the theory of relations were presented. The author formulated there the whole of syllogistic using the concepts of Boolean algebra augmented by a certain type of existential quantifier.

His follower at the chair was Zygmunt Zawirski (1882–1948) who studied philosophy, mathematics and physics at the Lvov University and received his Ph.D. in Lvov in 1906 and *Habilitation* at the Jagiellonian University in Cracow in 1924. Since 1928 he was professor at the University of Poznań. He worked in the methodology of physics as well as in the formal logic. In the latter he was interested mainly in possible applications of logic, in particular in interrelations between many-valued logics and the probability theory.

At the beginning of 1937 Zawirski moved to Cracow where he became ordinary professor at the Philosophical Faculty of the Jagiellonian University. The Ministry of Education asked all the relevant professors in Poland to suggest a candidate to fill the vacancy. Alfred Tarski (1901–1983) was unanimously recommended but was not appointed. The reason was the atmosphere in Poznań – always a stronghold of right-wing conservatism and dominated by the Catholic church, had, since Piłsudski's death in 1935, moved even further to the right and became outright fascistic and anti-Semitic. Because there was no way to appoint anyone else without making the reasons for denying Tarski the professorship patently clear, the position was eliminated.[3]

Before the war at the University in Poznań worked also a philosopher and logician Adam Wiegner (1889–1967). He obtained his *Habilitation* in 1934 and was active mainly after the Second World War – hence we will talk about him later.

One can see that before 1945 there were no real investigations in logic *sensu stricto* in Poznań. Logic was developed by philosophers interested mainly in the methodology. On the other hand mathematicians in Poznań were not interested in logic at all. During the Second World War the university was closed by the German authorities. It existed and worked, however, as the underground University of Western Territories (Uniwersytet Ziem Zachodnich).

The university was reopened in 1945.[4] And now the situation of logic has changed: philosophers who moved to Poznań (mainly from Lvov) were really interested in logic and scientifically active in this field and on the other hand ma-

---

3   Cf. Woleński (1995b) as well as Feferman and Feferman (2004, pp. 102–103).
4   In 1955 the university assumed the name of Adam Mickiewicz – after the greatest Polish Romantic poet of the first half of the 19th century. This name was proposed by professor Janusz Pajewski and this helped to avoid the proposal made by the authorities, namely the name of Joseph Stalin or of Julian Marchlewski (Polish communist).

thematicians, in particular Władysław Orlicz (a collaborator of Stefan Banach in Lvov) who re-launched the mathematical center in Poznań – though not working in logic – set a high value on it.

In 1945 at the Faculty of Mathematics and Natural Sciences the Chair of Theory and Methodology of Science has been founded – its head was professor Kazimierz Ajdukiewicz (who came to Poznań from Lvov via Cracow). In 1951 its name was changed to Chair of Logic at the Faculty of Mathematics, Physics and Chemistry. Members of the chair were (in various periods): Seweryna Łuszczewska-Romahnowa, Roman Suszko, Tadeusz Strumiłło, Franciszek Zeidler, Zbigniew Czerwiński, Andrzej Malewski.

Researches led at the chair were devoted to epistemology, formal logic and methodology. Beside the doctorate and *Habilitation* of R. Suszko (see below) two other procedures leading to *Habilitation* were carried out: by Henryk Mehlberg (on the basis of the dissertation *Essai sur la theorie causale du temps*) and by Maria Lutman (dissertation *O względności prawdy* [On the relativity of truth]).

Classes given by members of the chair were devoted to: main principles of logic and methodology (with exercises) (this course was obligatory for all students of the Faculty of Mathematics and Natural Sciences), mathematical logic (with exercises), logic (for students of other faculties), elements of set theory and mathematical logic, seminar on methodology as well as proseminars. Ajdukiewicz led also a seminar in logic for postgraduate and graduate students (see below).

In 1945 at the Faculty of Humanities, the Chair of Philosophy was reactivated. It was renamed in January 1951 to Chair of the History of Philosophy. Since 1947 its head was Adam Wiegner. In 1951 Wiegner became the head of the newly opened Chair of Logic which, since 1952, was located at the Philosophical-Historical Faculty. In this chair were active among others Jerzy Giedymin (since 1953) and Zbigniew Czerwiński (from 1951 till 1953).

An important rôle in the scholar life in Poznań after the war was played by Poznań Friends of Scholarship. Poznań logicians repeatedly delivered lectures at meetings of the Philosophical Committee of this society (often held as joint meetings of PTPN and the Polish Philosophical Society). Among the invited speakers one finds also: Stanisław Jaśkowski, Tadeusz Kotarbiński, Tadeusz Czeżowski, Maria Kokoszyńska-Lutmanowa, Izydora Dąmbska, Maria Ossowska.[5]

PTPN was also active in editing books and journals. In minutes of administrative meetings one finds some interesting information on editorial plans (as well as changes of them). For example, Ajdukiewicz intended to publish in October 1950 a work *Krytyczna analiza idealizmu* [Critical analysis of idealism] (10 printed sheets) (in the minute it was written: "the subject of the work is of great world-

---

[5] Let us add, as a curious detail, that on March 5, 1953, the day of the death of Joseph Stalin, a lecture by Rev. Mieczysław Dybowski on "Existentialism in psychology" was given (30 persons were present).

view significance"). In plans for the fourth quarter of 1951 a work under the same title is planned in the size of 5 printed sheets only. In 1952 Ajdukiewicz withdrew a work (planned for this year) under the title *Analiza klasycznej problematyki epistemologicznej* [An analysis of the classical problems of epistemology] – instead a paper by R. Suszko *O antynomiach logicznych* [On logical antinomies] (1.5 printed sheets) was planned. The latter was then moved in plans to 1953, next to 1954 (ultimately it was never published in Poznań; Suszko published „W sprawie antynomii kłamcy i semantyki języka naturalnego" [Concerning the antinomy of the liar and semantics of natural language] in Warsaw). Heated discussions were led concerning the question of publishing or rejecting a paper by T. Włodarczyk *Logika zdań u Dunsa Szkota i Pseudo-Szkota* [Propositional logic by Duns Scotus and Pseudo-Scotus].

Poznań logical center was active also in publishing journals. Note that the first volume of the new journal devoted to logic, namely *Studia Logica* has been prepared for publication in Poznań in 1953. The Editor-in-Chief was Kazimierz Ajdukiewicz, and Roman Suszko was the first secretary of the Editorial Board. Kazimierz Ajdukiewicz was also the Editor-in-Chief of the *Studia Philosophica*. The fourth, and at the same time the last, volume of this journal has appeared in 1951 (containing texts from 1949–1951).[6]

Below scientific researches and results of four leading Poznań logicians (Kazimierz Ajdukiewicz, Adam Wiegner, Seweryna Łuszczewska-Romahnowa and Roman Suszko) obtained in the considered period will be shortly presented and discussed.

2. Kazimierz Ajdukiewicz

The main figure among logicians active at the considered period in Poznań was doubtlessly Kazimierz Ajdukiewicz (December 12, 1890 – April 12, 1963). During his "Poznań time" he published more than thirty works (among them were his textbooks). He gained a world wide reputation already before the war. In Poznań he served as a rector of the University (1948–1952). He belonged of course to those whose opinion on the organization of scientific life in Poland after the war was of great importance.

Work and achievements of Ajdukiewicz were presented and discussed in many places.[7] We shall not attempt here to present them again, even in short. But on the other hand one should say some words about his works published

---

6   It is not true that *Studia Logica* was a continuation of *Studia Philosophica*. The ruling authorities simply closed *Studia Philosophica*, in the same way as they have done it to *Kwartalnik Filozoficzny*.

7   His works are presented, e.g., in papers by L. Borkowski and Z. Czerwiński in vol. XVI of *Studia Logica* – complete bibliography is provided there.

during his activity in Poznań in the considered period. As most important and representative we consider his following papers:

- "Logika i doświadczenie" [Logic and experience] (1947),
- "Zmiana i sprzeczność" [Change and contradition] (1948a),
- "Epistemologia i semiotyka" [Epistemology and semiotics] (1948b),
- "Metodologia i metanauka" [Methodology and metascience] (1948c),
- "On the notion of existence" (1951),
- "W sprawie artykułu prof. A. Schaffa o moich poglądach filozoficznych" [Concerning the paper by Professor A. Schaff on my philosophical views] (1953),
- "Klasyfikacja rozumowań" [Classification of reasonings] (1955).

In the period 1945–1955 some translations of important Ajdukiewicz's works written before the war were also published, for example *The Scientific World Perspective* and *Syntactic Connection*.

In the papers written in the considered period Ajdukiewicz repeatedly criticized severely and explicitly various idealistic trends in philosophy. He analysed the languages of the considered doctrines from the logical point of view and indicated various logical errors either in the construction of concepts or in reasonings.

Considering the idea maintained by some idealists according to which logic can be treated as a system of empirically verifiable hypotheses, Ajdukiewicz characterized strict empiricism as a certain *program* of developing science. He showed that such a program is in fact realizable but simultaneously stressed the inconformity of the hitherto development of science to the program of strict empiricism.

Analyzing paradoxes connected with notions of change and contradiction Ajdukiewicz attempted to explain misconceptions concerning the law of contradiction and to make precise some concepts involved in paradoxes.

In papers written in the period 1945–1955 Ajdukiewicz very often – for obvious reasons – undertook discussions with Marxists. And what is interesting, he not only defended his own philosophical views against attacks of opponents but also sometimes seemed to advance the latter some solutions in favor of their ideas.

In a paper on the classification of reasonings Ajdukiewicz critically analysed earlier proposals of Łukasiewicz and Czeżowski and presented his own classification. Critical remarks on it have been formulated by Wiegner.

As is known, Ajdukiewicz attached very great importance to the problem of teaching logic. He was an author of some excellent textbooks, in various papers expressed his opinion on the didactic of logic, organized meetings devoted to these problems. He spoke also in a discussion (important for the history of Polish logic and philosophy) on teaching logic which took place in the journal *Myśl Filozoficzna* [Philosophical Thought] in the fifties.

3. Adam Wiegner

In contrast to other logicians presented in this paper, Adam Wiegner (December 16, 1889 – September 28, 1967) was connected with Poznań during his whole life.[8] He studied philosophy at the Jagiellonian University in Cracow in 1908–1914 where he obtained his doctorate in 1923. His *Habilitation* took place at the University of Poznań in 1934. Since 1928 he was a member of the faculty of this university.

Main works of Wiegner (except his textbooks in logic) were written before the war. In the period considered here Wiegner did not publish a lot. Nevertheless he took an active part in scientific life. Let us quote as an example titles of some of his lectures delivered at that time: "Conference of German logicians in Berlin" (30 March 1954), "On paraphrasing subject statements into language" (18 October 1946), "On philosophical meaning of *Gestalttheorie*" (15 March 1948). He very often took part in discussions after talks presented on meetings of the Polish Philosophical Society or Poznań Friends of Scholarship.

Scientific works of Wiegner belong to various domains: history of philosophy, epistemology, ontology, psychology, philosophical foundations of physics and formal logic. Most important are his achievements in epistemology. His position was characterized by himself as "holistic empiricism". He claimed that what is directly given to us are certain wholes, "Gestalten", structures (together with causal relations). Elements of those data separated in the process of study are obtained by the process of abstraction. All scientific statements have a non-observational contents (of theoretical or hypothetical character). Hence Wiegner anticipated the thesis of hypothetism (though he himself attributed this to Avenarius).

Wiegner attempted to defend the principle of reciprocity between the contents and the extension of a notion. He claimed that sources of some of critics of this principle can be seen in terminological mistakes and in unsound assumptions concerning the concept of richness of the contents. He introduced a concept of a derived feature, the addition of which do not lead to the enlargement of the contents.

Wiegner was also interested in traditional logic – in some of his papers he considered the problem of a theoretical foundations of the traditional formal logic. He developed and improved the result of Ajdukiewicz which makes possible to obtain the principal part of the traditional formal logic in the system of the quantifier logic. This result of Wiegner was obtained earlier than a similar one by Ivo Thomas. Wiegner extended the system of axioms proposed by Ajdukiewicz (they ensured the non-emptiness of all considered names and the existence of three pairwise disjoint names) by adding an axiom

---

8   Information on his life and scientific activity can be found in Batóg (1968a, 1968b) and Nowakowa (2001). See also (Wiegner, 2005) for the selection of his works in the philosophy of science.

ensuring the non-universality of all considered names. He proposed to add to the rule of substituting terms with nonempty denotation for variables representing general names a rule of substituting terms with a non-universal denotation.

Wiegner's textbooks (cf. 1948 and 1952) were accurate and simultaneously clear and easy to understand. He proposed axiomatics for the propositional calculus which turned out to be of a didactic value. Primitive notions of this system are conjunction and negation and the axioms are the following:

(1) $p \supset p \cdot p$,
(2) $p \cdot q \supset p$,
(3) $p|q \supset q|p$,
(4) $(p \supset q) \supset (q|r \supset p|r)$,

where $\cdot, \supset$ and $|$ are symbols denoting conjunction, implication and Sheffer's stroke, resp. (the two latter being defined by negation and conjunction in the known way).

Wiegner carried out an analysis of important concepts of the philosophical logic such as: abstraction, generalization, idealization, concretization. Those analyzes influenced in a significant way the methodological reflection undertaken later in Poznań. In *Studia Logica* one can find short notes by Wiegner concerning for example attempts to specify logical terminology and some polemics with the classification of reasonings proposed by Ajdukiewicz.

4. Seweryna Łuszczewska-Romahnowa

Seweryna Łuszczewska-Romahnowa (August 10, 1904 – June 27, 1978) has published relatively little.[9] The main reason for that were, without any doubts, her dramatic experiences during the Second World War and the consequences these events had for her health afterwards.

One can recognize in her works the influence of her studies in Lvov: philosophical (under the guidance of Kazimierz Twardowski, Kazimierz Ajdukiewicz and Roman Ingarden) and mathematical (under the supervision of Hugo Steinhaus and Stefan Banach). She was awarded the degree of doctor of philosophy by Lvov University in 1932 on the basis of the dissertation "O wyrazach okazjonalnych" (On occasional terms; unpublished).

One of her earliest papers from the considered period "Wieloznaczność a język nauki" (Polysemy and the language of science; 1948) is devoted to the ambiguity of concepts used in the language of science. She argues that certain ambiguities in that language can not be avoided. However, scientific idiom remains

---

9   Information about Łuszczewska-Romahnowa's life and academic work can be found in Matóg (2001).

intelligible and serves as an efficient tool of communication. This is due – in Łuszczewska-Romahnowa's opinion – to certain structural as well as semantic properties of scientific texts.

The first volume of *Studia Logica* contained an extensive article by Łuszczewska-Romahnowa about a generalization of Venn's diagrams: "Analiza i uogólnienie metody sprawdzania formuł logicznych przy pomocy diagramów Venna" (Analysis and generalization of the method of verifying logical formulae by means of Venn's diagrams; 1953). The author presented there a method of checking the decidability of the first-order monadic predicate calculus.

In the period we are interested in, Łuszczewska-Romahnowa had also published a short note about natural classifications. Later on, she had also worked on multi-level classifications as well as on the distance functions connected with such classifications – cf. her "Classification as a kind of distance function" published in *Studia Logica* in 1961 and "A generalized theory of classifications I, II" written together with Tadeusz Batóg and published in 1965 in the same journal. According to those papers, any multi-level classification of a given set of objects is a linearly ordered hierarchy of partitions of that set. The ordering in question is that of "being a finer than" partition. The index of similarity of two objects is determined by the last level of the hierarchy at which these objects still remain elements of the same member of some partition in the hierarchy. It appears that the difference between the number of partitions involved and the index of two objects is the quasi-distance D between these objects (in the case where each member of the last partition is a one-element set it is simply the distance, i.e., a metric in the space of objects). The concept of naturalness of a multi-level classification is of a relative character: we say that a multi-level classification of a set of objects is natural with respect to an, *a priori* given, metric d, if for any objects x, y, u, v: if $D(x,y) < D(u,v)$, then $d(x,y) < d(u,v)$. Thus, natural multi-level classifications reflect similarity between objects given by an independent measure.

Łuszczewska-Romahnowa was also interested in methodological problems of the sciences and especially in those connected with 17th century. She had translated the important logical treatise *Logique de Port-Royal* of Nicole and Arnauld into Polish. It seems that she focused her attention exactly on those topics in the period we are describing.

Quite recently (2002) there appeared the special issue of *Kwartalnik Filozoficzny* (*Philosophical Quarterly*) vol. XIX, No 3/4, 1950, whose publication – at that time, exactly in 1950 – has been cancelled by the ruling political authorities. The volume includes the first part of a dissertation by Łuszczewska-Romahnowa *Rozważania o "metodzie" w filozofii francuskiej XVII wieku* [*Considerations concerning "the method" in French philosophy of the 17th century*]. The text ends with the words: *to be continued*. However, we do not know whether any continuation of this text exists. We have found, in the archive of the University of

Poznań, a review of Łuszczewska-Romahnowa's scientific achievements (written by Kazimierz Ajdukiewicz), where the reviewer mentioned – without title – her contribution sent for publication in the *Kwartalnik Filozoficzny* with a note about cancelation of the volume. In the published part one finds, among others, a classification of action programs. It is used in discussion of positions taken, e.g., by Pascal, Malebranche, Descartes and Leibniz as well as the views presented in the famous *Logique de Port-Royal*.

Seweryna Łuszczewska-Romahnowa was working at the University in Poznań up to her retirement in 1974. Some of her important works have been published after the period we are interested in. Let us shortly indicate the subjects touched upon in those works.

A few of Łuszczewska-Romahnowa's publications deal with argumentation theory – "Pewne pojęcie poprawnej inferencji i pragmatyczne pojęcie wynikania" (A certain concept of valid inference and a pragmatic concept of consequence) from 1962, "Z teorii racjonalnej dyskusji" (From the theory of rational discussion) from 1964. She had suggested a modernization of the classical theory of argumentation. She had been also working on a proper definition of the concept of pragmatic entailment.

Łuszczewska-Romahnowa was also interested in the problem of induction – her paper "Indukcja a prawdopodobieństwo" (Induction and probability) on that topic has appeared in 1957 in *Studia Logica*. It contains a critique of the probabilistic approach to the problem of induction and author's own suggested solution of this problem, which could be called a pragmatic one. In particular, she stresses that the goal of cognitive activity is not to arrive at absolute truth, but rather to get a better recognition of the environment and to find a smooth correlation between cognitive subject's actions and that environment.

5. Roman Suszko

Roman Suszko[10] (November 9, 1919 – June 3, 1979) was a student of physics at the University of Poznań in the years 1937–1939. He wrote his M.A. Thesis under the supervision of Zygmunt Zawirski in Cracow during the wartime. He has started his work at Ajdukiewicz's Chair of Theory and Methodology of Science in Poznań in 1946.

Suszko's first publications concerned logic without axioms (Suszko 1947a, 1948). He has shown how, given an axiomatic system T, one can eliminate its axioms and introduce a set of certain special finitistic inference rules so that the initial consequence relation of T is being preserved. Suszko has solved this problem in the case of propositional calculus and has pointed out to the

---

10 Some bio-bibliographical information on Suszko can be found in Omyła (1986, 2001) and Omyła and Zygmunt (1984).

limitations of the corresponding solution in the case of the predicate calculus with identity.

Another field of Suszko's interest at that time was the theory of definitions. He has developed an approach to this theory with an original criterion of extensional equivalence as replacing conditions of translatability and non-creativity. He has proposed a natural hierarchy of extensions-by-definitions of axiomatic systems. His main attention in (Suszko, 1949b) is devoted to such definitions of some functors which need an extra inference rule to be accepted in the system under consideration in order to introduce this definition in a correct, unique way.

The works (Suszko, 1949a and 1949b) were published together, as a separate booklet in the series edited by the Poznań Friends of Scholarship. They also correspond to the essential part of Suszko's Ph.D. Thesis submitted to the University of Poznań in Autumn 1948. The supervisor was Kazimierz Ajdukiewicz.

On November 19, 1951 Roman Suszko defended his *Habilitationsschrift* at the Faculty of Mathematics, Physics and Chemistry of the University of Poznań. The Polish text of this dissertation has appeared in print only recently in the above mentioned special issue of the *Kwartalnik Filozoficzny* under the title *Konstruowalne przedmioty i kanoniczne systemy aksjomatyczne* [*Constructible objects and canonic axiomatic systems*] [sent for publication on June 12, 1950]. Up to now, it was believed that the text of Suszko's *Habilitationsschrift* was published in *Studia Philosophica* in 1951 under the title *Canonic axiomatic systems* [sent for publication on November 25, 1950]. Actually, the latter is a genuine English translation of the former (there is a change of the title only). The reviewers of Suszko's dissertation were: Kazimierz Ajdukiewicz, Andrzej Mostowski and Władysław Orlicz.

Suszko's main aim in *Canonic axiomatic systems* is, according to his own words, "to precise and develop certain reasonings conveyed generally in connection with the so called Löwenheim–Skolem paradox". The distinguished feature of Suszko's approach, as compared with other (contemporary as well as later) resolutions of the paradox is that Suszko does not involve the Löwenheim–Skolem theorem at all. The author makes an essential use of his proposals developed earlier in the theory of definitions. The dissertation is the only one among Suszko's publications where he is directly addressing himself to set theoretical problems. He works in a system of set theory resembling those of Bernays and Gödel . In that system Cantor's theorem, stating the existence of an uncountable set, is of course provable. In his metatheorethical approach Suszko makes use of Tarski-style description of "morphology" of formal systems. The k-names introduced by Suszko correspond to categorematic names (here: closed terms obtained without the use of the descriptive operator) and the relation of k-designation holds between k-names and their extra-linguistic correlates, called *constructible* objects (sets). One obtains theorems concerning relative consistency of the investigated systems. If the universe of a given system consists entirely of constructible objects, then such a system is called *canonic*. The property of canonicity corresponds, according to Suszko, to Fraenkel's Axiom of Limitation (*Beschränkheitsaxiom*), which says that there are no other sets than these whose existence is postulated by the axioms (of a given axiomatic system of set theory). Suszko's metasystems are canonic. Constructible sets in canonic systems are k-designated by k-names and there exist only countably k-names. Thus we have a resolution of the paradox.

During his work in Poznań Suszko has also published a few other, minor articles. In particular, he reported about mathematical logic and foundational research in the Soviet Union (Suszko 1949c) and presented a critical discussion of logical positivism (Suszko 1952).

As it follows from the archives, Suszko has begun his work on *diachronic logic* already in Poznań – cf. also his first footnote to the text of (Suszko, 1957a) which (together with 1957b) is one of the first comprehensive applications of the Tarskian model theory to epistemological investigations. In the bibliography of this paper one finds also a reference to his work *Syntax and Model* (in preparation for print in *Studia Logica*). One may presume that what Suszko had in mind here are papers published a few years later in *Studia Logica* under the title *Syntactic structure and semantical reference* – part I in 1958, part II in 1960. Also the paper „W sprawie antynomii kłamcy i semantyki języka naturalnego" [Concerning the antinomy of the liar and semantics of natural language] published in Warsaw in 1957 has its roots in Suszko's investigations conducted in Poznań; cf. Suszko (1957c).

In 1953 Suszko left Poznań and moved to Warsaw.

6. Ajdukiewicz's Seminar

An important rôle in the scientific life in Poznań was played by the seminar led by Ajdukiewicz. We had interviewed some participants of the Ajdukiewicz's seminar. According to their memories, the problems discussed there were very diversified. Professor Jerzy Albrycht remembers among others discussion concerning Hume's philosophy. A vividly discussed topic was connected with logical fallacies in a natural language (common day) reasoning. Much effort had been also devoted to investigation of the logic of induction.

Some of the participants of the seminar changed later their interests from logic to other fields. Among them were Zbigniew Czerwiński, Dobiesław Bobrowski and Zbyszko Chojnicki.

Professor Zbigniew Czerwiński has written his M.A. thesis about logic of induction. He has also published some papers on this topic in *Studia Logica*. Some of Czerwiński's proposals appeared to be of importance to inferences conducted in economic sciences. At present Czerwiński is *professor emeritus* at the Academy of Economics in Poznań.

Professor Dobiesław Bobrowski's M.A. thesis has concerned propositional calculi without axioms. He remembers having many consultations with Roman Suszko on this topic. His further academic work was connected mainly with probability theory. At present he is *professor emeritus* at the Faculty of Mathematics and Computer Science at the Adam Mickiewicz University in Poznań.

Professor Zbyszko Chojnicki devoted his M.A. thesis to the logic of norms. At present he is *professor emeritus* at the Institute of Geography at the Adam Mickiewicz University in Poznań. Quite recently he has published a book devoted to the philosophy of science.

Andrzej Malewski was for a short time a teaching assistant at the Ajdukiewicz's Chair. He has published then a quite interesting textbook *ABC porządnego myślenia* [*ABC of well-ordered thinking*] which can be classified as belonging to logico-linguistic pragmatics. He has also collaborated with the late Professor Jerzy Topolski on the methodology of history.

7. Concluding Remarks

Above we have described the logical investigations carried out at the University of Poznań in the period 1945–1955. One can ask the following question: was there any continuation of the Lvov-Warsaw Logical School at the University of Poznań right after the Second World War? Unfortunately, the answer is far from obvious. Without any doubts one can find a stigma of that School in the post-war publications of Ajdukiewicz and Łuszczewska-Romahnowa. On the other hand, Suszko belonged to the new generation. Further, one has also to be fully aware of the political situation in Poland right after the war. One can grasp a touch of it, e.g., reading the discussion concerning the teaching of logic published in the fifties of the last century in *Myśl Filozoficzna* – the participants were, on the one side, prominent Polish logicians (a.o. Ajdukiewicz, Suszko, Grzegorczyk, Szaniawski, Przełęcki) and, on the other side, Marxists philosophers (headed by Adam Schaff). It is a real challenge for a historian of science to present a comprehensive report of the situation and ways of development of the academic life in Poland at the beginning of the post-war period. Some works already published are not free from emotional evaluations.

References

Ackermann, W. (1924/1925). Begründung des tertium non datur mittels der Hilbertschen Theorie der Widerspruchsfreiheit. *Mathematische Annalen* 93, 1–36.

Ackermann, W. (1940). Zur Widerspruchsfreiheit der Zahlentheorie. *Mathematische Annalen* 117, 162–194.

Addison, J. W. (1954). *On Some Points of the Theory of Recursive Functions*. Unpublished doctoral dissertation. University of Wisconsin.

Addison, J. W. (1958/1959). Separation Principles in the Hierarchies of the Classical and Effective Descriptive Set Theory. *Fundamenta Mathematicae* 46, 123–135.

Ajdukiewicz, K. (1934). Logistyczny antyirracjonalizm w Polsce [Logistical Antiirrationalism in Poland]. *Przegląd Filozoficzny* 37, 399–408. German translation: Der logistische Antiirrationalismus in Polen. *Erkenntnis* 5 (1935), 151–164.

Ajdukiewicz, K. (1947). Logika a doświadczenie [Logic and Experience]. *Przegląd Filozoficzny* XLIII, z. 1–4, 3–21.

Ajdukiewicz, K. (1948a). Zmiana i sprzeczność [Change and Contradition]. *Myśl Współczesna* 8/9, 35–52.

Ajdukiewicz, K. (1948b). Epistemologia i semiotyka [Epistemology and Semiotics]. *Przegląd Filozoficzny* XLIV, 336–347.

Ajdukiewicz, K. (1948c). Metodologia i metanauka [Methodology and Metascience]. *Życie Nauki* VI, 4–15.

Ajdukiewicz, K. (1951). On the Notion of Existence. *Studia Philosophica* IV, 7–22.

Ajdukiewicz, K. (1953). W sprawie artykułu prof. A. Schaffa o moich poglądach filozoficznych [Concerning the Paper by Professor A. Schaff on My Philosophical Views]. *Myśl Filozoficzna* 2(8), 292–334.

Ajdukiewicz, K. (1955). Klasyfikacja rozumowań [Classification of reasonings]. *Studia Logica* II, 278–300.

Anderson, R. and W. W. Bledsoe (1970). A Linear Format for Resolution with Merging and a New Technique for Establishing Completeness. *Journal of the Association for Computing Machines* 17, 525–534.

Andrews, P. (1976). Refutations by Matings. *IEEE Trans. on Computers* C-25, 801–807.

Andrews, P. (1981). Transforming Matings into Natural Deduction Proofs. *Proc. Fifth Conf. on Automated Deduction*, ed. by W. Bibel and R. Kowalski, pp. 281–292. Lecture Notes in Computer Science 87. Berlin-Heidelberg-New York: Springer-Verlag.

Andrews, P., D. A. Miller, E. L. Cohen and F. Pfenning (1984). Automating Higher-Order Logic. In: W. W. Bledsoe and D. W. Loveland (eds.) *Automated Theorem Proving. After 25 Years*, Contemporary Mathematics 29, pp. 169–192. Providence, Rhode Island: American Mathematical Society.

Apt, K. R. and W. Marek (1974). Second Order Arithmetic and Related Topics. *Annals of Mathematical Logic* 6, 177–239.

Arai, T. (1990). Derivability Conditions on Rosser's Provability Predicates. *Notre Dame Journal of Formal Logic* 4, 487–497.

Aristoteles (1960). *Aristotelis opera ex recensione Immanuelis Bekkeri edidit Academia Regia Borussica*. Editio altera. Edited by O. Gigon. Berlin: Walter de Gruyter.

Awodey, S. (1996). Structure in Mathematics and Logic: A Categorical Perspective. *Philosophia Mathematica* (3) Vol. 4, 209–237.

Awodey, S. (2004). An Answer to Hellman's Question: 'Does Category Theory Provide a Framework for Mathematical Structuralism'. *Philosophia Mathematica* (3) Vol. 12, 54–64.

Awodey, S. (2006). *Category theory*. Oxford: Oxford University Press.

Ax, J. (1968). Theory of Finite Fields. *Annals of Mathematics* 88, 239–271.

Ax, J. and S. Kochen (1965a). Diophantine Problems Over Local Fields. *American Journal of Mathematics* 87, 605–630.

Ax, J. and S. Kochen (1965b). Diophantine Problems Over Local Fields. II: A Complete Set of Axioms for p-adic Number Theory. *American Journal of Mathematics* 87, 631–648.

Ax, J. and S. Kochen (1966). Diophantine problems Over Local Fields. III: Decidable Fields. *Annals of Mathematics* 83, 437–456.

Bain, A. (1870). *Logic. Part first: Deduction, Part two: Induction*. London: Longmans, Green, Reader & Dyer.

Bain, A. (1878). *Logika* [Logic]. Tom I: *Dedukcya* [Deduction], Tom II: *Indukcya* [Induction]. Trans. Franciszek Krupiński. Warszawa: Skład główny w księgarni Gebethnera i Wolffa.

Banach, S. and S. Mazur (1937). Sur le fonctions calculables. *Annales de la Société Polonaise de Mathématique* 16, 223.

Barwise, J. (1980). Infinitary Logics. In: E. Agazzi (ed.) *Modern Logic – A Survey*, pp. 3–112. Dordrecht: D. Reidel Publishing Company.

Barzin, M. (1940). Sur la Portée du Théorème de M. Gödel'. *Académie Royale de Belgique, Bulletin de la Classe des Sciences*, Series 5, 26, 230–239.

Batóg, T. (1968a). Problematyka logiki tradycyjnej w pracach Adama Wiegnera [Problems of the Traditional Logic in Works by Adam Wiegner]. *Studia Logica* XXIII, 143–146.

Batóg, T. (1968b). Adam Wiegner. *Studia Filozoficzne* 1 (52), 236–238.

Batóg, T. (2001). Seweryna Łuszczewska-Romahnowa – Logic and Methodology of Science. In: W. Krajewski (ed.) *Polish Philosophers of Science and Nature in the 20th Century*, Poznań Studies in the Philosophy of the Sciences and the Humanities, vol. 74, pp. 113–119. Amsterdam-New York: Editions Rodopi.

Bedürftig, Th. and R. Murawski (2010). *Philosophie der Mathematik*. Berlin-New York: Walter de Gruyter.

Behmann H. (1921). Das Entscheidungsproblem der mathematischen Logik. *Jahresberichte der Deutschen Mathematiker-Vereinigung*, 2. Abteilung, 30.

Behmann H. (1922). Beiträge zur Algebra der Logik, insbesondere zum Entscheidungsproblem. *Mathematische Annalen* 86, 163–229.

Bell, J. L. (2006). Abstract and Variable Sets in Category Theory. In: *Advanced Studies in Mathematics and Logic: What is Category Theory?*, Polimetrica S.a.s., 9–16.

Benacerraf, P. (1967). God, the Devil, and Gödel . *The Monist* 51, 9–32.

Benacerraf, P. and H. Putnam, eds. (1964). *Philosophy of Mathematics. Selected Readings*. Englewood Cliffs, New Jersey: Prentice-Hall, Inc. Second edition: Cambridge: Cambridge University Press, 1983.

Bernays, P. (1934). Quelques points essentiels de la métamathématiques. *L'Enseignement mathématique* 34, 70–95.

Bernays, P. (1935a). Sur le platonisme dans les mathématiques. *L'Enseignement Mathématique* 34, 52–69.

Bernays, P. (1935b). Hilberts Untersuchungen über die Grundlagen der Arithmetik. In: D. Hilbert, *Gesammelte Abhandlungen*, Bd. 3, pp. 196–216. Berlin: Verlag von Julius Springer.

Bernays, P. (1937). A System of Axiomatic Set Theory. I. *Journal of Symbolic Logic* 2, 65–77.

Bernays, P. (1941). Sur les questions méthodologiques actuelles de la théorie hilbertienne de la démonstration. In: F. Gonseth (ed.), *Les entretiens de Zurich sur les fondements et la méthode des sciences mathématiques, 6–9 décembre 1938*, pp. 144–152. Zurich: Leemann. Discussion, pp. 153–161.

Bernays, P. (1967). Hilbert David. In: P. Edwards (ed.), *Encyclopedia of Philosophy*, vol.3, pp. 496–504. New York: Macmillan and Free Press.

Bernays, P. (1970). On the Original Gentzen Consistency Proof for Number Theory. In: J. Myhill et al. (eds.), *Intuitionism and Proof Theory*, pp. 409–417. Amsterdam: North-Holland Publishing Company.

Bernays P. and M. Schönfinkel (1928). Zum Entscheidungsproblem der mathematischen Logik. *Mathematische Annalen* 99, 342–372.

Black, M. (1948). The semantic definition of truth. *Analysis* 8.

Bläsius, K., N. Eininger, J. Siekmann, G. Smolka, A. Herold and C. Walther (1981). The Markgraf Karl refutation Procedure. In: *Proc. Seventh Intern. Joint Conf. on Artificial Intelligence*, pp. 511–518.

Bledsoe, W. W. (1983). Using Examples to Generate Instantiations for Set Variables. In: *Proc. Intern. Joint Conf. on Artificial Intelligence*.

Bledsoe, W. W. (1984). Some Automatic Proofs in Analysis. In: W. W. Bledsoe and D. W. Loveland (eds.) *Automated Theorem Proving. After 25 Years*, Contemporary Mathematics 29, pp. 89–118. Providence, Rhode Island: American Mathematical Society.

Boolos G. (1987). A Curious Inference. *Journal of Philosophical Logic* 16, 1–12.

Borkowski, L. (1965). Kazimierz Ajdukiewicz (1890–1963). I. *Studia Logica* XVI, 7–29.

Bowie, G. L. (1982). Lucas' Number Is Finally Up, *Journal of Philosophical Logic* 11, 279–285.

Boyer, R.S. (1971). *A Restriction of Resolution*. Ph. D. Thesis, University of Texas at Austin, Texas.

Boyer, R. S. and J. S. Moore (1975). Proving Theorems about LISP Functions. *Journal of the Association for Computing Machines* 22, 129–144.

Boyer, R. S. and J. S. Moore (1979). *A Computational Logic*. New York: Academic Press.

Boyer, R. S. and J. S. Moore (1981). A Verification Condition Generator for FORTRAN. In: R. S. Boyer and J. S. Moore (eds.) *The Correctness Problem in Computer Science*, London.

Börger E., E. Grädel and Y. Gurevich (1997). *The Classical Decision Problem*, Berlin: Springer Verlag.

Brouwer, L. E. J. (1907). *Over de Grondslagen der Wiskunde*. Amsterdam: Mass en van Suchtelen.

Brouwer, L. E. J. (1912). *Intuitionisme en formalisme*. Groningen: Noordhoff. Also in: *Wiskundig Tijdschrift* 9 (1912), 180–211. English translation: Intuitionism and Formalism, *Bulletin of the American Mathematical Society* 20 (1913), 81–96.

Brouwer, L. E. J. (1948). Consciousness, Philosophy and Mathematics. In: E. W. Beth, H. J. Pos and H. J. A. Hollak (eds.), *Library of the Tenth International Congress in Philosophy*, vol. 1, pp. 1235–1249. Amsterdam: North-Holland Publishing Company.

Buss, S. R. (1995). On Gödel's Theorem on Lengths of Proofs II: Lower Bounds for Recognizing k Symbol Provability. In: P. Clote and J. Remmel (eds.), *Feasible Mathematics II*, pp. 57–90. Basel: Birkhäuser.

Büchi, J. R. (1960). Weak Second-Order Arithmetic and Finite Automata. *Zeitschrift für mathematische Logik und Grundlagen der Mathematik* 6, 66–92.

Büchi, J. R. (1962). On a Decision Method in Restricted Second Order Arithmetic. In: E. Nagel et al. (eds.), *Logic, Methodology and Philosophy of Science . Proceedings of the 1960 Congress*, pp. 1–11. Stanford, California: Stanford University Press.

Cantor, G. (1895). Beiträge zur Begründung der transfiniten Mengenlehre. *Mathematische Annalen* 46, 481–512, 49, 207–246. Also in: Cantor (1932), pp. 281–351. English translation: *Contributions to the Founding of the Theory of Transfinite Numbers*, translated and edited by Ph.E.B. Jourdain, Dover Publications, Inc., New York 1955.

Cantor, G. (1932). *Gesammelte Abhandlungen mathematischen und philosophischen Inhalts mit erläuternden Anmerkungen sowie mit Ergänzungen aus dem Briefwechsel Cantor–Dedekind.* Edited by E. Zermelo. Berlin: Verlag von Julius Springer. Reprinted: Springer-Verlag, Berlin-Heidelberg-New York 1980.

Carnap, R. (1934). *Logische Syntax der Sprache*. Wien: Julius Springer.

Chaitin, G. (1974). Information-Theoretic Computational Complexity. *IEEE Transactions on Information Theory* IT-20, 10–15. Reprinted in: Th. Tymoczko (ed.), *New Directions in the Philosophy of Mathematics. An Anthology*, pp. 289–299. Boston-Basel-Stuttgart: Birkhäuser, 1985.

Chaitin, G. (1982). Gödel's Theorem and Information. *International Journal of Theoretical Physics* 21, 941–954. Reprinted in: Th. Tymoczko (ed.), *New Directions in the Philosophy of Mathematics. An Anthology*, pp. 300–311. Boston-Basel-Stuttgart: Birkhäuser, 1985.

Chang, C. L. (1970). The Unit Proof and the Input Proof in Theorem Proving. *Journal of the Association for Computing Machines* 17, 698–707.

Chang, C. L. and R. Lee (1973). *Symbolic Logic and Mechanical Theorem Proving*. New York-San Francisco-London: Academic Press.

Chihara, Ch. S. (1990). *Constructibility and Mathematical Existence*. Oxford: Clarendon Press.

Chinlund, T. J., M. Davis, G. Hineman and D. McIlroy (1964). *Theorem Proving by Matching*. Bell Laboratories.

Church, A. (1936). A Note on the Entscheidungsproblem. *Journal of Symbolic Logic* 1, 40–41. Correction: *ibid.*, 101–102.

Chwistek, L. (1912). Zasada sprzeczności w świetle nowszych badań Bertranda Russella [Principle of Contradiction in the Light of New Studies by Bertrand Russell]. *Rozprawy Akademii Umiejętności* 30, 270–334.

Chwistek, L. (1917). Trzy odczyty odnoszące się do pojęcia istnienia [Three Lectures Concerning the Concept of Existence]. *Przegląd Filozoficzny* 20, 122–151.

Chwistek, L. (1935). *Granice nauki. Zarys logiki i metodologii nauk ścisłych* [Limits of Science. An Outline of Logic and Methodology of Science]. Lwów-Warszawa: Książnica-Atlas. English translation: *The Limits of Science. An Outline of Logic and Methodology of Science*. New York-London 1948.

Cohen, P. J. (1969). Decision Procedures for Real and *p*-adic Fields. *Communications on Pure and Applied Mathematics* 22, 131–151.

Couturat, L. (1901). *La logique de Leibniz d'apres des documents inédits*. Paris: Alcan.

Couturat, L. (1905). *L'algèbre de la logique*. Paris: Gauthier-Villars.

Curry, H. B. (1951). *Outlines of a Formalist Philosophy of Mathematics*. Amsterdam: North-Holland Publishing Company.

Czermak J. (2002). *Abriss des ontologischen Arguments*. In: Köhler (2002), Band II, pp. 309–324.

Czerwiński, Z. (1965). Problematyka indukcji w pracach i działalności Kazimierza Ajdukiewicza [Problems of Induction in Works by Kazimierz Ajdukiewicz]. *Studia Logica* XVI, 31–38.

Davis, M. (1963). Eliminating the Irrelevant from Mechanical Proofs. *Proc. Symp. Applied Mathematics* 25, 15–30.

Davis, M., ed. (1965). *The Undecidable: Basic Papers on Undecidable Propositions, Unsolvable Problems, and Computable Functions*. Hewlett, N.Y.: Raven Press.

Davis, M., ed. (1994). *Solvability, Provability, Definability: The Collected Works of Emil L. Post*. Boston-Basel-Berlin: Birkhäuser.

Davis, M. and H. Putnam (1960). A Computing Procedure for Quantification Theory. *Journal of the Association for Computing Machines* 7, 201–215.

Dawson, J. W., Jr. (1984). Discussion on the Foundation of Mathematics. *History and Philosophy of Logic* 5, 111–129.

Dawson, J. W., Jr. (1985a). The Reception of Gödel's Incompleteness Theorems. In: P. D. Asquith and Ph. Kitcher (eds.), *PSA 1984: Proceedings of the 1984 Biennial Meeting of the Philosophy of Science Association*, vol. 2, pp. 253–271. Reprinted in: S. G. Shanker (ed.), *Gödel's Theorem in Focus*, pp. 74–95 (1988). London: Croom Helm.

Dawson, J. W., Jr. (1985b). Completing the Gödel–Zermelo Correspondence. *Historia Mathematica* 12, 66–70.

Dawson, J. W., Jr. (1997). *Logical Dilemmas. The Life and Work of Kurt Gödel*. Wellesley, Mass.: A.K. Peters.

Dawson, J. W., Jr. (1998). What Hath Gödel Wrought? *Synthese* 114, 3–12.

Dedekind, R. (1872). *Stetigkeit und irrationale Zahlen*. Braunschweig: Friedrich Vieweg und Sohn. 7th edition: 1967. English translation: *Continuity and Irrational Numbers*, in: *Essays on the Theory of Numbers*, translated by W. W. Beman, pp. 1–27. New York: Dover Publications, Inc., 1963.

Descartes, R. (1628). *Regulae ad directionem ingenii*. Reprinted: *Regulae ad directionem ingenii – Rules for the Direction of the Natural Intelligence*. A bilingual edition edited by G. Hefferman, Amsterdam-Atlanta, GA: Edition Rodopi, 1998.

Descartes, R. (1637). *Discours de la Methode. Pour bien conduire sa raison, chercher la verité dans la sciences. Plus la Dioptrique. Les Meteores. Et la Geometrie. Qui sont des essais de cete Methode*. Leyden: Ian Marie. Reprinted in: *Oeuvres de Descartes*. Edited by Charles Adam and Paul Tannery, vol. 6: *Discours de la Méthode, Essais*. Paris: Léopold Cerf, 1902.

Detlefsen, M. (1979). On Interpreting Gödel's Second Theorem. *Journal of Philosophical Logic* 18, 297–313.

Detlefsen, M. (1986). *Hilbert's Program. An Essay on Mathematical Instrumentalism*. Dordrecht-Boston-Lancaster-Tokyo: D. Reidel Publishing Company.

Detlefsen, M. (1990). On the Alleged Refutation of Hilbert's Program Using Gödel's First Incompleteness Theorem. *Journal of Philosophical Logic* 19, 343–377.

Dickstein, S. (1891). *Pojęcia i metody matematyki* [Concepts and Methods of Mathematics]. Tom 1, Część 1: Teorya działań, Warszawa: Wydawnictwo redakcyi "Prac matematycznofizycznych".

Dickstein, S. (1896). *Katalog dzieł i rękopisów Hoene-Wrońskiego. Catalogue des oeuvres imprimées et manuscriptes de Hoene Wroński*. Kraków: Nakładem Akademii Umiejętności. Also in: Dickstein S., *Hoene Wroński. Jego życie i prace*, 1896, pp. 239–351. Kraków: Nakładem Akademii Umiejętności.

Donner, J., A. Mostowski and A. Tarski (1978). The Elementary Theory of Well-Ordering. A Metamathematical Study. In: A. Macintyre, L. Pacholski and J. Paris (eds.), *Logic Colloquium'77*, pp. 1–54. Amsterdam: North-Holland Publishing Company.

Drake, F. R. (1989). On the Foundations of Mathematics in 1987. In: H.-D. Ebbinghaus *et al.* (eds.), *Logic Colloquium'87*, pp. 11–25. Amsterdam: Elsevier Science Publishers B.V.

Drossos, C. A. (2006). Sets, Categories and Structuralism. In: *Advanced Studies in Mathematics and Logic: What is Category Theory?*, Polimetrica S.a.s., 95–126.

Duda, R. (2004). On the Warsaw Interactions of Logic and Mathematics in the Years 1919–1939. *Annals of Pure and Applied Logic* 127, 289–301.

Dummett, M. (1978). Wittgenstein's Philosophy of Mathematics. In: M. Dummett, *Truth and Other Enigmas*, pp. 166–185. Cambridge, Mass.: Harvard University Press.

Ehrenfeucht, A. (1959). Decidability of the Theory of Linear Ordering Relation. *Notices of the American Mathematical Society* 6, 268–269.

Ehrenfeucht A. and Mycielski J. (1971). Abbreviating Proofs by Adding New Axioms. *Bulletin of the American Mathematical Society* 77, 366–367.

Eilenberg S., S. Mac Lane (1945). General Theory of Natural Equivalences. *Transactions of the American Mathematical Society* 58, 231–294.

Elgot, C. C. (1961). Decision Problems of Finite Automata Design and Related Arithmetics. Transactions of the American Mathematical Society 98, 21–52.

Euclid (1883–1886). *Euclidis Elementa*. vol. I–IV. Edidit I. L. Heiberg. Lipsiae: Teubner.

Euclid (1956). *The Thirteen Books of Euclid's Elements*. Translated from the text of Heiberg, with Introduction and Commentary by Sir Th. L. Heath. New York: Dover Publishers, Inc.

Euklid (1991). *Die Elemente. Buch I–XII*. Translated and edited by Clemens Thaer. Friedrich Vieweg und Sohn: Braunschweig and Wissenschaftliche Buchgesellschaft: Darmstadt (8th edition).

Feferman, S. (1960). Arithmetization of Metamathematics in a General Setting. *Fundamenta Mathematicae* 49, 35–92.

Feferman, S. (1962). Transfinite Recursive Progressions of Axiomatic Theories. *Journal of Symbolic Logic* 27, 259–316.

Feferman, S. (1964, 1968). Systems of Predicative Analysis. Part I: *Journal of Symbolic Logic* 129 (1964), 1–30; Part II: *Journal of Symbolic Logic* 133 (1968), 193–220.

Feferman, S. (1984). Kurt Gödel: Conviction and Causation. In: *Philosophia Naturalis*, a special issue, P. Weingartner *et al.* (eds.), *Philosophy of Science – History of Science. A Selection of Contributed Papers of the 7th International Congress of Logic, Methodology and Phi-

losophy of Science, Salzburg, 1983. Meisenheim/Glan: Verlag Anton Hain. Reprinted in: S. G. Shanker (ed.), *Gödel's Theorem in Focus*, pp. 96–114 (1988). London: Croom Helm.

Feferman, S. (1988). Hilbert's Program Revisited: Proof-Theoretical and Foundational Reductions. *Journal of Symbolic Logic* 153, 364–384.

Feferman A. and S. Feferman (2004). *Alfred Tarski. Life and Logic.* Cambridge: Cambridge University Press.

Field, H. (1980). *Science without Numbers.* Oxford: Basil Blackwell.

Field, H. (1989). *Realism, Mathematics, and Modality.* Oxford: Basil Blackwell.

Finsler, P. (1926). Formale Beweise und die Entscheidbarkeit. *Mathematische Zeitschrift* 25, 676–682. English translation: Formal Proofs and Undecidability, in: Heijenoort, J. van (ed.) (1967), pp. 440–445.

Fisher, M. J. and Rabin, M. O. (1974). Super Exponential Complexity of Presburger's Arithmetic. In: R. Karp (ed.) *Complexity of Computation, Proc. SIAM-AMS Sympos. Appl. Math.* 7, pp. 27–41.

Floyd, J. (1995). On Saying What You Really Want to Say: Wittgenstein, Gödel, and the Trisection of the Angle. In: J. Hintikka (ed.), *From Dedekind to Gödel. Essays on the Development of the Foundations of Mathematics,* pp. 373–425. Dordrecht: Kluwer Academic Publishers.

Frege, G. (1879). *Begriffsschrift, eine der arithmetischen nachgebildete Formelsprache des reinen Denkens.* Halle: Louis Nebert.

Frege, G. (1884). *Die Grundlagen der Arithmetik – Eine logisch-mathematische Untersuchung über den Begriff der Zahl.* Breslau: Wilhelm Koebner.

Frege, G. (1891). *Function und Begriff. Unveänderter Nachdruk eines Vortrages, Sitzung v. 9.1.1891 der Jenaischen Gesellschaft für Medizin und Naturwissenschaft.* Jena.

Frege, G. (1893). *Grundgesetze der Arithmetik – Begriffschriftlich abgeleitet,* vol. I. Jena: H. Pohle.

Frege, G. (1976). *Wissenschaftlicher Briefwechsel,* Hrsg. G. Gabriel, H. Hermes, F. Kambartel, Ch. Thiel, A. Veraart. Hamburg: Felix Meiner Verlag.

Friedman, H. (1975). Some Subsystems of Second Order Arithmetic and Their Use. In: *Proceedings of the International Congress of Mathematicians Vancouver 1974,* vol.1, Canadian Mathematical Congress, pp. 235–242.

Friedman, H. (1977). Personal communication to L. Harrington.

Friedman, H. (1981). On the Necessary Use of Abstract Set Theory. *Advances in Mathematics* 141, 209–280.

Gaifman, H. (2004). Non-Standard Models in a Broader Perspective, in: A. Enayat and R. Kossak (eds.) *Nonstandard Models of Arithmetic and Set Theory,* Contemporary Mathematics vol. 361, pp. 1–22. Providence, Rhode Island: American Mathematical Society.

Gawecki, B. J. (1958). *Wroński i o Wrońskim. Katalog dzieł filozoficznych Hoene-Wrońskiego oraz literatury dotyczącej jego osoby i filozofii* [Wroński and About Wroński. Catalogue of Philosophical Works by Hoene-Wroński and of Literature Concerning his Person and Philosophy]. Warszawa: PWN.

Gelernter, H. (1959). Realization of a Geometry Theorem-Proving Machine. *Proc. Intern. Conf. on Information Processing,* Paris: UNESCO House, pp. 273–282. Also in: Feigenbaum and Feldman (eds.) *Computers and Thought.* McGrow-Hill.

Gelernter, H., J. R. Hanson and D. W. Loveland (1960). Empirical Explorations of the Geometry-Theorem Proving Machine. *Proc. Western Joint Computer Conf.,* pp. 143–147. Also in: Feigenbaum and Feldman (eds.) *Computers and Thought.* McGrow-Hill.

Gentzen, G. (1936). Die Widerspruchsfreiheit der reinen Zahlentheorie. *Mathematische Annalen* 112, 493–565. English translation: The Consistency of Elementary Number Theory. In: M. E. Szabo (ed.), *The Collected Papers of Gerhard Gentzen*, pp. 132–213. Amsterdam: North-Holland Publishing Company.

Gentzen, G. (1938). Neue Fassung des Widerspruchsfreiheitsbeweises für die reine Zahlentheorie. *Forschung zur Logik und zur Grundlagen der exakten Wissenschaften, New Series* 4, 19–44. English translation: New Version of the Consistency Proof for Elementary Number Theory. In: M. E. Szabo (ed.), *The Collected Papers of Gerhard Gentzen*, pp. 252–286. Amsterdam: North-Holland Publishing Company.

Gentzen, G. (1969). On the Relation Between Intuitionistic and Classical Arithmetic. In: M. E. Szabo (ed.), *The Collected Papers of Gerhard Gentzen*, pp. 53–67. Amsterdam: North-Holland Publishing Company.

Gentzen, G. (1974). Der erste Widerspruchsfreiheitsbeweis für die klassische Zahlentheorie. *Archiv fuer math. Logik und Grundlagenforschungen* 116, 97–112.

Gilmore, P. C. (1959). A Program for the Production of Proofs for Theorems Derivable within the First Order Predicate Calculus from Axioms. *Proceedings of the International Conference on Information Processing*. Paris: UNESCO House.

Gilmore, P. C. (1960). A Proof Method for Quantification Theory: its Justification and Realization. *IBM Journal Research and Devel*, pp. 28–35.

Givant, S. R. (1991). A Portrait of Alfred Tarski. *The Mathematical Intelligencer* 13, No 3, 16–32.

Gödel, K. (1929). Über die Vollständigkeit des Logikkalküls. Doctoral dissertation, submitted in 1929. Published and translated in: Gödel (1986), pp. 60–101.

Gödel, K. (1930a). Die Vollständigkeit der Axiome des logischen Funktionenkalküls. *Monatshefte für Mathematik und Physik* 37, 349–360. Reprinted with English translation: The Completeness of the Axioms of the Functional Calculus of Logic, in: Gödel (1986), pp. 102–123.

Gödel, K. (1930b). Einige metamathematische Resultate über Entscheidungsdefinitheit und Widerspruchsfreiheit. *Anzeiger der Akademie der Wissenschaften in Wien* 67, 214–215. Reprinted with English translation: Some Metamathematical Results on Completeness and Consistency, in: Gödel (1986), pp. 140–143.

Gödel, K. (1931a). Über formal unentscheidbare Sätze der *Principia Mathematica* und verwandter Systeme. I. *Monatshefte für Mathematik und Physik* 38, 173–198. Reprinted with English translation: On Formally Undecidable Propositions of *Principia Mathematica* and Related Systems, in: Gödel (1986), pp. 144–195.

Gödel, K. (1931b). Diskussion zur Grundlegung der Mathematik. English translation: Discussion on Providing a Foundation for Mathematics. In: Gödel (1986), pp. 200–203.

Gödel, K. (1931?). Über unentscheidbare Sätze; first published (German text and English translation: On Undecidable Sentences) in: Gödel (1995), pp. 30–35.

Gödel, K. (1932). Über Vollständigkeit und Widerspruchsfreiheit. *Ergebnisse eines mathematischen Kolloquiums* 3, 12–13.

Gödel, K. (1933a). Zur intuitionistischen Arithmetik und Zahlentheorie. *Ergebnisse eines Mathematischen Kolloquiums* 14, 34–38. Reprinted with English translation: On Intuitionistic Arithmetic and Number Theory, in: Gödel (1986), pp. 286–295.

Gödel, K. (1933b). The Present Situation in the Foundations of Mathematics. First published in: Gödel (1995), pp. 45–53.

Gödel, K. (1933c). Zum Entscheidungsproblem des logischen Funktionenkalküls. *Monatshefte für Mathematik und Physik* 40, 433–443. Reprinted with English translation: On the Decision Problem for the Functional Calculus of Logic, in: Gödel (1986), pp. 306–326.

Gödel, K. (1934). *On Undecidable Propositions of Formal Mathematical Systems* (mimeographed lecture notes, taken by S. C. Kleene and J. B. Rosser), Princeton. Reprinted with revisions in: Davis (1965), pp. 39–74. Reprinted also in Gödel (1986), pp. 346–371.

Gödel, K. (1936). Über die Länge von Beweisen. *Ergibnisse eines mathematischen Kolloquiums* 7, 23–24. Reprinted with English translation: On the Length of Proofs, in: Gödel (1986), pp. 396–399.

Gödel, K. (1940). *The Consistency of the Axiom of Choice and of the Generalized Continuum Hypothesis with the Axioms of Set Theory*. Lecture notes taken by George W. Brown. Princeton, N.J.: Princeton University Press. Reprinted with additional notes in 1951 and with further notes in 1966. Reprinted in: Gödel (1990), pp. 33–101.

Gödel, K. (1944). Russell's Mathematical Logic. In: P. A. Schilpp (ed.), *The Philosophy of Bertrand Russell*, pp. 123–153. Evanston, Ill.: Northwestern University. Reprinted in: Gödel (1990), pp. 119–141.

Gödel, K. (1946). Remarks Before the Princeton Bicentennial Conference on Problems in Mathematics, 1–4. First published in: Davis (1965), pp. 84–88. Reprinted in: Gödel (1990), pp. 150–153.

Gödel, K. (1947). What is Cantor's Continuum Problem? *The American Mathematical Monthly* 54, 515–525. Second revised version in: P. Benacerraf and H. Putnam (eds.), *Philosophy of Mathematics. Selected Readings*, pp. 258–273 (1964). Englewood Cliffs, N.J.: Prentice-Hall, Inc. Reprinted in: Gödel (1990), pp. 176–187.

Gödel, K. (1951). Some Basic Theorems on the Foundations of Mathematics and Their Implications. First published in: Gödel (1995), pp. 304–323.

Gödel, K. (1958). Über eine bisher noch nicht benützte Erweiterung des finiten Standpunktes. *Dialectica* 12, 280–287. Reprinted with English translation: On a Hitherto Unutilized Extension of the Finitary Standpoint, in: Gödel (1990), pp. 240–251.

Gödel, K. (1961). The Modern Development of the Foundations of Mathematics in the Light of Philosophy. First published (German text and English translation) in: Gödel (1995), pp. 374–387.

Gödel, K. (1972). On an Extension of Finitary Mathematics Which Has Not Yet Been Used. Revised and expanded English version of Gödel (1958), to have appeared in *Dialectica*. First published in: Gödel (1990), pp. 271–280.

Gödel, K. (1986). *Collected Works*. Vol. I. Edited by S. Feferman, J. W. Dawson, Jr., S. C. Kleene, G. H. Moore, R. M. Solovay and J. van Heijenoort. New York: Oxford University Press and Oxford: Clarendon Press.

Gödel, K. (1990). *Collected Works*. Vol. II. Edited by S. Feferman, J. W. Dawson, Jr., S. C. Kleene, G. H. Moore, R. M. Solovay and J. van Heijenoort. New York and Oxford: Oxford University Press.

Gödel, K. (1995). *Collected Works*. Vol. III. Edited by S. Feferman, J. W. Dawson, Jr., W. Goldfarb, Ch. Parsons and R. M. Solovay. New York and Oxford: Oxford University Press.

Gödel, K. (2003a). *Collected Works*. Vol. IV. Edited by S. Feferman, J. W. Dawson, Jr., W. Goldfarb, Ch. Parsons and W. Sieg. Oxford: Clarendon Press.

Gödel, K. (2003b). *Collected Works.* Vol. V. Edited by S. Feferman, J.W. Dawson, Jr., W. Goldfarb, Ch. Parsons and W. Sieg. Oxford: Clarendon Press.

Goliński, S. (1894). *Historia gimnazjum przemyskiego. Sprawozdanie dyrekcyi c.k. gimnazjum w Przemyślu za rok szkolny 1894* [History of the Grammar-School in Przemyśl. A Report of the Board of Directors for the School Year 1894]. Przemyśl: (publisher's name not indicated).

Good, I. J. (1967). Human and Machine Logic. *British Journal for the Philosophy of Science* 18, 145–146.

Good, I. J. (1969). Gödel's Theorem is a Red Herring. *British Journal for the Philosophy of Science* 19, 357–358.

Goodstein, R. L. (1944). On the Restricted Ordinal Theorem. *Journal of Symbolic Logic* 19, 33–41.

Gould, W. E. (1966). A Matching Procedure for ω-Order Logic. *Air Force Cambridge Research Laboratories, Report 66-781-4.*

Grattan-Guinness, I. (1979). In Memoriam Kurt Gödel: His 1931 Correspondence with Zermelo on his Incompletability Theorem. *Historia Mathematica* 6, 294–304.

Grelling, K. (1937). Gibt es eine Gödelsche Antynomie?. *Theoria* 3, 297–306. Zusätze und Berichtigungen: *Theoria* 4 (1938), 68–69.

Grot, K. (ed.) (1971). *50 lat Uniwersytetu im. Adama Mickiewicza 1919–1969* [50 Years of Adam Mickiewicz University 1919–1969]. Poznań: Wydawnictwo Naukowe Uniwersytetu im. Adama Mickiewicza.

Grot, K. (ed.) (1972). *Dzieje Uniwersytetu im. Adama Mickiewicza 1919–1969* [History of Adam Mickiewicz University 1919–1969]. Poznań: Wydawnictwo Naukowe Uniwersytetu im. Adama Mickiewicza.

Grzegorczyk, A. (1953). Some Classes of Recursive Functions. *Rozprawy Matematyczne* IV, 1–46.

Grzegorczyk, A. (1955a). Computable Functionals. *Fundamenta Mathematicae* 42, 168–202.

Grzegorczyk, A. (1955b). Elementarily Definable Analysis. *Fundamenta Mathematicae* 41, 311–338.

Grzegorczyk, A. (1955c). On the Definition of Computable Functionals. *Fundamenta Mathematicae* 42, 232–239.

Grzegorczyk, A. (1957). On the Definitions of Computable Real Continuous Functions. *Fundamenta Mathematicae* 44, 61–71.

Grzegorczyk, A. (1959). Some Approaches to Constructive Analysis. In: A. Heyting (ed.), *Constructivity in Mathematics. Proceedings of the Colloquium held at Amsterdam, 1957*, pp. 43–61. Amsterdam: North-Holland Publishing Company.

Grzegorczyk, A. (1962). An Example of Two Weak Essentially Undecidable Theories F and F^*. *Bulletin de l'Académie Polonaise des Sciences, Série des sciences math., astr. et phys.*, 10, 5–9.

Grzegorczyk, A. (1962a). A Theory Without Recursive Models. *Bulletin de l'Académie Polonaise de Sciences, Série des sciences math., astr. et phys.* 10, 63–69.

Grzegorczyk, A. (1964). Recursive Objects in All Finite Types. *Fundamenta Mathematicae* 54, 73–93.

Guard, J. R., F. C. Oglesby, J. H. Bennett and L. G. Settle (1969). Semiautomated mathematics. *Journal of the Association for Computing Machines* 16, 49–62.

Guaspari, D. and R. Solovay (1974). Rosser Sentences. *Annals of Mathematical Logic* 116, 81–99.

Hájek, P. and P. Pudlák (1993). *Metamathematics of First-Order Arithmetic.* Berlin-Heidelberg-New York: Springer Verlag.

Hart, W. D., ed. (1996). *The Philosophy of Mathematics.* Oxford: Oxford University Press.

Hartmanis, J. (1989). Gödel, von Neumann and the $P = ?NP$ problem. *Bulletin of the European Association for Theoretical Computer Science* (EATCS) 38, 101–107.

Heijenoort, J. van, ed. (1967). *From Frege to Gödel. A Source Book in Mathematical Logic 1879–1931.* Cambridge, Mass.: Harvard University Press.

Heintz, J., M. F. Roy and P. Solerno (1989). Complexité du principe de Tarski–Seidenberg. *C. R. Acad. Sci. Paris*, Sér. Math. 309, 825–830.

Hellman, G. (1989). *Mathematics without Numbers. Towards a Modal-Structured Interpretation.* Oxford: Clarendon Press.

Hellman, G. (2003). Does Category Theory Provide a Framework for Mathematical Structuralism?. *Philosophia Mathematica* (3) Vol. 11, 129–157.

Helmer, O. (1937). Perelman versus Gödel. *Mind* 46, 58–60.

Hensel, G. and Putnam, H. (1969). Normal Models and the Field Σ_1^*. *Fundamenta Mathematicae* 64, 231–240.

Herbrand, J. (1928). Sur la théorie de la démonstration. *Comptes rendus hebdomadaires des séances de l'Academie des sciences (Paris)* 186, 1274–1276.

Herbrand J., (1929). Sur le problème fondamental des mathématiques. *Comptes rendus hebdomadaires des séances de l'Academie des sciences (Paris)* 186, 554–556, 720. English translation: On the Fundamental Problem of Mathematics, in: J. Herbrand, *Logical Writings*, ed. W. D. Goldfarb, pp. 41–43. Dordrecht: D. Reidel Publishing Company, 1971.

Herbrand, J. (1930). Recherches sur la théorie de la demonstration. *Travaux de la Sociéte des Sciences et des Lettres de Varsovie*, Classe III, Warszawa. English translation: Investigations in proof theory, in: J. Herbrand, *Logical Writings*, ed. W. D. Goldfarb, pp. 46–188, 272–276. Dordrecht: D. Reidel Publishing Company, 1971.

Herbrand, J. (1931a). Sur le problème fondamental de la logique mathématique. *Sprawozdania z Posiedzeń Towarzystwa Naukowego Warszawskiego*, Wydział III, 24, 12–56. English translation: *On the Fundamental Problem of Mathematical Logic*, in: Herbrand J., *Logical Writings*, ed. W. D. Goldfarb, pp. 215–259. Dordrecht: D. Reidel Publishing Company, 1971.

Herbrand, J. (1931b). Sur la non-contradiction de l'arithmétique. *Journal für reine und angewandte Mathematik* 166, 1–8. English translation: On the Consistency of Arithmetic, in: Heijenoort (1967), pp. 620–628.

Heyting, A. (1957). Some Remarks on Intuitionism. In: A. Heyting (ed.), *Constructivity in Mathematics. Proceedings of the Colloquium held in Amsterdam, 1957*, pp. 69–71. Amsterdam: North-Holland Publishing Company.

Hewitt, C. (1971). *Description and Theoretical Analysis(Using Schemata) of PLANNER: a Language for Proving Theorems and Manipulating Models in a Robot*, Ph.D. Thesis, MIT.

Hilbert, D. (1899). *Grundlagen der Geometrie. Festschrift zur Feier der Enthüllung des Gauss–Weber–Denkmals*, pp. 3–92. Leipzig: B.G. Teubner.

Hilbert, D. (1900). Über den Zahlbegriff. *Jahresbericht der Deutschen Mathematikervereinigung* 8, 180–184.

Hilbert, D. (1901) Mathematische Probleme. *Archiv der Mathematik und Physik* 1, 44–63 and 213–237. Reprinted in: Hilbert (1935), pp. 290–329. English translation: Mathematical Problems. *Bulletin of the American Mathematical Society* 8 (1901–2), 437–479. Also in: F. Browder (ed.), *Mathematical Developments Arising from Hilbert's Problems*. Proceedings of the Symposia in Pure Mathematics 28, pp. 1–34 (1976). Providence, R.I.: American Mathematical Society.

Hilbert, D. (1902/1903). Über den Satz von der Gleichheit der Basiswinkel im gleichschenkligen Dreieck. *Proceedings of the London Mathematical Society* 35, 50–68.

Hilbert, D. (1903). *Grundlagen der Geometrie*, second edition. Leipzig: Teubner.

Hilbert, D. (1905a). Logische Principien des mathematischen Denkens. Lecture notes by Ernst Hellinger, Mathematisches Institut, Georg-August-Universität Göttingen, Sommer-Semester 1905. Unpublished manuscript.

Hilbert, D. (1905b). Über die Grundlagen der Logik und der Arithmetik'. In: A. Krazer (ed.), *Verhandlungen des dritten Internationalen Mathematiker-Kongresses in Heidelberg vom 8. bis 13. August 1904*, pp. 174–185. Leipzig: Teubner. English translation: On the Foundations of Logic and Arithmetic. In: Heijenoort (1967), pp. 129–138.

Hilbert, D. (1917/1918). Prinzipien der Mathematik. Lecture notes by Paul Bernays. Mathematisches Institut, Georg-August-Universität Göttingen, Wintersemester 1917–1918. Unpublished typescript.

Hilbert, D. (1918). Axiomatisches Denken. *Mathematische Annalen* 78, 405–415. Reprinted in: Hilbert (1935), pp. 146–177.

Hilbert, D. (1926). Über das Unendliche. *Mathematische Annalen* 95, 161–190. English translation: On the Infinite, translated by S. Bauer-Mengelberg. In: Heijenoort (1967), pp. 369–392.

Hilbert, D. (1927). Die Grundlagen der Mathematik. *Abhandlungen aus dem mathematischen Seminar der Hamburgischen Universität* 6, 65–85. Reprinted in: D. Hilbert, *Grundlagen der Geometrie*, 7th edition, Leipzig 1930. English translation: The Foundations of Mathematics. In: Heijenoort (1967), pp. 464–479.

Hilbert, D. (1929). Probleme der Grundlegung der Mathematik. In: *Atti del Congresso Internationale dei Matematici, Bologna 3–10 September 1928*, Bologna, vol. I, pp. 135–141. Also in: *Mathematische Annalen* 102, 1–9.

Hilbert, D. (1930). Naturerkennen und Logik. *Naturwissenschaften* 18, 959–963. Reprinted in: Hilbert (1935), pp. 378–387.

Hilbert, D. (1931). Die Grundlegung der elementaren Zahlentheorie. *Mathematische Annalen* 104, 485–494. Reprinted in: Hilbert (1935), pp. 192–195.

Hilbert, D. (1935). *Gesammelte Abhandlungen*, Bd. 3. Berlin: Verlag von Julius Springer.

Hilbert, D. and W. Ackermann (1928). *Grundzüge der theoretischen Logik*. Berlin: Verlag von Julius Springer. English translation of the second edition: *Principles of Mathematical Logic*. 1950. New York: Chelsea Publishing Company.

Hilbert, D. and P. Bernays (1934, 1939). *Grundlagen der Mathematik*. Band I 1934, Band II 1939. Berlin: Springer-Verlag.

Hintikka, J., ed. (1969). *The Philosophy of Mathematics*. Oxford: Oxford University Press.

Hlodovskii, I. (1959). A New Proof of the Consistency of Arithmetic. *Usp. Mat. Nauk* 14, No 6, 105–140 (in Russian).

Hofstadter, D. R. (1979). *Gödel, Escher, Bach: An Eternal Golden Briad*. New York: Basic Books.

Huet, G. D. (1975). A Unification Algorithm for Typed λ-Calculus. *Theoretical Computer Science*, 27–57.

Ingarden, R. (1938). Działalność naukowa Kazimierza Twardowskiego [Scientific Activity of Kazimierz Twardowski]. In: *Kazimierz Twardowski: nauczyciel – uczony – obywatel* [Kazimierz Twardowski: Teacher – Scholar – Citizen]. Lwów. Also in: *Z badań nad filozofią współczesną* [From Studies on the Contemporary Philosophy]. Warszawa 1963. This same in English translation: The Scientific Activity of Kazimierz Twardowski. *Studia Philosophica* 3 (1948), 17–30.

Isaacson, D. (1987). Arithmetical Truth and Hidden Higher-Order Concepts. In: The Paris Logic Group (eds.), *Logic Colloquium'85*, pp. 147–169. Amsterdam: Elsevier Science Publishers B.V. (North-Holland).

Isaacson, D. (1992). Some Considerations on Arithmetical Truth and the ω-Rule. In: M. Detlefsen (ed.), *Proof, Logic and Formalization*, pp. 94–138. London and New York: Routledge.

Janiczak, A. (1950). A Remark Concerning Decidability of Complete Theories. *Journal of Symbolic Logic* 15, 277–279.

Janiczak, A. (1953). Undecidability of Some Simple Formalized Theories. *Fundamenta Mathematicae* 40, 131–139.

Janiczak, A. (1955). Some Remarks on Partially Recursive Functions. *Colloquium Mathematicum* 3, 37–38.

Janiszewski, Z. (1915a). Logistyka [Logistics]. In: *Poradnik dla samouków. Wskazówki metodyczne dla studjujących poszczególne nauki*. Wydanie nowe, tom I. Wydawnictwo A. Heflera i St. Michalskiego. Warszawa, 449–461.

Janiszewski, Z. (1915b). Zagadnienia filozoficzne matematyki [Philosophical Problems of Mathematics]. In: *Poradnik dla samouków. Wskazówki metodyczne dla studjujących poszczególne nauki*, tom I, pp. 462–489. Warszawa: Wydawnictwo A. Heflicha i St. Michalskiego.

Janiszewski, Z. (1916). O realizmie i idealizmie w matematyce [On Realism and Idealism in Mathematics]. *Przegląd Filozoficzny* 19, 161–170. French translation: 'Sur le réalisme et l'idéalisme en mathématiques'. In: Z. Janiszewski, *Oeuvres choisies*, rédigées par K. Borsuk et al., pp. 309–317 (1962). Warszawa: Państwowe Wydawnictwo Naukowe.

Janiszewski, Z. (1917). O potrzebach matematyki w Polsce [On the needs of mathematics in Poland]. In: *Nauka polska, jej potrzeby, organizacja i rozwój* 1, 11–18. Reprinted in: *Roczniki Polskiego Towarzystwa Matematycznego, Seria II: Wiadomości Matematyczne* 7 (1963), 3–8. English translation published in: S. M. G. Kuzawa, *Modern Mathematics: The Genesis of a School in Poland*, New Haven 1968.

Jensen, D. C. and T. Pietrzykowski (1976). Mechanizing ω-Order Type Theory through Unification. *Theoretical Computer Science*, 123–171.

Jeroslov, R. (1975). Experimental Logics and Δ_2^0-Theories. *Journal of Philosophical Logic* 14, 253–267.

Jevons, W. S. (1870). *Elementary Lessons in Logic: Deductive and Inductive*. London: Macmillan and co.

Jevons, W. S. (1886). *Logika. Objaśniona figurami i pytaniami* [Logic. Explained by Figures and Questions]. Trans. Henryk Wernic. Warszawa: Nakładem Księgarni Teodora Paprockiego i S-ki.

Jordan, Z. (1945). *The Development of Mathematical Logic and of Logical Positivism in Poland between the Two Wars*. London: Oxford University Press.

Kahr, A., E. Moore and H. Wang (1962). Entscheidungsproblem Reduced to the ∀∃∀ Case. *Proceedings of the National Academy of Sciences USA* 48, 365–377.

Kalmár, L. (1932). Zum Entscheidungsproblem der mathematischen Logik. In: *Verhandlungen des Internationalen Mathematischen Kongresses*, vol. II, pp. 337–338. Zürich.

Kalmár, L. (1936). Zurückführung des Entscheidungsproblems auf den Fall von Formeln mit einer einzigen, binären, Funktionsvariablen. *Compositio Mathematica* 4, 137–144.

Kalmár, L. (1943). Egyszerü példa eldönthetetlen aritmetikai problémára. *Matematikai és Fizikai Lapok* 50, 1–23.

Kant, I. (1781). *Critik der reinen Vernunft*. Riga: Johann Friedrich Hartknoch.

Kaye, R. (1991). *Models of Peano Arithmetic*. Oxford: Clarendon Press.

Ketelsen, Ch. (1994). *Die Gödelschen Unvollständigkeitssätze. Zur Geschichte ihrer Entstehung und Rezeption*. Stuttgart: Franz Steiner Verlag.

Kirby, L. and J. Paris (1977). Initial Segments of Models of Peano's Axioms. In: A. Lachlan, M. Srebrny and A. Zarach (eds.), *Set Theory and Hierarchy Theory V (Bierutowice, Poland, 1976)*, LNM 619, pp. 211–226. Berlin-Heidelberg-New York: Springer-Verlag.

Kirby, L. and J. Paris (1982). Accessible Independence Results for Peano Arithmetic. *Bulletin of London Mathematical Society* 14, 285–293.

Kitcher, Ph. (1976). Hilbert's Epistemology. *Philosophy of Science* 143, 99–115.

Kitcher, Ph. (1983). *The Nature of Mathematical Knowledge*. New York-Oxford: Oxford University Press.

Kleene, S. C. (1943). Recursive Predicates and Quantifiers. *Transactions of the American Mathematical Society* 53, No 1, 41–73. Reprinted in: Davis (1965), pp. 255–287.

Kleene, S. C. (1955). Arithmetical Predicates and Function Quantifiers. *Transactions of the American Mathematical Society* 79, 312–340.

Kleene, S. C. (1959). Recursive Functionals and Quantifiers of Finite Types. I. *Transactions of the American Mathematical Society* 91, 1–52.

Kleinert, E. (2002). Zur Ontologie der mathematischen Gegenstände. pp. 53–69. In: E. Kleinert, *Beiträge zur Philosophie der Mathematik*. Leipzig: Leipziger Universitätsverlag.

Knaster, B. (1960). Zygmunt Janiszewski (w 40-lecie śmierci) [Zygmunt Janiszewski (at the 40th anniversary of his death)], *Roczniki Polskiego Towarzystwa Matematycznego. Seria II: Wiadomości Matematyczne* 4, 1–9.

Knuth, D. E. and P. B. Bendix (1970). Simple Word Problems in Universal Algebra. In: Leech (ed.) *Combinatorial Problems in Abstract Algebras*, pp. 263–270. Pergamon-New York.

Kotarbińska, J. (1984). Głos w dyskusji [A Contribution to Discussion], *Studia Filozoficzne* 5(222), 69–93.

Kotarbiński, T. (1959). *La logique en Pologne, son originalité et les influences etrangères*. Roma: Accademia Polacca di Scienze e Lettere.

Kotarbiński, T. (1967). Notes on the Development of Formal Logic in Poland in the Years 1900–39. In: S. McCall (ed.), *Polish Logic 1920–1939*. Oxford: Clarendon Press.

Kowalski, R. (1975). A Proof Procedure Using Connection Graph. *Journal of the Association for Computing Machines* 22.

Kowalski, R. and D. Kuehner (1971). Linear Resolution with Selection Function. *Artificial Intelligence* 2, 227–260.

Köhler, E. (1991). Gödel und der Wiener Kreis. In: P. Kruntorad (ed.), *Jour Fixe der Vernunft*, pp. 127–158. Wien: Hölder-Pichler-Tempsky. Also in: E. Köhler *et al.* (eds.) (2002) Vol 1, pp. 83–108.

Köhler, E., P. Weibel, M. Stöltzner, B. Buldt, C. Klein and W. Depauli-Schimanovich-Göttig (2002). *Kurt Gödel. Wahrheit und Beweisbarkeit.* Vol 1–2. Wien: öbt & hpt.

Kotlarski, H. (1986). Bounded Induction and Satisfaction Classes. *Zeitschrift für Mathematische Logik und Grundlagen der Mathematik* 32, 531–544.

Kotlarski, H. and Z. Ratajczyk (1990a). Inductive Full Satisfaction Classes. *Annals of Pure and Applied Logic* 47, 199–223.

Kotlarski, H. and Z. Ratajczyk (1990b). More on Induction in the Language with a Full Satisfaction Class. *Zeitschrift für Mathematische Logik und Grundlagen der Mathematik* 36, 441–454.

Krajewski, S. (1976). Non-Standard Satisfaction Classes. In: W. Marek, M. Srebrny and A. Zarach (eds.), S*et Theory and Hierarchy Theory*, Proc. Bierutowice Conf. 1975, Lecture Notes in Mathematics 537, pp. 121–144. Berlin-Heidelberg-New York: Springer Verlag.

Krajewski, S. (1983). Philosophical Consequences of Gödel's Theorems. *Bulletin of the Section of Logic* 12, 157–164.

Krajewski, S. (1993). Did Gödel Prove That we Are Not Machines?. In: Z. W. Wolkowski (ed.), *First International Symposium on Gödel's Theorems*, pp. 39–49. Singapore-New Jersey-London-Hong Kong: World Scientific.

Krajewski, S. (2003). *Twierdzenie Gödla i jego interpretacje filozoficzne. Od mechanicyzmu do postmodernizmu* [Gödel's Theorems and Their Philosophical Interpretations. From Mechanicism to Postmodernism]. Warszawa: Wydawnictwo Instytutu Filozofii i Socjologii PAN.

Kreisel, G. (1953). Note on Arithmetic Models for Consistent Formulae of the Predicate Calculus. II. In: *Proceedings of the XIth International Congress of Philosophy*, vol. XIV, pp. 39–49. Amsterdam-Louvain.

Kreisel, G. (1958). Hilbert's Programme. *Dialectica* 112, 346–372. Revised with Postscript in: P. Benacerraf and H. Putnam (eds.) (1964), pp. 157–180.

Kreisel, G. (1959). Interpretation of Analysis by Means of Constructive Functionals of Finite Types. In: A. Heyting (ed.), *Constructivity in Mathematics. Proceedings of the Colloquium held at Amsterdam, 1957*, pp. 101–128. Amsterdam: North-Holland Publishing Company.

Kreisel, G. (1968). A Survey of Proof Theory. *Journal of Symbolic Logic* 133, 321–388.

Kreisel, G. (1976). What Have We Learned from Hilbert's Second Problem? In: F. Browder (ed.), *Mathematical Developments Arising from Hilbert Problems*. Proceedings of the Symposia in Pure Mathematics 28, pp. 93–130. Providence, Rhode Island: American Mathematical Society.

Kreisel, G. (1980). Kurt Gödel 1906–1978. *Biographical Memoires of Fellows of the Royal Society* 26, 149–224.

Kreisel, G. and G. Takeuti (1974). Formally Self-Referential Propositions for Cut-Free Analysis and Related Systems. *Dissertationes Mathematicae* 118, 4–50.

Kronecker, L. (1887). Über den Zahlbegriff. *Journal für die reine und angewandte Mathematik* 101, 337–355.

Kuczyński, J. (1938). O twierdzeniu Gödla [On Gödel's Theorem]. *Kwartalnik Filozoficzny* 14, 74–80.

Kuehner, D. G. (1971). A Note on the Relation between Resolution and Maslov's Inverse Method. In: B. Meltzer and D. Michie (eds.) *Machine Intelligence* 6, pp. 73–76. New York.

Kuratowski, K. and Mostowski, A. (1952). *Teoria mnogości* [Set Theory]. Warszawa-Wrocław: Polskie Towarzystwo Matematyczne. Second edition: Państwowe Wydawnictwo Naukowe, Warszawa 1966; English translation: *Set Theory*. Warszawa: Polish Scientific Publishers and Amsterdam: North-Holand Publishing Company, 1967.

Ladrière, J. (1957). *Les limitations internes des formalismes. Étude sur la signification du théorème de Gödel et des théorèmes apparenté dans la théorie des fondements des mathématiques.* Louvain: E. Nauwelaerts and Paris: Gauthier-Villars.

Lakatos, I. (1963a). Proofs and Refutations (I). *British Journal for the Philosophy of Science* 14, 1–25.

Lakatos, I. (1963b). Proofs and Refutations (II). *British Journal for the Philosophy of Science* 14, 120–139.

Lakatos, I. (1963c). Proofs and Refutations (III). *British Journal for the Philosophy of Science* 14, 221–245.

Lakatos, I. (1964). Proofs and Refutations (IV). *British Journal for the Philosophy of Science* 14, 296–342.

Lakatos, I. (1967). A Renaissance of the Empiricism in the Recent Philosophy of Mathematics? In: I. Lakatos (ed.), *Problems in the Philosophy of Mathematics*, pp. 199–202. Amsterdam: North-Holland Publishing Company. Extended version in: I. Lakatos, *Philosophical Papers*. Volume 2: *Mathematics, Science and Epistemology*, J. Worall and G. Currie (eds.), pp. 24–42 (1978). Cambridge-London-New York-Melbourne: Cambridge University Press.

Lakatos, I. (1976). *Proofs and Refutations. The Logic of Mathematical Discovery.* Cambridge: Cambridge University Press.

Langford, C. H. (1927). Some Theorems on Deducibility. *Annals of Mathematics* (2nd series) 28, 16–40; Theorems on Deducibility. (Second Paper), *ibidem*, 459–471.

Läuchli, H., and J. Leonard (1966). On the Elementary Theory of Linear Order. *Fundamenta Mathematicae* 59, 109–116.

La Mettrie, J. O. de (1748). *L'homme-machine*. Critical edition, A. Vartanian (ed.), Paris 1960.

Landry, E. (1999). Category Theory: The Language of Mathematics. *Philosophy of Science* 66, 3: supplement, S14–S27.

Landry, E. (2001). Landry E., 2001, Logicism, Structuralism and Objectivity. *Topoi* 20, 79–95.

Landry, E. and J.-P. Marquis (2005). Categories in Context: Historical, Foundational, and Philosophical. *Philosophia Mathematica* (3) Vol. 13, 1–43.

Lebesgue, H. (1922). Á propos d'une nouvelle revue mathématique: *Fundamenta Mathematicae*. *Bulletin des Sciences Mathématiques* 46.

Leibniz, G. W. (1875–1890). *Philosophische Schriften*. Edited by C. I. Gerhardt. 7 volumes. Berlin: Weidmannsche Buchhandlung.

Liard, L. (1884). *Logique*. Paris: Masson.

Liard, L. (1886). *Logika* [Logic]. Trans. not indicated. Warszawa: Nakładem Redakcyi "Prawdy".

Livesey, M. and J. Siekmann (1976). Unification of A+C-terms (bags) and A+C+I-terms (Sets). *Universität Karlsruhe, Interner Bericht* Nr.5/76, Karlsruhe.

Löb, M. H. (1955). Solution of a Problem of Leon Henkin. *Journal of Symbolic Logic* 120, 115–118.

Löb, M. H. and S. S. Wainer (1970a). Hierarchies of Number-Theoretic Functions. I. *Archiv für Mathematische Logik und Grundlagenforschung* 13, 39–51.

Löb, M. H. and S. S. Wainer (1970b). Hierarchies of Number-Theoretic Functions. II. *Archiv für Mathematische Logik und Grundlagenforschung* 13, 97–113.

Lorenzen, P. (1951). Algebraische und logistische Untersuchungen über freie Verbände. *Journal of Symbolic Logic* 16, 81–106.

Loveland, D. W. (1968). Mechanical Theorem Proving by Model Elimination. *Journal of the Association for Computing Machines* 15, 236–251.

Loveland, D. W. (1969). A Simplified Format for the Model Elimination Procedure. *Journal of the Association for Computing Machines* 16, 349–363.

Loveland, D. W. (1970). A Linear Format for Resolution. In: *Proc. IRIA Symp. on Automatic Demonstration*, Versailles, France 1968, LNM 125, pp. 147–162. Berlin-New York: Springer Verlag.

Loveland, D. W. (1972). A Unifying View of some Linear Herbrand Procedures. *Journal of the Association for Computing Machines* 19, 366-384.

Loveland, D. W. (1978). *Automated Theorem Proving: A Logical Basis*. Amsterdam-New York-Oxford: North-Holland Publishing Company.

Loveland, D. W. (1984). Automated Theorem-Proving: A Quarter-Century Review. In: W. W. Bledsoe and D. W. Loveland (eds.) *Automated Theorem Proving. After 25 Years*, *Contemporary Mathematics*, 29, pp. 1–46. , Providence, Rhode Island: American Mathematical Society.

Löwenheim, L. (1915). Über Möglichkeiten im Relativkalkül. *Mathematische Annalen* 76, 447–470.

Lucas, J. R. (1961). Minds, Machines and Gödel. *Philosophy* 36, 112–127. Reprinted in: A. R. Anderson (ed.), *Minds and Machines*. 1964. Englewood Cliffs: Prentice Hall.

Lucas, J. R. (1967). Human and Machine Logic. A Rejoinder. *British Journal for the Philosophy of Science* 19, 155–156.

Lucas, J. R. (1968). Satan Stultified: A Rejoinder to Paul Benacerraf. *The Monist* 52, 145–158.

Luckham, D. (1970). Refinements in Resolution Theory. In: *Proc. IRIA Symp. on Automatic Demonstration*, Versailles, France 1968, LNM 125, pp. 163–190. Berlin-New York: Springer Verlag.

Łukasiewicz, J. (1907). Logika a psychologia [Logic and Psychology]. *Przegląd Filozoficzny* 10, 489–491.

Łukasiewicz, J. (1910). *O zasadzie sprzeczności u Arystotelesa: studyum krytyczne* [On the Principle of Contradition by Aristotle. A Critical Study]. Kraków: Akademia Umiejętności.

Łukasiewicz, J. (1912). O twórczości w nauce [Creative Elements in Science]. In: *Księga pamiątkowa ku uczczeniu 250 rocznicy założenia Uniwersytetu Lwowskiego*, pp. 1–15. Lwów. English translation: Creative elements in science. In: Łukasiewicz (1970), pp. 1–15.

Łukasiewicz, J. (1916). O pojęciu wielkości (Z powodu dzieła Stanisława Zaremby) [On the Concept of Magnitude. (In Connection with Stanisław Zaremba's Work)]. *Przegląd Filozoficzny* 19, 1–70. English translation: On the Concept of Magnitude (In connection with Stanisław Zaremba's *Theoretical Arithmetic*). In: Łukasiewicz (1970), pp. 16–83.

Łukasiewicz, J. (1936). Logistyka a filozofia [Logistic and Philosophy]. *Przegląd Filozoficzny* 39, 115–131. English translation: Logistic and philosophy. In: Łukasiewicz (1970), pp. 218–235.

Łukasiewicz, J. (1937). W obronie logistyki [In Defence of Logistic]. In: *La pensée catholoque et la logique moderne*, pp. 7–13. Kraków: Wydawnictwo Wydziału Teologicznego UJ. English translation: In Defence of Logistic. In: Łukasiewicz (1970), pp. 236–249.

Łukasiewicz, J. (1952). On the Intuitionistic Theory of Deduction. *Indagationes Mathematicae* 14, 202–212. Reprinted in: Łukasiewicz (1970), pp. 325–340.

Łukasiewicz, J. (1970). *Selected Works*. Edited by L. Borkowski. Amsterdam-London: North-Holland Publishing Company and Warszawa: Państwowe Wydawnictwo Naukowe (Polish Scientific Publishers).

Łuszczewska-Romahnowa, S. (1948). Wieloznaczność a język nauki [Polysemy and the language of science]. *Kwartalnik Filozoficzny* 1, 47–58.

Łuszczewska-Romahnowa, S. (1950). Rozważania o „metodzie" w filozofii francuskiej XVII wieku [Considerations on the "Method" in French Philosophy of the 17th Century]. *Kwartalnik Filozoficzny* XIX, no 3/4, 171–205. First published in 2002 by Polska Akademia Umiejętności and Uniwersytet Jagielloński, Kraków.

Łuszczewska-Romahnowa, S. (1953). Analiza i uogólnienie metody sprawdzania formuł logicznych przy pomocy diagramów Venna [Analysis and Generalization of the Method of Verifying Logical Formulae by Means of Venn's Diagrams]. *Studia Logica* I, 185–213.

Łuszczewska-Romahnowa, S. (1957). Indukcja a prawdopodobieństwo [Induction and Probability]. *Studia Logica* V, 71–90.

Łuszczewska-Romahnowa, S. (1961). Classification as a Kind of Distance Function. Natural Classifications. *Studia Logica* XII, 41–81.

Łuszczewska-Romahnowa, S. (1962). Pewne pojęcie poprawnej inferencji i pragmatyczne pojęcie wynikania [A Concept of a Correct Inference and Pragmatic Concept of Consequence]. *Studia Logica* XIII, 203–208.

Łuszczewska-Romahnowa, S. (1964). Z teorii racjonalnej dyskusji [On the Theory of Rational Discussion]. In: *Rozprawy logiczne: Księga pamiątkowa ku czci profesora Kazimierza Ajdukiewicza*, pp. 103–112. Warszawa: Państwowe Wydawnictwo Naukowe.

Łuszczewska-Romahnowa, S. and T. Batóg (1965a). A Generalized Theory of Classifications. I. *Studia Logica* XVI, 53–74.

Łuszczewska-Romahnowa, S. and T. Batóg (1965b). A Generalized Theory of Classifications. II. *Studia Logica* XVII, 7–30.

Mackie, J. L. (1973). *Truth, Probability and Paradox*. Oxford: Oxford University Press.

Mac Lane, S. (1996). Structure in Mathematics. *Philosophia Mathematica* (3) Vol. 4, 174–183.

MacLarty, C. (2004). Exploring Categorical Structuralism. *Philosophia Mathematica* (3) Vol. 12, 37–53.

Maddy, P. (1980). Perception and Mathematical Intuition. *Philosophical Review* 89, 163–196.

Maddy, P. (1990a). *Realism in Mathematics*. Oxford: Clarendon Press.

Maddy, P. (1990b). Physicalistic Platonism. In: A. D. Irvine (ed.), *Physicalism in Mathematics*, pp. 259–289. Dordrecht: Kluwer Academic Publishers.

Mancosu, P. (1998). Hilbert and Bernays on Metamathematics. In: P. Mancosu, *From Brouwer To Hilbert. The Debate on the Foundations of Mathematics in the 1920s*, pp. 149–188. New York-Oxford: Oxford University Press.

Mancosu, P. (1999). Between Vienna and Berlin: The Immediate Reception of Gödel's Incompleteness Theorems. *History and Philosophy of Logic* (Series III) 20, 33–45.

Mancosu, P. (2005). Harvard 1940–1941: Tarski, Carnap and Quine on a Finitistic Language of Mathematics and Science, *History and Philosophy of Logic* 26, 327–357.

Marciszewski W. and R. Murawski (1995). *Mechanization of Reasoning in a Historical Perspective*, Amsterdam/Atlanta, GA: Editions Rodopi.

Marczewski, E. (1948). *Rozwój matematyki w Polsce* [Development of Mathematics in Poland]. Nakładem Polskiej Akademii Umiejętności z zasiłku Prezydium Rady Ministrów. Skład Główny w Księgarni Gebethnera i Wolffa, Warszawa-Kraków-Łódź-Poznań-Zakopane.

Marquis, J.-P. (1997). Category Theory and Structuralism in Mathematics: Syntactical Considerations. In: E. Agazzi and G. Darvas (eds.), *Philosophy of Mathematics Today*, pp. 123–136. Oxford: Clarendon Press.

Maslov, S.Ju. (1964). An Inverse Method of Establishing Deducibility in Classical Predicate Calculus. *Dokl. Akad. Nauk SSR*, pp. 17–20.

Maslov, S.Ju. (1971). Proof-Search Strategies for Methods of the Resolution Type. B. Meltzer and D. Michie (eds.) *Machine Intelligence*, 6, pp. 77–90. New York.

Mazur, S. (1963). Computable Analysis. Ed. by A. Grzegorczyk and H. Rasiowa. *Rozprawy Matematyczne* XXXIII, 1–111.

Mazurkiewicz, S. (1923). Teoria mnogości w stosunku do innych działów matematyki [Set Theory in Relation to Other Domains of Mathematics]. In: *Poradnik dla samouków*, Tom III: *Matematyka. Uzupełnienia do tomu pierwszego*. Wydawnictwo A. Heflera i St. Michalskiego. Warszawa, 89–98.

Meltzer, B. (1966). Theorem-Proving for Computers: Some Results on Resolution and Renaming. *Computer Journal* 8, 341–343.

Mendelson, E. (1964). *Introduction to Mathematical Logic*. Princeton-Toronto-New York-London: D. Van Nostrand Company, Inc.

Menger, K. (1978). *Selected Papers in Logic and Foundations, Didactics, Economics*. Dordrecht-Boston-London: D. Reidel Publishing Company.

Meyer, A. R. (1975). The Inherent Complexity of Theories of Ordered Sets. In: *Proceedings of the International Congress of Mathematicians, Vancouver 1974*, vol. 2, pp. 477–482.

Miller, D. A., E. L. Cohen and P. B. Andrews (1982). A Look at TPS. In: D. W. Loveland (ed.) *Proc. Sixth Conf. on Automated Deduction*. Lecture Notes in Computer Science 138, pp. 60–69. Berlin-Heidelberg-New York: Springer-Verlag.

Moore, G. H. (1980). Beyond First-Order Logic: The Historical Interplay Between Mathematical Logic and Axiomatic Set Theory. *History and Philosophy of Logic* 1, 95–137.

Moore, G. H. (1991). Sixty Years After Gödel. *The Mathematical Intelligencer* 13, 6–11.

Morris, J. B. (1969). E-resolution: Extension of Resolution to Include the Equality. In: *Proc. Intern. Joint Conf. Artificial Intelligence*. Washington D.C., pp. 287–294.

Mostowski, A. (1946). O zdaniach nierozstrzygalnych w sformalizowanych systemach matematyki [On Sentences Undecidable in Formalized Systems of Mathematics], *Kwartalnik Filozoficzny* 16, 223–277.

Mostowski, A. (1947). On Definable Sets of Positive Integers. *Fundamenta Mathematicae* 34, 81–112.

Mostowski, A. (1948a). *Logika matematyczna. Kurs uniwersytecki* [Mathematical Logic. University Course]. Monografie Matematyczne. Warszawa-Wrocław.

Mostowski, A. (1948b). On a Set of Integers Not Definable by Means of One-Quantifier Predicate. *Annales de la Société Polonaise de Mathématique* 21, 114–119.

Mostowski, A. (1949). Sur l'interpretation géométrique et topologique des notions logiques. In: *Actes du X-ème Congres International de Philosophie (Amsterdam 11–18 août 1948)*, pp. 610–617. Amsterdam.

Mostowski, A. (1949–1950). A Classification of Logical Systems. *Studia Philosophica* 4, 237–274 (published in 1951); reprinted in: Mostowski 1979, vol. II, pp. 154–191.

Mostowski, A. (1952). *Sentences Undecidable in Formalized Arithmetic. An Exposition of the Theory of Kurt Gödel.* Amsterdam: North-Holland Publishing Company.

Mostowski, A. (1953a). O tzw. konstruktywnych pogladach w dziedzinie podstaw matematyki [On So Called Constructive Views in the Foundations of Mathematics]. *Myśl Filozoficzna* 1 (7), 230–241.

Mostowski, A. (1953b). On a System of Axioms Which Has No Recursively Enumerable Arithmetical Model. *Fundamenta Mathematicae* 40, 56–61.

Mostowski, A. (1954). Podstawy matematyki na VIII Zjeździe Matematyków Polskich [Foundations of Mathematics at the 8th Congress of Polish Mathematicians], *Myśl Filozoficzna* 2 (12), 328–330.

Mostowski, A. (1955a). Współczesny stan badań nad podstawami matematyki [The Present State of Investigations on the Foundations of Mathematics]. *Prace Matematyczne* 1, 13–55.

Mostowski, A. (in collaboration with A. Grzegorczyk, S. Jaśkowski, J. Łoś, S. Mazur, H. Rasiowa and R. Sikorski) (1955b). The Present State of Investigations on the Foundations of Mathematics. *Rozprawy Matematyczne* 9, 1–48.

Mostowski, A. (1955c). Examples of Sets Definable by Means of Two and Three Quantifiers. *Fundamenta Mathematicae* 42, 259–270.

Mostowski, A. (1955d). A Formula With No Recursively Enumerable Model. *Fundamenta Mathematicae* 42, 125–140.

Mostowski, A. (1957a). *Sentences Undecidable in Formalized Arithmetic. An Exposition of the Theory of Kurt Gödel.* Amsterdam: North-Holland Publishing Company.

Mostowski, A. (1957b). On Recursive Models of Formalized Arithmetic. *Bulletin de l'Académie Polonaise de Sciences (Cl. III)*, 5, 705–710.

Mostowski, A. (1959). On Various Degrees of Constructivism. In: A. Heyting (ed.), *Constructivity in Mathematics. Proceedings of the Colloquium held in Amsterdam, 1957*, vol. II, pp. 178–194. Amsterdam: North-Holland Publishing Company. Reprinted in: Mostowski 1979, pp. 359–375.

Mostowski, A. (1964). Widerspruchsfreiheit und Unabhängigkeit der Kontinuumhypothese. *Elemente der Mathematik* 19, 121–125.

Mostowski, A. (1965). Thirty Years of Foundational Studies. Lectures on the Development of Mathematical Logic and the Study of the Foundations of Mathematics in 1930–1964. *Societas Philosophical Fennica*, Helsinki. Reprinted in: Mostowski 1979, vol. I, pp. 1–76.

Mostowski, A. (1967a). Recent Results in Set Theory. In: I. Lakatos (ed.), *Problems in the Philosophy of Mathematics*, pp. 82–96 and 105–108. Amsterdam: North-Holland Publishing Company.

Mostowski, A. (1967b). O niektórych nowych wynikach meta-matematycznych dotyczacych teorii mnogości [On Some New Metamathematical Results Concerning Set Theory]. *Studia Logica* 20, 99–116.

Mostowski, A. (1967c). Tarski, Alfred. In: P. Edwards (ed.), *The Encyclopedia of Philosophy*, vol. 8, pp. 77–81. New York: Macmillan.

Mostowski, A. (1968). Niesprzeczność i niezależność hipotezy kontinuum [Consistency and Independence of the Continuum Hypothesis]. *Roczniki Polskiego Towarzystwa Matematycznego, Seria II: Wiadomości Matematyczne* 10, 175–182.

Mostowski, A. (1969). *Constructible Sets with Applications*. Warszawa: PWN-Polish Scientific Publishers and Amsterdam: North-Holland Publishing Company.

Mostowski, A. (1972a). Matematyka a logika. Refleksje przy lekturze książki A. Grzegorczyka „Zarys arytmetyki teoretycznej" wraz z próba recenzji [Mathematics vs. Logic. Impressions of the Study of "Outline of Theoretical Arithmetic" by A. Grzegorczyk with an Attempt of a Review]. *Roczniki Polskiego Towarzystwa Matematycznego, Seria II: Wiadomości Matematyczne* 15, 79–89.

Mostowski, A. (1972b). Sets. In: *Scientific Thought. Some Underlying Concepts, Methods and Procedures*, pp. 1–34. The Hague: Mouton/Unesco.

Mostowski, A. (1975). Travaux de W. Sierpiński sur la théorie des ensembles et ses applications. In: W. Sierpiński, *Œuvres choisies*, Tome II, PWN-Éditions Scientifiques de Pologne, Warszawa 1975, 9–13.

Mostowski, A. (1979). *Foundational Studies. Selected Works*, vols. I–II. Amsterdam: North-Holland Publishing Company.

Murawski, R. (1976a). On Expandability of Models of Peano Arithmetic. I. *Studia Logica* 35, 409–419.

Murawski, R. (1976b). On Expandability of Models of Peano Arithmetic. II. *Studia Logica* 35, 421–431.

Murawski, R. (1977). On Expandability of Models of Peano Arithmetic. III. *Studia Logica* 36, 181–188.

Murawski, R. (1984a). Expandability of Models of Arithmetic. In: G. Wechsung (ed.), *Proceedings of Frege Conference 1984*, pp. 87–93. Berlin: Akademie-Verlag.

Murawski, R. (1984b). G. Cantora filozofia teorii mnogości. *Studia Filozoficzne* 11–12, 75–88. English translation: G. Cantor's Philosophy of Set Theory. In: Murawski (2010), pp. 15–28.

Murawski, R. (1984c). Matematyczna niezupełność arytmetyki [Mathematical Incompleteness of Arithmetic]. *Roczniki Polskiego Towarzystwa Matematycznego, Seria II: Wiadomości Matematyczne* 26, 47–58.

Murawski, R. (1985). Giuseppe Peano – pioneer and promoter of symbolic logic. Komunikaty i Rozprawy Instytutu Matematyki UAM, Poznań. Reprinted in: Murawski (2010), pp. 169–182.

Murawski, R., ed. (1986). *Filozofia matematyki. Antologia tekstów klasycznych* [Philosophy of Mathematics. Anthology of Classical Texts]. Poznań: Wydawnictwo Naukowe UAM. Second revised edition: 1994. Third revised edition: 2003.

Murawski, R. (1987). Generalizations and Strengthenings of Gödel's Incompleteness Theorem. In: J. Srzednicki (ed.), *Initiatives in Logic*, pp. 84–100. Dordrecht-Boston-Lancaster: Martinus Nijhof Publishers.

Murawski, R. (1993). Rozwój programu Hilberta [The Development of Hilbert's Programme]. *Roczniki Polskiego Towarzystwa Matematycznego, Seria II: Wiadomości Matematyczne* 30, 51–72.

Murawski, R. (1994). Hilbert's Program: Incompleteness Theorems vs. Partial Realizations. In: J. Woleński (ed.), *Philosophical Logic in Poland*, pp. 103–127. Dordrecht-Boston-London: Kluwer Academic Publishers. Also in this volume, pp. 83–100.

Murawski, R. (1996). Contribution of Polish Logicians to Decidability Theory. *Modern Logic* 6, 37–66. Reprinted in: Murawski (2010), pp. 211–231.

Murawski, R. (1997a). Gödel's Incompleteness Theorems and Computer Science. *Foundations of Science* 2 (1997), 123–135. Also in this volume, pp. 127–135.

Murawski, R. (1997b). Satisfaction Classes – a Survey. In: R. Murawski and J. Pogonowski (eds.), *Euphony and Logos*, pp. 259–281. Amsterdam-Atlanta, GA: Editions Rodopi.

Murawski, R. (1998). Undefinability of Truth. The Problem of the Priority: Tarski vs Gödel. *History and Philosophy of Logic* 19, 153–160. Also in this volume, pp. 177–185.

Murawski, R. (1999a). Undefinability vs. Definability of Satisfaction and Truth. In: J. Woleński and E. Köhler (eds.), *Alfred Tarski and the Vienna Circle*, pp. 203–215. Dordrecht-Boston-London: Kluwer Academic Publishers.

Murawski, R. (1999b). *Recursive Functions and Metamathematics. Problems of Completeness and decidability, Gödel's Theorems*. Dordrecht-Boston-London: Kluwer Academic Publishers.

Murawski, R. (2002a). On the Distinction Proof-Truth in Mathematics. In: P. Gärdenfors, J. Woleński and K. Kijania-Placek (eds.), *In the Scope of Logic, Methodology and Philosophy of Science*, vol. I, pp. 287–303. Dordrecht: Kluwer Academic Publishers. Also in this volume, pp. 101–113

Murawski, R. (2002b). Leibniz's and Kant's Philosophical Ideas and the Development of Hilbert's Programme. *Logique et Analyse* 179–180, 421–437. Reprinted in Murawski (2010), pp. 29–39.

Murawski, R. (2003). Reactions to the Discovery of the Incompleteness Phenomenon. In: J. Hintikka *et al.* (eds.), *Philosophy and Logic. In Search of the Polish Tradition*. Dordrecht: Kluwer Academic Publishers, pp. 213–227. Also in this volume, pp. 115–126.

Murawski, R. (2004a). Church's Thesis and Its Epistemological Status. *Annales UMCS Informatica AI* 2, 57–70. Reprinted in: Murawski (2010), pp. 123–134.

Murawski, R. (2004b). Philosophical Reflection on Mathematics in Poland in the Interwar Period. *Annals of Pure and Applied Logic* 127, 325–337. Also in this volume, pp. 215–226.

Murawski, R. (2008). *Filozofia matematyki. Zarys dziejów* [Philosophy of Mathematics. An Outline of History]. Third edition. Poznań: Wydawnictwo Naukowe Uniwersytetu im. Adama Mickiewicza.

Murawski, R. (2010). *Essays in the Philosophy and History of Logic and Mathematics*. Amsterdam-New York, NY: Editions Rodopi.

Nagel, E. and J. R. Newman (1958). *Gödel's Proof*. London: Routledge & Kegan Paul, Ltd.

Nerode A. and L. A. Harrington (1984). The Work of Harvey Friedman. *Notices of the American Mathematical Society* 31, 563–566.

Neumann, J. von (1927), Zur Hilbertschen Beweistheorie. *Mathematische Zeitschrift* 26, 1–46. Reprinted in: Neumann (1961), pp. 256–302.

Neumann, J. von (1961). *Collected Works*. Vol. I: *Logic, Theory of Sets and Quantum Mechanics*. Oxford-London-New York-Paris: Pergamon Press.

Neumann, J. von (1966). *Theory of Self-Reproducing Automata*. A. W. Burks (ed.). Urbana: University of Illinois.

Nevins, A. J. (1974). A Human Oriented Logic for Automatic Theorem Proving. *Journal of the Association for Computing Machines* 21, 606–621.

Nevins, A. J. (1975a). Plane Geometry Theorem Proving Using Forward Chaining. *Artificial Intelligence* 6, 1–23.

Nevins, A.J. (1975b). A Relaxation Approach to Splitting in an Automatic Theorem Prover. *Artificial Intelligence* 6, 25–39.

Newell, A., J. C. Shaw and H. A. Simon (1956). Empirical Explorations of the Logic Theory Machine: A Case Study in Heuristics. In: *Proc. Western Joint Computer Conf.* 15, 218–239. Also in: Feigenbaum and Feldman (eds.), *Computers and Thought*. McGrow-Hill 1963.

Nicolaus Cusanus (1514). *Liber de mente*. In: *Nicolai Cusae Cardinalis Opera*. Parisiis.

Nicolaus Cusanus (1514). *De docta ignorantia*. In: *Nicolai Cusae Cardinalis Opera*. Parisiis.

Niebergall, K.-G. (1996). *Zur Metamathematik nichtaxiomatisierbarer Theorien*. Centrum für Informations- und Sprachverarbeitung Ludwig-Maximilians-Universität München. Bericht 96–87.

Norton, L. M. (1966). *ADEPT – a Heuristic Program for Proving Theorems of Group Theory*, Ph. D. Thesis, MIT.

Nowakowa, I. (2001). Adam Wiegner – Nonstandard Empiricism. In: W. Krajewski (ed.), *Polish Philosophers of Science and Nature in the 20th Century*, Poznań Studies in the Philosophy of the Sciences and the Humanities, vol. 74, pp. 79–87. Amsterdam-New York: Editions Rodopi.

Omyła, M. (1986). O życiu i twórczości Romana Suszki [On the Life and Work of Roman Suszko]. *Studia Semiotyczne* XIV–XV, 13–22.

Omyła, M., ed. (2001). *Logiczne idee Romana Suszki* [Logical Ideas of Roman Suszko]. Materiały XLV Konferencji Historii Logiki (Kraków 1999). Warszawa: Wydział Filozofii i Socjologii Uniwersytetu Warszawskiego.

Omyła M. and J. Zygmunt (1984). Roman Suszko (1919–1979): A Bibliography of the Published Work with an Outline of His Logical Investigations. *Studia Logica* XLIII, 421–441.

Parikh R. (1971). Existence and Feasibility in Arithmetic. *The Journal of Symbolic Logic* 36, 494–508.

Paris, J. (1978). Some Independence Results for Peano Arithmetic. *Journal of Symbolic Logic* 143, 725–731.

Paris, J. and L. Harrington (1977). A Mathematical Incompleteness in Peano Arithmetic. In: Barwise, J. (ed.), *Handbook of Mathematical Logic*, pp. 1133–1142. Amsterdam: North-Holland Publishing Company.

Pasch, M. (1882). *Vorlesungen über neuere Geometrie*. Leipzig und Berlin: B.G. Teubner.

Peckhaus, V. (1990). *Hilbertprogramm und Kritische Philosophie. Das Göttinger Model interdisziplinärer Zusammenarbeit zwischen Mathematik und Philosophie*. Göttingen: Vandenhoeck & Ruprecht.

Penrose, R. (1989). *The Emperor's New Mind. Concerning Computers, Minds, and the Laws of Physics*. Oxford: Oxford University Press.

Penrose, R. (1994). *Shadows of the Mind: a Search for the Missing Science of Consciousness*. Oxford-New York-Melbourne: Oxford University Press.

Péter, R. (1951). *Rekursive Funktionen*. Budapest: Akadémiai Kiadó.

Piątkiewicz, S. (1888). Algebra w logice [Algebra in Logic]. In: *Sprawozdania dyrektora c.k. IV gimnazyum we Lwowie za rok szkolny 1888*, pp. 1–52. Lwów: Nakładem Funduszu Naukowego.

Plotkin, G. D. (1972). Building-in Equational Theories. In: B. Meltzer and D. Michie (eds.) *Machine Intelligence* 7, pp. 73–90. New York.

Poincaré, H. (1902). *La science et l'hypotèse*. Paris: Flammarion.

Poincaré, H. (1908). *Science et méthode*, Flammarion, Paris.

Peano, G. (1889). *Arithmetices principia nova methodo exposita*. Torino: Bocca.

Perelman, Ch. (1936). L'Antinomie de M. Gödel. *Academie Royale de Belgique, Bulletin de la Classe des Sciences*, Series 5, 22, 730–736.

Post, E. L. (1944). Recursively Enumerable Sets of Positive Integers and Their Decision Problem. *Bulletin of the American Mathematical Society*, 50, 284–316. Reprinted in Davis (1994), pp. 461–494.

Post, E. L. (1965). Absolutely Unsolvable Problems and Relatively Undecidable Propositions. Account of an Anticipation. In: Davis (1965), pp. 340–433.

Prawitz, D. (1981). Philosophical Aspects of Proof Theory. In: G. Fløistad (ed.), *Contemporary Philosophy. A New Survey*, pp. 235–277. The Hague-Boston-London: Martinus Nijhoff Publishers.

Prawitz, D. (1960). An Improved Proof Procedure. *Theoria* 26, 102–139.

Presburger, M. (1930). Über die Vollständigkeit eines gewisses Systems der Arithmetik ganzer Zahlen, in welchem die Addition als einzige Operation hervortritt. In: *Sprawozdanie z I Kongresu Matematyków Krajów Słowiańskich – Comptus Rendus, I Congres des Math. des Pays Slaves*, pp. 92–101, 395. Warszawa.

Proclus (1873). *Procli Diadochi in Primum Elementorum Librum Comentarii*. Edited by G. Friedlein. Leipzig: B.G. Teubner. Reprinted: G. Olms, Hildesheim 1967.

Putnam, H. (1965). Trial and Error Predicates and the Solution to a Problem of Mostowski. *Journal of Symbolic Logic* 30, 49–57.

Putnam, H. (1967). Mathematics without Foundations. *Journal of Philosophy* 64, 5–22. Revised version in: H. Putnam, *Mathematics, Matter and Method. Philosophical Papers*, vol. I. Cambridge-London-New York-Melbourne: Cambridge University Press, 1975, pp. 43–59.

Putnam, H. (1975). What Is Mathematical Truth? In: H. Putnam, *Mathematics, Matter and Method. Philosophical Papers*, vol. I, pp. 60–78. Cambridge-London-New York-Melbourne: Cambridge University Press. Reprinted in: Th. Tymoczko (ed.), *New Directions in the Philosophy of Mathematics. An Anthology*. Boston-Basel-Stuttgart: Birkhäuser, 1985, pp. 50–65.

Quine, W. V. O. (1951a). Two Dogmas of Empiricism. *Philosophical Review* 60 (1), 20–43. Also in: W. V. O. Quine, *From a Logical Point of View*. Cambridge, Mass.: Harvard University Press, 1964, pp. 20–46.

Quine, W. V. O. (1951b). On Carnap's Views on Ontology. *Philosophical Studies* 2, 65–72.

Quine, W. V. O. (1953). On What There Is. In: Quine W. V. O., *From a Logical Point of View*, pp. 1–19. Cambridge, Mass.: Harvard University Press.

Ramsey, F. P. (1929). On a Problem of Formal Logic. *Proc. London Math. Soc.* 1(2) 30, 264–286.

Reid, C. (1970). *Hilbert*. Berlin-Heidelberg-New York: Springer-Verlag.

Reiter, R. (1971). Two Results on Ordering for Resolution with Merging and Linear Format. *Journal of the Association for Computing Machines* 18, 630–646.

Resnik, M. D. (1974). On the Philosophical Significance of Consistency Proofs. *Journal of Philosophical Logic* 13, 133–147.

Resnik, M. D. (1981). Mathematics as a Science of Patterns: Ontology and Reference. *Noûs* 15, 529–550.

Resnik, M. D. (1982). Mathematics as a Science of Patterns: Epistemology. *Noûs* 16, 95–105.

Resnik, M. D., ed. (1995). *Mathematical Objects and Mathematical Knowledge*. Aldershot-Brookfield, USA-Singapore-Sydney: Dartmouth.

Robinson, R. M. (1951). Undecidable Rings. *Transactions of the American Mathematical Society* 70, 137–159.

Robinson, J. (1949). Definability and Decision Problems in Arithmetic. *Journal of Symbolic Logic* 14, 98–114. Reprinted in: *The Collected Works of Julia Robinson*, ed. by S. Feferman, pp. 7–23 (1996). Providence, RI: American Mathematical Society.

Robinson, J. A. (1965a). A Machine-Oriented Logic Based on the Resolution Principle. *Journal of the Association for Computing Machines* 12, 23–41.

Robinson, J. A. (1965b). Automated Deduction with Hyperresolution. *International Journal Computational Mathematics* 1, 227–234.

Rodríguez-Consuegra, F. A. (1993). Russell, Gödel and Logicism. In: Czermak, J. (ed.) *Philosophy of Mathematics. Proceedings of the 15th International Wittgenstein-Symposium*, Part 1, pp. 233–242. Wien: Verlag Hölder-Pichler-Tempsky.

Rogers, H., Jr. (1967). *Theory of Recursive Functions and Effective Computability*. New York-St.Luis-San Francisco-Toronto-London-Sydney: Mc-Graw Hill.

Rosser, J. B. (1936). Extensions of Some Theorems of Gödel and Church. *Journal of Symbolic Logic* 11, 8791.

Rosser, J. B. (1937). Gödel Theorems for Non-Constructive Logics. *Journal of Symbolic Logic* 2, 129–137.

Rosser, B. (1939). An Informal Exposition of Proofs of Gödel's Theorems and Church's Thesis. *Journal of Symbolic Logic* 4, 53–60.

Rowe, D. E. (1989). Klein, Hilbert, and the Göttingen Mathematical Tradition. *Osiris* (2) 5, 186–213.

Rucker, R. (1982). *Infinity and the Mind*, Boston-Basel-Stuttgart: Birkhäuser.

Russell, B. (1901). Mathematics and Metaphysicians. *The International Monthly* 4, 83–101. Reprinted in: *Mysticism and Logic and Other Essays*, pp. 74–96. London: George Allen & Unwin Ltd., 1949 (8th edition).

Russell, B. (1903). *The Principles of Mathematics*. Cambridge: The University Press.

Russell, B. (1919). *Introduction to Mathematical Philosophy*. London: George Allen & Unwin, Ltd., and New York: The Macmillan Co.

Russell, B. (1937). On Verification. *Proceedings of the Aristotelian Society* 38, 1–20.

Russell, B. (1959). *My Philosophical Development*. London: George Allen & Unwin Ltd.

Schilpp, P., ed. (1944). *The Philosophy of Bertrand Russell*. Evanston, Ill.: Northwestern University.

Schröder, E. (1877). *Der Operationskreis des Logikkalkuls*. Leipzig: Teubner.

Schröder, E. (1895). *Vorlesungen über die Algebra der Logik (Exakte Logik).* Vol. 3: *Algebra und Logik der Relative.* Leipzig.

Schütte, K. (1951). Beweistheoretische Erfassung der unendlichen Induktion in der Zahlentheorie. *Mathematische Annalen* 122, 369–389.

Schütte, K. (1960). *Beweistheorie.* Berlin: Springer-Verlag. English translation: *Proof Theory,* Berlin-Heidelberg-New York: Springer-Verlag, 1977.

Shanker, S. G. (1988). Wittgenstein's Remarks on the Significance of Gödel's Theorem. In: S.G. Shanker (ed.) *Gödel's Theorem in Focus,* pp. 155–256. London: Croom Helm.

Shapiro, S. (1989). Structure and Ontology. *Philosophical Topics* 17, 145–171.

Shapiro, S. (1991). *Foundations without Foundationalism.* Oxford: Clarendon Press.

Shapiro, S. (2005). Categories, Structures, and Frege–Hilbert Controversy: The Status of Metamathematics. *Philosophia Mathematica* (3) 13, 61–77.

Shoenfield, J. R., (1967). *Mathematical Logic.* Reading, Mass.: Addison-Wesley.

Shostak, R. E. (1976). Refutation Graphs. *Artificial Intelligence* 7.

Sibert, E. E. (1969). A Machine Oriented Logic Incorporating the Equality Relation. In: B. Meltzer and D. Michie (eds.)*Machine Intelligence* 4, pp. 103–134. New York.

Sickel, S. (1976). Interconnectivity Graphs. *IEEE Trans. on Computers* C-25.

Sieg, W. (1985). Fragments of Arithmetic. *Annals of Pure and Applied Logic* 128, 33–71.

Sieg, W. (1988). Hilbert's Program Sixty Years Later. *Journal of Symbolic Logic* 53, 338–348.

Sierpiński, W. (1909). Pojęcie odpowiedniości w matematyce [The Concept of Correspondence in Mathematics]. *Przegląd Filozoficzny* 12, 8–19.

Sierpiński, W. (1912). *Zarys teorii mnogości* [An Outline of Set Theory]. Skład Główny w Księgarni E. Wendego i S-ki, Warszawa.

Sierpiński, W. (1918). L'axiome de M. Zermelo et son role dans la théorie des ensembles et l'analyse, *Bulletin International de l'Académie des Sciences et des Lettres de Cracovie,* classe de Sciences mathématiques et naturelles, Série A: sciences mathématiques, 97–152. Rreprinted in: W. Sierpiński, *Oeuvres choisies,* vol. 2, pp. 208–255. Warszawa: Państwowe Wydawnictwo Naukowe, 1975.

Sierpiński, W. (1923). *Zarys teorii mnogości* [An Outline of Set Theory]. Warszawa: Wydawnictwo Kasy im. Mianowskiego.

Sierpiński, W. (1965). *Cardinal and Ordinal Numbers.* Warszawa: Polish Scientific Publishers.

Sigmund, K. (1995). A Philosopher's Mathematician: Hans Hahn and the Vienna Circle. *The Mathematical Intelligencer* 17, 16–29.

Simpson, S. G. (1985a). Friedman's Research on Subsystems of Second Order Arithmetic. In: L. Harrington *et al.* (eds.), *Harvey Friedman's Research in the Foundations of Mathematics,* pp. 137–159. Amsterdam: North-Holland Publishing Company.

Simpson, S. G. (1985b). Reverse Mathematics. *Proceedings of Symposia in Pure Mathematics* 142, 461–471.

Simpson, S. G. (1987). Subsystems of Z and Reverse Mathematics. Appendix to: G. Takeuti, *Proof Theory,* pp. 432–446, 1987. Amsterdam: North-Holland Publishing Company.

Simpson, S. G. (1988a). Partial Realizations of Hilbert's Program. *Journal of Symbolic Logic* 153, 349–363.

Simpson, S. G. (1988b). Ordinal Numbers and the Hilbert's Basis Theorem. *Journal of Symbolic Logic* 153, 961–974.

References

Simpson, S. G. (1998). *Subsystems of Second Order Arithmetic.* New York: Springer-Verlag, Inc.

Skolem, Th. (1919). Untersuchungen über die Axiome des Klassenkalküls und über Produktions- und Summationsprobleme, welche gewisse Klassen von Aussagen betreffen. *Videnskapsselskapets skrifter, I, Matematik-naturvidenskabeling klasse*, no. 3, 30–37.

Skolem, Th. (1930). Über einige Satzfunktionen in der Arithmetik. *Skriften utgit av videnskasselskapet i Kristina*, I Klasse, No 7, Oslo. Also in: T. Skolem, *Selected Works in Logic*, ed. by J.E. Fenstad, Oslo 1970, pp. 281–306.

Skolimowski, H. (1967). *Polish Analytical Philosophy: a Survey and Comparison with British Analytical Philosophy*, London: Routledge and Kegan Paul.

Slagle, J. R. (1967). Automated Theorem Proving with Renamable and Semantic Resolution. *Journal of the Association for Computing Machines* 14, 687–697.

Sleszyński, J. (1923). O znaczeniu logiki dla matematyki [On the Meaning of Logic for Mathematics]. In: *Poradnik dla samouków*, tom III: *Matematyka. Uzupełnienia do tomu pierwszego*, pp. 39–52. Warszawa: Wydawnictwo A. Heflicha i St. Michalskiego.

Smart, J. J. C. (1961). Gödel's Theorem, Church's Thesis, and Mechanism. *Synthese* 13, 105–110.

Smoryński, C. (1977). The Incompleteness Theorems. In: J. Barwise (ed.), *Handbook of Mathematical Logic*, pp. 821–865. Amsterdam: North-Holland Publishing Company.

Smoryński, C. (1981). Fifty Years of Self-Reference in Arithmetic. *Notre Dame Journal of Formal Logic* 122, 357–374.

Smoryński, C. (1985). *Self-Reference and Modal Logic.* New York-Berlin-Heidelberg-Tokyo: Springer-Verlag.

Smoryński, C. (1988). Hilbert's Programme. *CWI Quarterly* 1, 3–59.

Specker, E. (1949). Nichtkonstruktiv beweisbare Sätze der Analysis. *Journal of Symbolic Logic* 14, 145–158.

Stamm, E. (1909). O aprjoryczności matematyki [On the Apriority of Mathematics]. *Przegląd Filozoficzny* 12, 504–514.

Stamm, E. (1910). Czem jest i czem będzie matematyka? [What is and What Will be Mathematics]. *Wiadomości Matematyczne* 14, 181–196.

Stamm, E. (1911a). Logiczne podstawy nauk matematycznych [Logical Foundations of Mathematical Sciences]. *Przegląd Filozoficzny* 14, 251–274.

Stamm, E. (1911b, 1912). Zasady algebry logiki [Principles of the Algebra of Logic]. *Wiadomości Matematyczne* 15 (1911), 1–87 and 16 (1912), 1–31.

Statman R. (1978). Bounds for Proof-Search and Speed up in the Predicate Calculus, *Annals of Mathematical Logic* 15, 225–287.

Steinhaus, H. (1921). Zygmunt Janiszewski – wspomnienie pośmiertne [Zygmunt Janiszewski – An Obituary]. *Przegląd Filozoficzny* 22, 113–117.

Stickel, M. E. (1981). A Complete Unification Algorithm for Associative-Commutative Functions. In: *Proc. Fourth Intern. Joint Conf. on Artificial Intelligence*, Tbilisi, USSR. Also in: *Journal of the Ass. for Computing Machines* 28, 423–434.

Stickel, M. E. (1985). Automated Deduction by Theory Resolution. *Journal of Automated Reasoning* 1, 333–356.

Suszko, R. (1947a). O zdaniach tautologicznych [On Tautological Propositions]. *Sprawozdania Poznańskiego Towarzystwa Przyjaciół Nauk* 14, 159–160.

Suszko, R. (1947b). Z teorii definicji [On Theory of Definitions], *Sprawozdania Poznańskiego Towarzystwa Przyjaciół Nauk* 14, 160–161.

Suszko, R. (1948). W sprawie logiki bez aksjomatów [Concerning Logic Without Axioms]. *Kwartalnik Filozoficzny* XVII, 199–205.

Suszko, R. (1949a). O analitycznych aksjomatach i logicznych regułach wnioskowania [On Analytical Axioms and Logical Rules of Inference]. *Poznańskie Towarzystwo Przyjaciół Nauk. Prace Komisji Filozoficznej* VII, nr 5, 1–30.

Suszko, R. (1949b). Z teorii definicji [On Theory of Definitions]. *Poznańskie Towarzystwo Przyjaciół Nauk. Prace Komisji Filozoficznej* VII, nr 5, 31–59.

Suszko, R. (1949c). Logika matematyczna i teoria podstaw matematyki w ZSRR [Mathematical Logic and the Theory of the Foundations of Mathematics in the Soviet Union]. Myśl Współczesna, nr 12 (43), 390–396.

Suszko, R. (1950). Konstruowalne przedmioty i kanoniczne systemy aksjomatyczne [Constructible Objects and Canonic Axiomatic Systems]. *Kwartalnik Filozoficzny*, vol. XIX, no 3/4, 331–359. First published in 2002 by Polska Akademia Umiejętności and Uniwersytet Jagielloński, Kraków.

Suszko, R. (1951). Canonic Axiomatic Systems. *Studia Philosophica* IV, 301–330.

Suszko, R. (1952). Aksjomat, analityczność i flexioryzm [Axiom, Analyticity and Apriorism]. *Myśl Filozoficzna* nr 4 (6), 129–161.

Suszko, R. (1957a). Logika formalna a niektóre zagadnienia teorii poznania. Diachroniczna logika formalna [Formal Logic and Certain Problems of Epistemology. Diachronic Formal Logic]. *Myśl Filozoficzna* nr 2 (28), 27–56.

Suszko, R. (1957b). Logika formalna a niektóre zagadnienia teorii poznania (Diachroniczna logika formalna) [Formal Logic and Certain Problems of Epistemology (Diachronic Formal Logic)]. Myśl Filozoficzna nr 3 (29), 34–67.

Suszko, R. (1957c). W sprawie antynomii kłamcy i semantyki języka naturalnego [Concerning the Antinomy of the Liar and Semantics of Natural Language]. *Zeszyty Naukowe Wydziału Filozoficznego Uniwersytetu Warszawskiego* nr 3, Warszawa: Państwowe Wydawnictwo Naukowe.

Suszko, R. (1958). Syntactic Structure and Semantical Reference. I. *Studia Logica* VIII, 213–244.

Suszko, R. (1960). Syntactic Structure and Semantical Reference. II. *Studia Logica* IX, 185–216.

Szmielew, W. (1949a). Decision Problem in Group Theory. In: *Proceedings of the Xth International Congress of Philosophy*, vol. 1, Fascicule 2, pp. 763–766. Amsterdam.

Szmielew, W. (1949b). Arithmetical Classes and Types of Abelian Groups. *Bulletin of the American Mathematical Society* 55, 65 and 1192.

Tait, W. W. (1981). Finitism. *Journal of Philosophy* 178, 524–546.

Takeuti, G. (1975). *Proof Theory*. Amsterdam: North-Holland Publishing Company.

Tarski, A. (1930). Fundamentale Begriffe der Methodologie der deduktiven Wissenschaften. I. *Monatshefte für Mathematik und Physik* 37, 361–404. English translation: Fundamental Concepts of the Methodology of the Deductive Sciences. In: Tarski (1956a), pp. 60–109.

Tarski, A. (1933a). *Pojęcie prawdy w językach nauk dedukcyjnych* [The Notion of Truth in Languages of Deductive Sciences]. Warszawa: Nakładem Towarzystwa Naukowego Warszawskiego. English translation: Tarski (1956a).

Tarski, A. (1933b). Einige Betrachtungen über die Begriffe der ω-Widerspruchsfreiheit und der ω-Vollständigkeit. *Monatshefte für Mathematik und Physik* 40, 97–112. English translation: Some Observations on the Concept of ω-Consistency and ω-Completeness. In: Tarski (1956a), pp. 279–295.

Tarski, A. (1936). Der Wahrheitsbegriff in den formalisierten Sprachen. *Studia Philosophica* 1, 261–405 (offprints dated 1935).

Tarski, A. (1944). The Semantic Conception of Truth and the Foundations of Semantics. *Philosophy and Phenomenological Research* 4, 341–375.

Tarski, A. (1949a). Arithmetical Classes and Types of Boolean Algebras. *Bulletin of the American Mathematical Society* 55, 64 and 1192.

Tarski, A. (1949b) Arithmetical Classes and Types of Algebraically Closed and Real-Closed Fields. *Bulletin of the American Mathematical Society* 55, 64 and 1192.

Tarski, A. (1949c). Undecidability of the Theories of Lattices and Projective Geometries. *Journal of Symbolic Logic* 14, 77–78.

Tarski, A. (1951). *A Decision Method for Elementary Algebra and Geometry*. 2nd revised ed. Berkeley and Los Angeles: University of California Press.

Tarski, A. (1953). A General Method in Proofs of Undecidability. In: Tarski, Mostowski and Robinson (1953), pp. 1–35.

Tarski, A. (1953a). Undecidability of the Elementary Theory of Groups. In: Tarski. Mostowski and Robinson (1953), pp. 75–87.

Tarski, A. (1954). Contributions to the Discussion of P. Bernays, '*Zur Beurteilung der Situation in der beweistheoretischen Forschung*'. *Revue Internationale de Philosophie* 8, 16–20. Reprinted in: A. Tarski, *Collected Papers*, vol. IV, pp. 711–714. Basel: Birkhäuser, 1986.

Tarski, A. (1956a). The Concept of Truth in Formalized Languages. In: Tarski (1956b), pp. 152–278.

Tarski, A. (1956b). *Logic, Semantics, Metamathematics. Papers From 1923 To 1938*. Oxford: Clarendon Press. Second edition with corrections and emendations, ed. by J. Corcoran, Indianapolis, IN: Hackett, 1983.

Tarski, A. (1965). The Concept of Truth in Formalized Languages. In: *Logic, Semantics, Metamathematics. Papers From 1923 To 1938*, pp. 152–278. Oxford: Clarendon Press.

Tarski, A. (1995). Some Current Problems in Metamathematics. Ed. by J. Tarski and J. Woleński. *History and Philosophy of Logic* 16, 159–168.

Tarski, A., A. Mostowski and R. M. Robinson (1953). *Undecidable Theories*. Amsterdam: North-Holland Publishing Company.

Thiel, Ch., ed. (1982). *Erkenntnistheoretische Grundlagen der Mathematik*. Hildesheim: Gerstenberg Verlag.

Turing, A. (1936–1937). On Computable Numbers, with an Application to the Entscheidungsproblem. *Proceedings of the London Mathematical Society*, Series 2, 42, pp. 230–265.

Turing, A. (1950). Computing Machinery and Intelligence. *Mind* 59, 433–460. Reprinted in: A.R. Anderson (ed.), *Minds and Machines*, pp. 4–42 (1964). Englewood Cliffs: Prentice Hall.

Tymoczko, Th., ed. (1985). *New Directions in the Philosophy of Mathematics*. Boston-Basel-Stuttgart: Birkhäuser.

Visser, A. (1989). Peano's Smart Children: A Provability Logical Study of Systems with Built-In Consistency. *Notre Dame Journal of Formal Logic* 130, 161–196.

Wainer, S. S. (1970). A Classification of the Ordinal Recursive Functions. *Archiv für Mathematische Logik und Grundlagenforschung* 13, 136–153.

Wang, Hao (1960a). Toward Mechanical Mathematics. *IBM Journal Research and Development*, 2–22. Also in: *Logic, Computers and Sets*, Chelsea, New York, 1970.

Wang, Hao (1960b). Proving Theorems by Pattern Recognition. Part I. *Commun. Assoc. Comput. Mach.* 3, 220–234.

Wang, Hao (1961). Proving Theorems by Pattern Recognition. Part II. *Bell System Technical Journal* 40, 1–41.

Wang, Hao (1974). *From Mathematics to Philosophy*. London: Routledge and Kegan Paul.

Wang, Hao (1981). Some Facts About K. Gödel. *Journal of Symbolic Logic* 46, 653–659.

Wang, Hao (1987). *Reflections on Kurt Gödel*. Cambridge, Mass.: The MIT Press.

Wang, Hao (1991a). Imagined Discussions with Gödel and with Wittgenstein. Address to the Kurt Gödel Society Meeting at Kirchberg am Wechsel, Austria.

Wang, Hao (1991b). To and From Philosophy – Discussions with Gödel and Wittgenstein. *Synthese* 88 (2), 229–277.

Wang, Hao (1996). *A Logical Journey. From Gödel to Philosophy*. Cambridge, Mass., and London, England: The MIT Press.

Weyl, H. (1918). *Das Kontinuum. Kritische Untersuchungen über die Grundlagen der Analysis*. Leipzig: Veit.

Whitehead, A. N. and Russell, B. (1910–1913). *Principia Mathematica*. Vol. I: 1910. Vol. II: 1912. Vol. III: 1913. Cambridge: Cambridge University Press.

Wiegner, A. (1948). *Elementy logiki formalnej* [Elements of Formal Logic]. Poznań: Księgarnia Akademicka.

Wiegner, A. (1952). *Zarys logiki formalnej* [An Outline of Formal Logic]. Warszawa: Państwowe Wydawnictwo Naukowe.

Wilder, R. L. (1968). *Evolution of Mathematical Concepts. An Elementary Study*. New York: John Wiley & Sons.

Wilder, R. L. (1981). *Mathematics as a Cultural System*. Oxford: Pergamon Press.

Wilkosz, W. (1938). O definicji przez abstrakcję [On Definitions by Abstracting]. *Kwartalnik Filozoficzny* 14, 1–13.

Wilkosz, W. (1939). Znaczenie logiki matematycznej dla matematyki i innych nauk ścisłych [The Meaning of Mathematical Logic for Mathematics and Other Sciences]. *Przegląd Filozoficzny* 39, 343–346.

Winker, S., L. Wos and E. Lusk (1981). Semigroups, Antiautomorphisms and Involutions: A Computer Solution to an Open Problem, I. *Mathematics of Computation*, 533–545.

Wittgenstein, L. (1953). *Philosophical Investigations*. Oxford: Basil Blackwell.

Wittgenstein, L. (1956). *Remarks on the Foundations of Mathematics*. Oxford: Basil Blackwell.

Woleński, J. (1989). *Logic and Philosophy in the Lvov-Warsaw School*, Dordrecht: Kluwer Academic Publishers.

Woleński, J. (1991). Gödel, Tarski and the Undefinability of Truth. In: *Yearbook 1991 of the Kurt Gödel Society* (Jahrbuch 1991 der Kurt-Gödel-Gesellschaft), pp. 97–108. Wien: Kurt

Gödel Gesellschaft. Reprinted in: J. Woleński, *Essays in the History of Logic and Logical Philosophy*, pp. 134–138. Kraków: Jagiellonian University Press, 1999.

Woleński, J. (1993). Tarski as a Philosopher. In: F. Coniglione, R. Poli and J. Woleński (eds.), *Polish Scientific Philosophy: The Lvov-Warsaw School*, pp. 319–338. Amsterdam-Atlanta, GA: Editions Rodopi.

Woleński, J. (1995a). On Tarski's Background. In: J. Hintikka (ed.), *Essays on the Development of the Foundations of Mathematics*, pp. 331–341. Dordrecht: Kluwer Academic Publishers.

Woleński, J. (1995b). Mathematical Logic in Poland 1900–1939: People, Circles, Institutions, Ideas. *Modern Logic* 5, 363–405.

Wos, L., G. A. Robinson, D. F. Carson and L. Shalla, (1967). The Concept of Demodulation in Theorem Proving. *Journal of the Association for Computing Machines* 14, 698–709.

Wos, L., D. F. Carson and G. A. Robinson (1964). The Unit Preference Strategy in Theorem Proving. *AFIPS Conf. Proc. 26*, Washington D.C., pp. 615–621.

Wos, L., G. A. Robinson and D. F. Carson (1965). Efficiency and Completeness of the Set of Support Strategy in Theorem Proving. *Journal of the Association for Computing Machines* 12, 536–541.

Wos, L. and G. A. Robinson (1970). Paramodulation and Set-of-Support. In: *Proc. IRIA Symp. on Automatic Demonstration*, Versailles, France 1968, LNM 125, pp. 276–310. Berlin-New York: Springer Verlag.

Yates, R., B. Raphael and T. Hart (1970). Resolution Graphs. *Artificial Intelligence* 1, 257–289.

Zaremba, S. (1911). Pogląd na te kierunki w badaniach matematycznych, które mają znaczenie teoretyczno-poznawcze [Remarks on Those Trends in Mathematical Investigations which have Epistemological Meaning] *Wiadomości Matematyczne* 15, 217–223.

Zaremba, S. (1912). *Arytmetyka teoretyczna* [Theoretical Arithmetic]. Kraków: Polska Akademia Umiejętności.

Zaremba, S. (1926). *La logique en mathématiques*. Paris: Gauthier-Villars.

Zaremba, S. (1938). Uwagi o metodzie w matematyce i fizyce [Remarks on the Methods of Mathematics and Physics]. *Przegląd Filozoficzny* 41, 31–36.

Zermelo, E. (1932). Über Stufen der Quantifikation und die Logik des Unendlichen. *Jahresbericht der Deutschen Mathematiker-Vereinigung* 41 (2), 85–88. Reprinted with English translation: On Levels of Quantification and the Logic of the Infinite, in: Zermelo (2010), pp. 542–549.

Zermelo, E. (2010). *Collected Works*. Vol. I: *Set Theory, Miscellanea*. Edited by H.-D. Ebbinghaus and A. Kanamori. Berlin and Heidelberg: Springer-Verlag.

Editorial Note

Mathematical Knowledge. In: I. Niiniluoto, M. Sintonen and J. Woleński (eds.), *Handbook of Epistemology*, pp. 571–606 (2004). Dordrecht-Boston-London: Kluwer Academic Publishers.
© 2004 Kluwer Academic Publishers.
Reprinted with kind permission of Springer Science and Business Media.

On the Power and Weaknesses of the Axiomatic Method – published (in Polish) under the title: O potędze i słabościach metody aksjomatycznej i ich konsekwencjach dla filozofii matematyki. In: S. Ziemiański SJ (red.), *Philosophia vitam alere*. pp. 455–465. Kraków: Ignatianum – WAM (2005).
Translated and published with the permission of the publisher.

Remarks on the Mathematical Universe – published under the title: Did Leibniz and Newton Discover or Create the Calculus?, *Fundamenta Informaticae* 81 (2007), 249–256.
© 2007 Polish Mathematical Society.
Reprinted with kind permission of Zarząd Główny Polskiego Towarzystwa Matematycznego.

Structuralism and Category Theory in the Contemporary Philosophy of Mathematics. *Logique et Analyse* 204 (2008), 365–373 (published as a joint paper with I. Bondecka-Krzykowska).
Reprinted with kind permission of the editor of the journal *Logique et Analyse*.

Hilbert's Program: Incompleteness Theorems vs. Partial Realizations. In: J. Woleński (ed.), *Philosophical Logic in Poland*, pp. 103–127 (1994). Dordrecht–Boston–London: Kluwer Academic Publishers.
© 1994 Kluwer Academic Publishers.
Reprinted with kind permission of Springer Science and Business Media.

On the Distinction Proof-Truth in Mathematics. In: Gärdenfors P. *et al.* (eds.), *In the Scope of Logic, Methodology and Philosophy of Science*, pp. 287–303 (2002). Dordrecht–Boston–London: Kluwer Academic Publishers.
© 1994 Kluwer Academic Publishers.
Reprinted with kind permission of Springer Science and Business Media.

Reactions to the Discovery of the Incompleteness Phenomenon. In: J. Hintikka *et al.* (eds.), *Philosophy and Logic. In Search of the Polish Tradition*, pp. 213–227 (2003). Dordrecht–Boston–London: Kluwer Academic Publishers.
© 2003 Kluwer Academic Publishers.
Reprinted with kind permission of Springer Science and Business Media.

Gödel's Incompleteness Theorems and Computer Science. *Foundations of Science* 2 (1997), 123–135.
© 1997 Kluwer Academic Publishers.
Reprinted with kind permission of Springer Science and Business Media.

The Present State of Mechanized Deduction, and the Present Knowledge of Its Limitations, *Studies in Logic, Grammar and Rhetoric* 9 (22) (2006), 31–60.

Reprinted with kind permission of Wydawnictwo Uniwersytetu w Białymstoku (Publishing House of the University in Białystok).

On Proofs of the Consistency of Arithmetic, *Studies in Logic, Grammar and Rhetoric* 4(17) (2001), 41–50.

Reprinted with kind permission of Wydawnictwo Uniwersytetu w Białymstoku (Publishing House of the University in Białystok).

Decidability vs. Undecidability. Logico-Philosophico-Historical Remarks, *Annales Universitatis Mariae Curie-Skłodowska, Sectio AI, Informatica* 3 (2005), 105–117.

Reprinted with kind permission of Wydawnictwo Uniwersytetu Marii Curie-Skłodowskiej (Publishing House of Maria Curie-Skłodowska University).

Undefinability of Truth. The Problem of the Priority: Tarski vs. Gödel. *History and Philosophy of Logic* 19 (1998), 153–160.

© 1998 Taylor & Francis Ltd.

Reprinted with kind permission of Taylor & Francis, Taylor & Francis Group plc http://www.tandf.co.uk/journals.

Troubles with (the Concept of) Truth in Mathematics, *Logic and Logical Philosophy* 15 (2006), 285–303.

© 2002 by Nicolaus Copernicus University Toruń.

Reprinted with kind permission of Wydawnictwo Naukowe Uniwersytetu Mikołaja Kopernika (Scientific Press of Nicolaus Copernicus University).

The Philosophy of Hoene-Wroński, *Organon* 35 (2006), 143–150.

Reprinted with kind permission of Instytut Historii Nauki Polskiej Akademii Nauk (Institute for the History of Science of Polish Academy of Sciences).

Philosophical Reflection on Mathematics in Poland in the Interwar Period, *Annals of Pure and Applied Logic* 127 (2004), 325–337.

© 2004 by Elsevier.

Reprinted with permission from Elsevier.

Philosophy of Mathematics in the Warsaw Mathematical School, *Axiomathes* 20 (2010), 279–293.

Reprinted with kind permission of Springer Science and Business Media.

Andrzej Mostowski on the Foundations and Philosophy of Mathematics. In: A. Ehrenfeucht, V.W. Marek and M. Srebrny (eds.), *Andrzej Mostowski and Foundational Studies*, pp. 324–337 (2008). Amsterdam/Berlin/Oxford/Tokyo/Washington, DC: IOS Press (published as a joint paper with J. Woleński).

© 2008 The authors and IOS Press.

Reprinted with kind permission of IOS Press BV.

Editorial Note

Stanisław Piątkiewicz and the Beginnings of Mathematical Logic in Poland. *Historia Mathematica* 23 (1996), 68–73 (published as a joint paper with T. Batóg).
© 1996 by Elsevier.
Reprinted with permission from Elsevier.

Contribution of Polish Logicians to Recursion Theory. In: K. Kijania-Placek and J. Woleński (eds.), *The Lvov-Warsaw School and Contemporary Philosophy*, pp. 265–282 (1998). Dordrecht-Boston-London: Kluwer Academic Publishers.
© 1998 Kluwer Academic Publishers.
Reprinted with kind permission of Springer Science and Business Media.

Logical Investigations at the University of Poznań in 1945–1955. In: P. Bernhard und V. Peckhaus (eds.), *Methodisches Denken im Kontext*, pp. 239–254 (2008). Paderborn: mentis Verlag (published as a joint paper with J. Pogonowski).
© 2008 mentis Verlag GmbH.
Reprinted with kind permission of mentis Verlag GmbH.

Index of Names

Ackermann W. 8, 39, 43, 87, 105, 115, 159, 160, 164, 166, 170, 295, 306
Addison J. W. 268, 274, 295
Agazzi E. 296, 313
Ajdukiewicz K. 221, 222, 261, 283, 285–289, 291–295, 297, 299
Albrycht J. 293
Anderson A. R. 311
Anderson R. 295
Andrews P. 152, 157, 295
Andrews P. B. 152, 313
Apt K. R. 96, 295
Arai T. 89, 295
Archimedes 19
Aristotle 17–20, 25–27, 53, 54, 63, 68, 101, 296, 311
Artin E. 100
Arzela C. 99
Ascoli G. 99
Asquith P. D. 299
Avenarius R. 288
Awodey S. 75–77, 79, 296
Ax, J. 172, 296

Bach J. S. 306
Bain A. 262, 296
Baire R. L. 33, 268
Balas Y. 104–106, 181, 183
Balzac H. 211
Banach S. 60, 98, 250, 269, 277–280, 289, 296
Banachiewicz T. 233
Barwise J. 121, 296, 317, 321
Barzin M. 124, 296
Batóg T. 9, 261, 288–290, 296, 312
Bauer-Mengelberg S. 306
Bedürftig Th. 296
Behmann H. 165, 170, 296
Bell J. L. 296
Beman W. W. 299
Benacerraf P. 128, 296, 303, 309, 311
Bendix P. B. 151, 308
Bennett J. H. 150, 304
Bernays P. 41, 43, 48, 63, 86, 88, 90, 94, 117, 119, 120, 161, 164, 179, 281,
292, 297, 306, 312, 323
Berry G. D. 42
Beth E. W. 139, 298
Bibel W. 295
Bishop E. 37
Black M. 187, 297
Bledsoe W. W. 148, 150, 151, 157, 295, 297, 311
Bläsius K. 152, 297
Bobrowski D. 293
Bolyai J. 42, 56
Bolzano B. 99, 263, 278
Bondecka-Krzykowska I. 9, 73
Boole G. 27, 42, 55, 262–264
Boolos G. 138, 155, 156, 297
Borel E. 33, 98, 268, 270
Borkowski L. 286, 297, 312
Bourbaki N. 48, 73, 75
Bowie G. L. 297
Boyer R. S. 149, 151, 297
Boy-Żeleński T. 211
Braun J. 212
Brouwer L. E. J. 32, 33, 35, 36, 67, 83, 84, 87, 243, 276, 297, 298, 312
Browder F. 309
Brown G. W. 303
Bukaty A. 212
Buldt B. 309
Burali-Forti C. 26, 42
Burks A. W. 182, 316
Buss S. R. 154, 298
Börger E. 174, 297
Büchi J. R. 174, 298

Cantor G. 20, 26, 30, 33, 42, 60, 64, 65, 83, 196, 229, 233, 256, 270, 292, 298, 315
Cardan G. 208
Carnap R. 107, 118, 126, 128, 182, 184, 244, 273, 298, 313, 318
Carson D. 140
Carson D. F. 149, 325
Carson G. A. 148
Cauchy A. 42, 61, 98, 99
Chaitin G. 51, 134, 298

Chang C. L. 144, 149, 298
Chihara Ch. S. 45, 298
Chinlund T. J. 140, 298
Chojnicki Z. 293, 294
Chomicz P. 212
Church A. 40, 43, 123, 137, 152, 157, 169, 170, 174, 267, 298, 316, 319, 321
Chwistek L. 31, 217, 218, 220, 222, 223, 243, 264, 298
Cieszkowski A. C. 205
Clote P. 298
Cohen E. L. 152, 157, 171, 295, 313
Cohen P. J. 44, 61, 62, 254, 255, 299
Colmerauer A. 149
Coniglione F. 325
Corcoran J. 323
Courant R. 119
Couturat L. 23, 264, 299
Currie G. 310
Curry H. B. 39, 299
Cusanus N. 65, 66
Czermak J. 299, 319
Czerwiński Z. 285, 286, 293, 299
Czeżowski T. 217, 285, 287

Darboux J. G. 279
Darvas G. 313
Davis M. 123, 133, 137–140, 173, 181, 182, 298, 299, 303, 318
Davis Ph. J. 51
Dawson J. W., Jr. 117, 119, 121, 122, 125, 179, 299, 303
Dąbrowski H. 206
Dąmbska I. 285
Dedekind R. 29, 42, 48, 66, 83, 85, 298, 299
Depauli-Schimanovich-Göttig W. 309
Descartes R. 20–22, 54, 127, 167, 299
Detlefsen M. 84, 86, 89, 299, 300, 307
Deuticke F. 117
Dickstein S. 212, 265, 300
Dirichlet J. 92
Domaradzki C. 212
Donner J. 171, 300
Drake F. R. 97, 300
Drossos C. A. 300
Duda R. 240, 300
Dummett M. 124, 300

Dybowski M. 285

Ebbinghaus H.-D. 300, 325
Edwards P. 297, 315
Ehrenfeucht A. 155, 172, 300
Eilenberg S. 76, 300
Eininger N. 152, 297
Elgot C. C. 300
Enayat A. 301
Encausse G. (Papus) 212
Enriques F. 28
Escher P. 306
Euclid 18–20, 27, 53–56, 64, 69, 101, 300
Eudoxus 19

Feferman A. 284, 301
Feferman S. 88, 89, 94, 95, 106, 117, 183, 195, 284, 300, 301, 303
Feigenbaum E. A. 301, 317
Feigl H. 118
Feldman J. 301, 317
Fermat P. 116, 208
Fichte J. G. 205
Field H. 45, 301
Finsler P. 122, 301
Fisher M. J. 154, 173, 301
Fløistad G. 318
Floyd J. 125, 301
Fraenkel A. 44, 60–62, 99, 156, 293
Frege G. 29–31, 39, 42, 43, 55, 83–85, 102, 263, 301, 305, 320
Friedlein G. 318
Friedman H. 42, 95, 97, 99, 100, 156, 301, 316, 320

Gabriel G. 301
Gaifman H. 200, 301
Gałecki Ł. 9
Gauss C. F. 33, 42, 56, 66
Gawecki B. J. 212, 301
Gelernter H. 137, 139, 301
Gentzen G. 41, 43, 94, 100, 121, 139, 163, 164, 251, 297, 302
Gerhardt C. I. 310
Gibb J. W. 107, 132
Giedymin J. 285
Gilmore P. C. 137, 139, 302
Givant S. R. 179, 302

Index of Names

Goldbach C. 69, 116
Goldfarb W. D. 303, 305
Goliński S. 262, 304
Gołuchowski J. 205
Gonseth F. 297
Good I. J. 128, 304
Goodstein R. L. 92, 93, 304
Gordan P. 85
Gould W. E. 152, 304
Grassmann R. 262
Grattan-Guinness I. 179, 304
Grelling K. 42, 124, 304
Grot K. 283, 304
Grädel E. 297
Grzegorczyk A. 250, 268–270, 275–277, 279–281, 294, 304, 313–315
Guard J. R. 150, 152, 304
Guaspari D. 89, 305
Gurevich Y. 297
Gärdenfors P. 316
Gödel K. 7, 8, 16, 28, 32, 40–44, 46, 51, 56–59, 61, 62, 65, 69, 70, 87, 88, 90, 91, 93–95, 98, 100, 101, 104, 106, 107, 111, 114–124, 126–135, 153–155, 157, 161–163, 166, 169, 170, 173–175, 177–185, 188, 189, 191, 194, 249, 251, 254, 256, 267, 272, 273, 275, 276, 280, 292, 296, 298–306, 308, 309, 312, 313, 315, 316, 318–320, 324

Hahn H. 98, 115–117, 320
Hamilton W. 262
Hanson H. J. 301
Harrington L. A. 40, 57, 91–93, 99, 156, 301, 316, 317, 320
Hart T. 148, 325
Hart W. D. 305
Hartmanis J. 154, 305
Hausdorff F. 234
Heath Th. L. 300
Hefferman G. 299
Hegel G. 205, 210
Heiberg I. L. 300
Heijenoort J. van 90, 117, 122, 301, 303, 305, 306
Heine E. 61, 98
Heintz J. 154, 305

Hellman G. 49, 77, 296, 305
Helmer O. 124, 305
Henkin L. 121, 125, 310
Hensel G. 281, 305
Herbart J. F. 263
Herbrand J. 119, 137, 139, 140, 147, 162, 163, 167, 169, 172, 251, 305
Hermes H. 301
Herold A. 152, 297
Hersh R. 51
Hewitt C. 150, 305
Heyting A. 36, 41, 184, 249, 250, 304, 305, 309, 314
Hilbert D. 7, 8, 25, 28, 37–43, 48, 55, 71, 78, 84–91, 94, 95, 97, 99, 100, 102–105, 114, 117–121, 133, 153, 159–166, 173, 174, 185, 191, 220, 243, 245, 267, 297, 299, 301, 305, 306, 308, 309, 312, 315–321
Hineman G. 298
Hintikka J. 301, 306, 316, 325
Hlodovskii I. 164, 306
Hoene-Wroński J. M. 8, 205–213, 215, 300, 301
Hofstadter D. R. 128, 306
Hollak H. J. A. 298
Horn A. 142
Huet G. D. 152, 307
Hume D. 293
Husserl E. 220
Hájek P. 193, 305

Ingarden R. 261, 289, 307
Irvine A. D. 312
Isaacson D. 113, 307

Janiczak A. 171, 173, 270, 307
Janiszewski Z. 216, 219, 227–234, 236–239, 307, 308, 321
Jankowski J. 212
Jastrzębiec-Kozłowski C. 212
Jaśkowski S. 43, 285, 314
Jensen D. C. 152, 307
Jeroslov R. 307
Jevons W. S. 262, 263, 307
Jordan Z. 261, 264, 307
Jurek S. 262

Kahr A. 174, 308
Kalmár L. 170, 173, 174, 267, 269, 308
Kambartel F. 301
Kanamori A. 325
Kant I. 16, 24–26, 29, 33, 35, 38, 66, 67, 85, 205, 206, 308, 316
Karp R. 121, 301
Kaye R. 109, 193, 308
Kent C. F. 91
Kepler J. 208
Ketelsen Ch. 122, 308
Kijania-Placek K. 316
Kirby L. 40, 57, 92, 93, 100, 308
Kitcher Ph. 45, 51, 86, 299, 308
Kleene S. C. 169, 182, 257, 267–271, 274, 275, 280, 303, 308
Klein C. 309, 319
Klein F. 34, 255
Kleinert E. 56, 308
Knaster B. 232, 308
Knuth D. E. 151, 308
Kochen S. 172, 296
Kokoszyńska-Lutmanowa M. 285
Kolankowski J. 262
Kołakowski L. 17
Kossak R. 199, 301
Kostrzewska-Kratochwilowa S. 262
Kościuszko T. 206
Kotarbińska J. 245, 308
Kotarbiński T. 71, 219, 220, 224, 247, 261, 285, 308
Kotlarski H. 112, 196–198, 309
Kowalski R. 148, 152, 295, 308
Kozłowski W. M. 283, 284
Krajewski S. 111, 199, 309
Krajewski W. 129, 131, 317
Krasiński Z. 205
Kreisel G. 86, 87, 89, 91, 94, 95, 117, 132, 275, 281, 309
Kremer J. 205
Kronecker L. 33, 83, 84, 309
Krupiński F. 296
Kruskal J. 156
Kuczyński J. 124, 309
Kuehner D. G. 148, 149, 308, 310
Kuratowski K. 217, 247, 253, 310
Kuzawa S. M. G. 307
Kwietniewski S. 230

Köhler E. 115, 184, 299, 309, 316
König D. 97, 98
La Mettrie J. O. de 127, 310
Lachlan A. 308
Ladrière J. 122, 310
Lakatos I. 47, 48, 310, 314
Landau E. 234
Landry E. 310
Langford C. H. 171, 310
Lebesgue H. L. 33, 236, 268, 310
Lee Bowie G. 131
Lee R. 144, 298
Leibniz G. W. 20, 22–24, 26, 27, 29, 50, 54, 55, 63, 71, 120, 158, 167, 168, 209, 262, 310, 316
Leo XII 210
Leonard J. 172, 310
Leśniewski S. 71, 218, 220, 222–224, 239, 240, 243, 247
Levy A. 91
Liard L. 310
Libelt K. 205, 211
Lindenbaum A. 240, 280
Livesey M. 151, 310
Lobachevsky N. I. 42, 56
Locke J. 29
Lorenzen P. 164, 311
Loveland D. W. 138, 140, 148, 295, 297, 301, 311, 313
Lubrański J. 283
Lucas J. R. 7, 127–131, 297, 311
Luckham D. 148, 311
Lullus R. 167
Lusk E. 151, 157, 324
Lutosławski W. 212
Luzin N. N. 33, 268
Läuchli H. 172, 310
Lévi E. 212
Löb M. H. 88, 161, 270, 310, 311
Löwenheim L. 16, 43, 59, 105, 165, 170, 171, 184, 280, 292, 311

Łoś J. 314
Łukasiewicz J. 9, 216, 217, 219–222, 230, 239, 240, 244, 261, 264, 287, 311, 312
Łukomski L. 212

Index of Names

Łuszczewska-Romahnowa S. 285, 286, 289–291, 294, 296, 312
Łuszczewski J. 212
Mac Lane S. 76, 300, 312
Macintyre A. 300
Mackie J. L. 187, 312
MacLarty C. 312
Maddy P. 46, 312
Madej A. 212
Mazur S. 314
Malewski A. 285, 294
Mancosu P. 115, 118, 159, 244, 312, 313
Marchlewski J. 284
Marciszewski W. 137, 141, 313
Marczewski E. 313
Marek W. 96, 295, 309
Markov A. 267
Marquis J.-P. 310, 313
Maslov J. Ju. 149
Maslov S. Ju. 313
Matiyasevich Y. 173
Mazur S. 250, 269, 277–280, 296, 313
Mazurkiewicz S. 227, 230, 234, 236–240, 313
McCall S. 308
McIlroy D. 298
Meinong A. 220
Meltzer B. 149, 310, 313, 318, 320
Mendelson E. 127, 188, 313
Menger K. 117, 313
Meyer A. R. 154, 313
Michie D. 310, 313, 318, 320
Mickiewicz A. 205, 284
Mill J. S. 25, 26, 68
Miller D. A. 152, 157, 295, 313
Minsky M. 139, 150
Mittag-Leffler C. 65
Mleczko P. 9
Moore E. 308
Moore G. H. 119, 121, 303, 313
Moore J. S. 151, 174, 297
Morgan A. De 27, 42, 55, 262, 264
Morris J. B. 149, 313
Mostowski A. 8, 91, 155, 162, 171, 189, 219, 221, 233, 240, 245–257, 267–274, 276, 280, 281, 292, 300, 310, 313–315, 318, 323

Murawski R. 40, 42, 55, 57, 60, 83, 93, 96, 106, 109, 111, 115, 120, 121, 127, 137, 141, 162, 169, 171, 179, 188, 191–193, 245, 296, 313, 315, 316
Mycielski J. 155, 300

Nagel E. 131, 316
Natorp P. 33
Nerode A. 156, 316
Neumann J. von 8, 90, 118, 119, 126, 153, 159–162, 168, 182, 305, 316
Nevins A. J. 150, 317
Newell A. 137–139, 317
Newman J. R. 131, 316
Newton I. 20, 26, 55, 63, 71, 209
Nicolaus Cusanus 317
Niebergall K.-G. 110, 193, 194, 317
Niementowski J. 212
Norton L. M. 150, 317
Nowakowa I. 288, 317

Oglesby F. C. 150, 304
Omyła M. 291, 317
Orlicz W. 285, 292
Ossowska M. 285
Overbeek R. 151

Pacholski L. 300
Pajewski J. 284
Parikh R. 155, 317
Paris J. 40, 57, 91–93, 100, 300, 308, 317
Parsons Ch. 303
Pascal B. 21, 22
Pasch M. 28, 42, 317
Peano G. 28, 39, 42, 55, 69, 89, 90, 93, 94, 96, 98, 99, 108, 110–113, 130, 162, 173, 175, 189, 191–193, 198, 200, 201, 231, 238, 265, 315, 317, 318, 324
Peckhaus V. 84, 317
Peirce Ch. S. 27, 55
Penrose R. 128, 317
Pepis J. 174
Perelman Ch. 124, 318
Pfenning F. 157, 295
Piątkiewicz S. 8, 261–265, 318
Pieri M. 28
Pietrzykowski T. 152, 307

Piłsudski J. 284
Plato 17, 18, 26, 27, 53, 54, 62–64, 68, 69, 101
Plotkin G. D. 151, 318
Ploucquet G. 262
Plummer D. F. 119
Pogonowski J. 9, 283, 316
Poincaré H. 18, 33–35, 67, 83, 215, 217, 227, 231, 255, 318
Poli R. 325
Popper K. 47, 48
Pos H. J. 298
Post E. L. 122, 123, 252, 267, 272, 299, 318
Prawitz D. 137, 140, 318
Presburger M. 138, 154, 171–173, 318
Proclus 18, 20, 318
Przełęcki M. 294
Pseudo-Scotus 286
Pudlák P. 193, 305
Putnam H. 45, 135, 137, 139, 140, 173, 281, 296, 299, 303, 305, 309, 318
Puzyna J. 233, 234
Péter R. 267, 269, 317

Quine W. V. O. 45, 46, 175, 244, 313, 318

Rabin M. O. 154, 173, 301
Ramsey F. P. 31, 92, 167, 281, 318
Raphael B. 148, 325
Rasiowa H. 277, 313, 314
Ratajczyk Z. 99, 112, 197, 198, 309
Reichenbach H. 118
Reid C. 84, 119, 120, 318
Reiter R. 148, 319
Remmel J. 298
Resnik M. D. 49, 86, 90, 319
Riemann B. 69, 98
Robinson A. 26, 125, 139
Robinson D. F. 148
Robinson G. 140
Robinson G. A. 149, 325
Robinson J. 173, 319
Robinson J. A. 137, 140, 141, 143–145, 148, 149, 319
Robinson R. M. 173, 319, 323
Rodríguez-Consuegra F. A. 125, 319
Rogers H., Jr. 189, 319

Rosser J. B. 88, 89, 91, 117, 162, 182, 303, 319
Roussel P. 149
Rowe D. E. 102, 319
Roy M. F. 305
Rucker R. 129, 319
Russell B. 27, 28, 30, 31, 39, 42, 43, 46, 55, 56, 71, 83, 84, 124, 125, 138, 238, 244, 264, 298, 303, 319, 324
Ruziewicz S. 234

Schaff A. 287, 294, 295
Schelling F. 205, 206
Schilpp P. A. 303, 319
Schröder E. 27, 55, 165, 262–264, 319, 320
Schönfinkel M. 167, 297
Schütte K. 94, 164, 320
Scotus D. 59, 286
Seidenberg A. 305
Settle L. G. 150, 304
Shalla L. 149, 325
Shanker S. G. 125, 299, 301, 320
Shapiro S. 49, 79, 320
Shaw J. C. 137–139, 317
Sheffer H. M. 289
Shepherdson J. 91
Shoenfield J. R. 188, 189, 320
Shostak R. E. 152, 320
Sibert E. E. 149, 320
Sickel S. 152, 320
Sieg W. 90, 100, 119, 303, 320
Siekmann J. 151, 152, 297, 310
Sierpiński W. 215, 216, 218, 227–230, 233, 234, 240, 254, 270, 315, 320
Sigmund K. 115, 320
Sikorski R. 314
Simon H. A. 137–139, 317
Simpson S. G. 42, 94, 95, 99, 320, 321
Skolem T. 16, 42, 43, 59, 86, 95, 105, 137, 170–172, 184, 280, 292
Skolem Th. 321
Skolimowski H. 261, 321
Slagle J. R. 321
Sleszyński J. 216, 217, 321
Słowacki J. 205
Smart J. J. C. 7, 127, 128, 321
Smith B. 151
Smolka G. 152, 297

Index of Names

Smoryński C. 84, 86, 87, 89–91, 127, 321
Solerno P. 305
Solovay R. M. 89, 303, 305
Souslin M. 272
Specker E. 279, 321
Spinoza 55
Srebrny M. 308, 309
Stalin J. 284, 285
Stamm E. 9, 218, 264, 321
Statman R. 155, 321
Steinhaus H. 240, 289, 321
Stickel M. E. 151, 321
Strumiłło T. 285
Stöltzner M. 309
Suszko R. 285, 286, 291–294, 317, 321, 322
Swart E. R. 51
Szabo M. E. 302
Szaniawski J. 294
Szmielew W. 171, 322

Śpiewak P. 17

Tait W. W. 86, 322
Takeuti G. 89, 94, 309, 320, 322
Tarski A. 8, 16, 32, 43, 59, 60, 71, 106, 111, 121, 126, 169, 171–173, 176–179, 181–185, 187–189, 191, 192, 199, 219, 221, 224, 225, 240, 244, 246, 247, 249, 251, 284, 300–302, 305, 313, 315, 316, 322–324
Thiel Ch. 301, 323
Thomas I. 288
Towiański A. 205
Trentowski B. 205
Turing A. 58, 127, 133, 134, 137, 152, 154, 157, 169, 175, 267, 323
Twardowski K. 215, 220, 239, 243, 261, 289, 307
Tymoczko Th. 298, 318, 323

Ujejski J. 211

Vartanian A. 310

Venn J. 27, 290
Veraart A. 301
Veronese G. 28
Visser A. 89, 324

Wainer S. S. 270, 311, 324
Waismann F. 118
Wallis J. 208
Walther C. 152, 297
Wang Hao 104–107, 115, 120, 125, 133, 134, 137, 139, 174, 180, 181, 183, 308, 324
Weibel P. 309
Weierstrass K. 29, 33, 42, 83, 99, 278
Weyl H. 83, 84, 123, 276, 324
Whitehead A. N. 30, 31, 43, 138, 238, 264, 324
Wiegner A. 284–286, 288, 289, 296, 317, 324
Wilder R. L. 48, 324
Wilkosz W. 218, 324
Winker S. 151, 157, 324
Wittgenstein L. 46, 47, 124–126, 320, 324
Włodarczyk T. 286
Woleński J. 9, 106, 121, 183, 191, 243, 261, 283, 284, 316, 323–325
Wolkowski Z. W. 309
Worall J. 310
Wos L. 140, 148, 149, 151, 157, 324, 325
Wójtowicz J. 262

Yates R. 148, 325

Zarach A. 308, 309
Zaremba S. 217, 230, 234, 311, 325
Zawirski Z. 284, 291
Zeidler F. 285
Zermelo E. 42, 44, 60–62, 99, 108, 156, 179, 180, 228, 298, 299, 304, 320, 325
Zygmunt III 283
Zygmunt J. 291, 317

Żorawski K. 234

Polish Contemporary Philosophy and Philosophical Humanities
Edited by Jan Hartman

Vol. 1 Roman Murawski: Logos and Máthēma. Studies in the Philosophy of Mathematics and History of Logic. 2011.

www.peterlang.de

 www.ingramcontent.com/pod-product-compliance
Ingram Content Group UK Ltd.
Pitfield, Milton Keynes, MK11 3LW, UK
UKHW021830210426
5322IPUK00004B/110